Intelligent Systems Reference Library

Volume 93

The aim of this series is to publish a Reference Library, including novel advances and developments in all aspects of Intelligent Systems in an easily accessible and well structured form. The series includes reference works, handbooks, compendia, textbooks, well-structured monographs, dictionaries, and encyclopedias. It contains well integrated knowledge and current information in the field of Intelligent Systems. The series covers the theory, applications, and design methods of Intelligent Systems. Virtually all disciplines such as engineering, computer science, avionics, business, e-commerce, environment, healthcare, physics and life science are included. The list of topics spans all the areas of modern intelligent systems such as: Ambient intelligence, Computational intelligence, Social intelligence, Computational neuroscience, Artificial life, Virtual society, Cognitive systems, DNA and immunity-based systems, e-Learning and teaching, Human-centred computing and Machine ethics, Intelligent control, Intelligent data analysis, Knowledge-based paradigms, Knowledge management, Intelligent agents, Intelligent decision making, Intelligent network security, Interactive entertainment, Learning paradigms, Recommender systems, Robotics and Mechatronics including human-machine teaming, Self-organizing and adaptive systems, Soft computing including Neural systems, Fuzzy systems, Evolutionary computing and the Fusion of these paradigms, Perception and Vision, Web intelligence and Multimedia.

Indexed by SCOPUS, DBLP, zbMATH, SCImago.

All books published in the series are submitted for consideration in Web of Science.

Spyros Papadimitriou · Maria Virvou

Artificial Intelligence—Based Games as Novel Holistic Educational Environments to Teach 21st Century Skills

Springer

Spyros Papadimitriou
Department of Informatics
University of Piraeus
Piraeus, Greece

Maria Virvou
Department of Informatics
University of Piraeus
Piraeus, Greece

ISSN 1868-4394 ISSN 1868-4408 (electronic)
Intelligent Systems Reference Library
ISBN 978-3-031-77463-8 ISBN 978-3-031-77464-5 (eBook)
https://doi.org/10.1007/978-3-031-77464-5

This Springer imprint is published by the registered company Springer Nature Switzerland AG
The registered company address is: Gewerbestrasse 11, 6330 Cham, Switzerland

Spyros Papadimitriou dedicates to his wife, Efi, and their boys, Stelios and Fanis.

Maria Virvou dedicates to her family.

Foreword

In the pursuit of education, humanity has continually sought innovative methods to transfer knowledge and foster the development of essential skills to the younger generations. From the ancient teachings of Aristotle to the modern advancements in technology, the essence of education remains stable to the principles of ethos, pathos, and logos. It is within this dynamic intersection of classical wisdom and contemporary innovation that *Artificial Intelligence—Based Games as Novel Holistic Educational Environments to Teach 21st Century Skills* emerges as an inspiration of transformative pedagogy.

Aristotle, the eminent philosopher of antiquity, integrated the significance of ethos, pathos, and logos as foundational elements in effective communication and persuasion. Ethos, rooted in the credibility and character of the communicator, establishes trust and authority. The authors of this book employ the narrative and plot of a prosocial computer adventure game for education to cultivate ethos in students.

Pathos, appealing to the emotions of the audience, evokes empathy and engagement. The authors of this book engage and evoke empathy to players who are also students through the immersive environment of a computer game. Logos, the logical appeal based on reason and evidence, fosters critical thinking and understanding. The authors of this book achieve the teaching of hard skills in programming and other domains but also cultivate problem-solving skills by a dynamically adapted scenario to the needs of students.

In this innovative exploration, the authors seamlessly weave together the timeless principles of ethos, pathos, and logos with the state-of-the-art capabilities of Artificial Intelligence (AI) and immersive gaming environments. Through careful research and innovative design, they reveal a paradigm shift in educational methodology, transcending traditional boundaries to create holistic learning environments that cultivate 21st-century skills. Indeed, at this century of the fourth industrial revolution, the strength of Artificial Intelligence and other technologies are very good for progressing the world's knowledge but at the same time they pose challenges on trustworthiness, robustness, ethics and human oversight, which render skills like problem solving, critical thinking, ethical attitudes more needed than ever for the success of new generations and the smooth establishment of technological breakthroughs into the human society.

To this end, the integration of AI-based games into educational settings presents new challenges and complexities. Yet, within these challenges lie unparalleled opportunities for growth and discovery. By employing the power of technology, educators can cultivate fertile teaching grounds where learners are empowered to explore, collaborate, and innovate.

This inspiring work not only illuminates the transformative potential of AI-based educational games but also highlights the enduring relevance of classical education in an era defined by rapid change and uncertainty. Just as Aristotle's teachings have stood the test of time, so too does this innovative approach to education revive the enduring goal for knowledge and enlightenment through holistic approaches to benefit human learning. In closing, *Artificial Intelligence—based Games as Novel Holistic Educational Environments to Teach 21st Century Skills* invites educators, researchers, and policymakers alike to embrace the synergies between tradition and innovation, ethos and AI, pathos and immersive experiences, logos and critical inquiry for the purposes of education that unites with the countless possibilities of the digital age.

Piraeus, Greece
September 2024

Prof. Dr. George A. Tsihrintzis

Preface

In an era marked by rapid technological advancement, the field of education is undergoing a profound transformation. As society embraces the empowering potential of Artificial Intelligence (AI) and the limitless access to information it provides, educators are tasked with equipping students with the skills necessary to be successful in this evolving landscape. The term "21st-century skills" has emerged to encapsulate both technical proficiencies and essential soft skills required for success in today's world.

Educational games have attracted significant research attention as a powerful tool for addressing educational needs, offering immersive experiences that engage students in meaningful learn-and-play activities. Inspired by Aristotle's rhetoric, these games combine the persuasive power of ethos, pathos, and logos to create environments where learning becomes an enjoyable pursuit, teaching reasoning and ethical skills. By seamlessly combining entertainment with education, they promote not only knowledge acquisition but also the development of essential skills such as problem-solving, communication, and collaboration.

In light of these principles, our research has examined educational games to provide individualized educational platforms for the benefit of learners. Thus, we have developed a novel suite of educational games and components titled "EPATHLO" (Ethos PATHos LOgos). This title, derived from the Greek word meaning a prize won in a contest, emphasizes how learning through these gaming environments can be perceived as a valuable reward, akin to the recognition attributed to victorious athletes in contests like the Olympic Games.

EPATHLO addresses a common issue found in educational games, the lack of robust adaptive capabilities. Existing games often fail to automatically tailor content to the users' needs. And when they do, they tend to prioritize educational aspects while neglecting the entertaining factor essential for engaging gameplay. Therefore, it is essential for educational games to dynamically adapt their content at both the instructional and game levels.

Our book outlines the innovative holistic educational gaming environments of EPATHLO, which utilize AI for dynamic personalization of learning content and gaming plots. These environments aim to teach a broad spectrum of skills to students, adapting educational objectives and game scenarios based on individual student knowledge and preferences.

The EPATHLO Suite exemplifies these principles, integrating distinct components to maximize educational benefits while elevating user enjoyment and motivation levels. The suite comprises the EPATHLO inter-game individualization module to host educational material, build student models, and provide tailored exercises across different gaming environments. This is an innovative mechanism for play-and-learn environments, abstracting AI-based personalization processes for use across different game platforms and categories. Meanwhile, the EPATHLO Fuzzy Module employs fuzzy reasoning to estimate student knowledge levels. The suite's game, Fuzzy EPATHLO, interfaces with these components, receiving exercises and updating scenarios based on student performance, thereby offering a seamless and personalized learning experience. The EPATHLO Author is a user-friendly environment that allows teachers to create their own educational games by providing their teaching content that benefits from all the reasoning and individualizing capabilities of the EPATHLO suite. Finally, AI ethics in terms of students' privacy issues are implemented adapted to the European regulation for privacy, GDPR, after research that resulted to the EPATHLO Privacy implementation.

In view of the above, this book presents a novel approach to educational gaming, emphasizing adaptivity, personalization, and engagement to promote comprehensive skill development in students. Each component of the EPATHLO suite represents a valuable step forward in the development of personalized educational games for teaching 21st-century skills. Their innovative features and functionalities contribute to the ongoing evolution of educational technology, offering novel holistic solutions to enhance learning experiences.

Piraeus, Greece
September 2024

Dr. Spyros Papadimitriou
Prof. Dr. Maria Virvou

Contents

Acronyms

AHP	Analytic Hierarchy Process
AI	Artificial Intelligence
AR	Augmented Reality
ATAG	Authoring Tool Accessibility Guidelines
CBM	Cognitive-Based Model
ChatGPT	Chat Generative Pre-Trained Transformer
DDA	Dynamic Difficulty Adjustment
DL	Deep Learning
DPD	Data Protection Directive
DPIA	Data Protection Impact Assessment
EFL	English as Foreign Language
EPATHLO	Ethos, Pathos, Logos
FAHP	Fuzzy Analytic Hierarchy Process
FFMS	Fuzzy Finite State Machine
FIS	Fuzzy Inference System
FLS	Fuzzy Logic System
FSM	Finite State Machine
FSP	Fuzzy Student Profile
FuzKSD	Fuzzy Knowledge State Definer
FuzzEG	Fuzzy Educational Game
GDPR	General Data Protection Regulation
HTML	HyperText Markup Language
ICAI	Intelligent Computer Assisted Instruction
ICT	Information and Computing Technologies
IDN	Interactive Digital Narratives
IEG	Intelligent Educational Games
IHS	Intelligent Help System
IoT	Internet of Things
ITS	Intelligent Tutoring System
JSON	JavaScript Object Notation
LLM	Large Language Model

MF	Membership Function
ML	Machine Learning
NLP	Natural Language Processing
NPC	Non-Playable Character
OSS	Open-Source Software
P2P	Peer-to-Peer
PDA	Personal Digital Assistant
REST	Representational State Transfer
SQL	Structured Query Language
STEM	Science, Technology, Engineering, and Mathematics
TSK	Takagi-Sugeno-Kang
UML	Unified Modeling Language
USDP	Unified Software Development Process
VR	Virtual Reality
WCAG	Web Content Accessibility Guidelines
WWW	World Wide Web
XR	Extented Reality

List of Figures

List of Tables

Chapter 1
21st Century Skills Via Prosocial Educational Games with Responsible Artificial Intelligence: An Overview

Abstract In modern educational settings, the integration of educational games and responsible Artificial Intelligence (AI) presents a promising avenue for developing digital skills and soft skills such as problem-solving, multidisciplinarity, interdisciplinarity, collaborativity, critical thinking, and ethics among learners, required for success in the 21st century. This chapter provides an overview of the intersection between these elements, exploring the synergistic relationship between prosocial educational games, AI technologies, including Intelligent Tutoring Systems (ITSs) and student modeling for personalization and adaptivity, and the cultivation of these important skills. Additionally, it examines the role of prosocial games in fostering empathy, teamwork, and other positive social behaviors. Beginning with an examination of the evolution of educational games and AI in education, the chapter examines the various ways in which these technologies can enhance learning experiences and promote skill development. Furthermore, it explores the ethical considerations and challenges associated with the integration of AI in educational games, emphasizing the importance of responsible AI design and implementation, including adherence to GDPR regulations for data protection. Through case studies and examples, this chapter illustrates the potential benefits and limitations of leveraging AI-powered educational games to develop digital skills and soft skills. It also discusses how fuzzy logic constitutes a transparent AI technology that can enhance the educational experience. By providing insights into the transformative role of educational games with responsible AI, this chapter aims to inform educators, researchers, and policymakers about the opportunities and considerations in harnessing these technologies to develop essential skills for the 21st century.

Keywords Educational software · Educational games · Artificial Intelligence in education · Fuzzy logic in education · User experience · AI-empowered software · User modeling · Human-AI interaction · e-learning games · 21st-century skills · Human-centred Artificial Intelligence

S. Papadimitriou and M. Virvou, *Artificial Intelligence—Based Games as Novel Holistic Educational Environments to Teach 21st Century Skills*, Intelligent Systems Reference Library 93, https://doi.org/10.1007/978-3-031-77464-5_1

1.1 The Motivation for the Research

The rapid growth of Information and Computing Technologies (ICT) has a critical impact on what has been labelled the 21st century skills that students should acquire to become competent in the new society challenges and careers [113]. Moreover, advancements in Artificial Intelligence (AI) impact decision making [105], by providing sophisticated AI-empowered software tools [104] featuring novel methodologies to engineering smart software systems [131], smart software applications in cyber-physical systems [130] and multimedia systems and services [98]. Such advancements have further highlighted the importance of these skills. Moreover, AI's role in Large Language Models (LLMs) [69, 124–126, 132] and ITSs is reshaping curricula and skill requirements by emphasizing methodologies, offering teaching tools, and stressing the need for soft skills such as critical thinking and ethical behavior, affecting teamwork, collaboration, communication, problem-solving, adaptability, and emotional intelligence, while also addressing technological limitations.

ITSs employ AI to deliver personalized learning experiences, adapting to individual student needs and preferences. These systems enhance education by providing tailored instruction and feedback, improving learning outcomes and engagement. For example, in [123] an ITS is developed which is integrated with social networks to improve mathematics learning, demonstrating the benefits of social interactivity. Cognitive modeling plays an important role, as seen in the work of Virvou and Manos [117], who individualized cognitive models to match students' memory capabilities. Cognitive modeling has been fused with a decision theory in [42]. Affective user modeling, discussed by Katsionis and Virvou [44], allows ITS to respond to students' emotional states, further personalizing the learning experience. Game-based learning systems like VIRGE, developed by Virvou and Katsionis [115], increase motivation by integrating educational content with interactive gameplay. Moreover, error diagnosis systems such as those by Virvou and Maras [118] help identify and correct student misconceptions, providing timely feedback. During the COVID-19 pandemic, Chrysafiadi et al. [23] used fuzzy logic to evaluate e-learning effectiveness, highlighting ITS's adaptability. Additionally, enhancing collaborative learning through ITS, as studied by Olsen et al. [68], demonstrates their potential to foster peer interaction and knowledge sharing. The research performed by Tourtoglou and Virvou [97] applies simulated annealing to optimize learner groups in UML education, enhancing both technical and soft skills. This approach promotes collaboration, communication, and teamwork, improving students' interpersonal skills alongside their technical learning.

In addition to the above ITS features, authoring tools are software applications that enable educators to create educational content and interactive learning experiences. Moundridou and Virvou [66] discuss tools for developing ITSs, which provide personalized and adaptive instruction over the web. These ITS tools allow educators to create interactive, scalable content tailored to individual learner needs. Virvou et al. [127] introduce VR-INTEGATE, an authoring tool for creating Virtual

Reality (VR) educational games. VR-INTEGATE facilitates the design of immersive and active learning environments that enhance engagement and motivation. The INMA tool exemplifies the integration of education and entertainment by using a knowledge-based approach to improve music perception and understanding through advanced educational methods [129]. These tools exemplify how technology can improve education by enabling the development of personalized, interactive learning experiences.

The widespread use of computers and AI is fundamentally changing the nature of professions, with AI, Machine Learning (ML), and Deep Learning (DL) [102] becoming pervasive in both local and global applications [34, 70, 99, 101, 103]. AI already enhances user experiences across various domains, including critical sectors such as healthcare (e.g. [112]), emergency help (e.g. [18]), environment (e.g. [54]), education (e.g. [92]), and tourism (e.g. [51]).

While AI promises to alleviating cognitive burdens by automating tasks and problem-solving, individuals must cultivate additional essential skills to navigate the dynamic demands of modern society. In response to this need, the concept of 21st-century skills has emerged within the field of education, offering a framework to equip learners with the competencies necessary for success in an interconnected world [9, 31, 40] and data mining for many domains [3].

These 21st-century skills encompass both technical proficiencies and a diverse range of competencies and soft skills. Technical skills include digital literacy and proficiency in AI, given the rapid integration of computer and AI-based technologies, which extend human-computer interaction to involve human-AI interaction as well [100, 111]. Desired competencies and soft skills include, but are not limited to, problem-solving [50], collaboration [5], critical thinking [6], adaptability [25], and interdisciplinary and multidisciplinary competence [52]. These skills empower individuals to navigate uncertainty, embrace innovation, and effectively engage with diverse perspectives and ideas.

Moreover, it is worth noting that required skills extend beyond traditional academic knowledge, emphasizing the importance of emotional intelligence [134], social awareness [47], and ethical decision-making [11] in personal and professional contexts. These skills empower children to navigate social interactions, build meaningful relationships, and resolve conflicts peacefully. Additionally, they lay the groundwork for effective communication, collaboration, and ethical behavior, preparing children to succeed not only academically but also in their future careers and personal lives. By integrating these skills into education from an early age, we can equip children with the tools they need to adapt, thrive, and contribute positively to an ever-evolving society.

In addition to their importance in personal and social contexts, ethical decision-making skills are also vital for the ethical and responsible development of technology. As advancements in technology continue to shape our world, ethical considerations surrounding issues such as data privacy, AI, and algorithmic bias become increasingly complex. Children who develop ethical reasoning abilities are better equipped to understand the impact of technology on society and make informed decisions that prioritize ethical principles such as fairness, accountability, and respect for human rights.

By fostering ethical skills alongside technical expertise, we can ensure that the next generation of innovators and creators contribute to the development of technology that benefits humanity while minimizing potential harm.

As such, integrating educational games and responsible AI into educational settings offers a promising avenue for cultivating these competencies among learners. Responsible AI refers to the development and deployment of AI systems in a manner that prioritizes ethical considerations, transparency, fairness, accountability, and the minimization of potential harms to individuals and society [88]. By combining interactive experiences with AI-driven personalization and adaptivity [110], we can create one-to-one tailored environments [45, 59, 78, 139]. Educators can empower students to develop both digital skills and soft skills necessary for succeeding in the 21st century. Additionally, employing prosocial games can further enhance the development of empathy, teamwork, positive social behaviors, and ethical decision-making, fostering a comprehensive skill set that prepares learners to contribute meaningfully to society.

Computer games have become a popular form of entertainment enjoyed by people of all ages. The continuous evolution of technology has significantly contributed to the advancement of computer-based entertainment, allowing users to access a wide variety of computer and mobile games [4, 26]. These games provide immersive environments with captivating graphics, sounds, music, and animation, constantly challenging and stimulating the players' motivation [138]. The appeal of games has led researchers to explore their potential as alternative educational tools. Educational digital games offer an entertaining experience that presents numerous challenges and incentives, effectively maintaining high levels of student engagement while facilitating learning in various subjects [21]. Therefore, these games serve a dual purpose, helping students master the subject matter while enjoying the process [35, 107].

Educational digital games are widely accepted as effective learning tools due to several features [29, 48, 89]. Firstly, they create an interactive and experiential learning environment, engaging students in problem-solving activities with immediate feedback on their performance. Secondly, they provide added motivation, particularly in classroom settings where students may have lost interest, by sparking curiosity and enthusiasm to continue playing and grasp the learning objectives. Additionally, students can make mistakes without fear of criticism, encouraging them to exert additional effort [83]. Lastly, the enjoyable nature of educational games deepens students' understanding of the subject matter and fosters the development of skills such as creativity, critical thinking, collaboration, communication, persistence, confidence, determination, and innovation [17, 96, 106].

Nonetheless, every student possesses distinct personality traits, individual needs, and preferences [41]. Each person interprets and processes information differently during the learning and reasoning process [109]. Similarly, when it comes to games, individuals have specific preferences regarding game categories, features, and devices used [114]. Teachers must recognize each student's unique characteristics in education and adjust their teaching approach accordingly. By doing so, educators can offer tailored guidance to help students overcome any challenges they may face. However, providing personalized instruction can be challenging in a classroom where multiple students are present, often resulting in uniform teaching methods for all students.

AI plays a crucial role in enhancing educational performance, particularly in the context of personalized learning. This is evident in the well-established technology of ITSs, a category of educational applications that leverage intelligent techniques to provide individualized guidance through adaptivity and personalization. They evaluate students' needs and autonomously tailor the content to meet those needs, delivering personalized learning instructions and support.

ITSs interpret students' responses as they interact, assess their performance, infer their level of knowledge, and deliver appropriate educational material [20, 24, 34, 99, 119]. These functions depend on the student modeling component integrated into the ITS architecture [19]. ITSs provide personalized instructions and guidance to address students' deficiencies or weaknesses in comprehending and mastering the study material. Several studies have shown that the guidance offered by ITSs to students produces almost as many positive effects as human guidance [103, 135].

The incorporation of adaptivity in educational games [37, 60, 72, 93] is crucial for customizing educational content to suit the individual needs of the student, just like Intelligent Help Systems (IHS) [81] and ITSs [49]. However, the challenges are more complex for educational games as they need to engage the user as both a student and a player [36, 116]. Ensuring a balance between entertainment and learning is essential; otherwise, the edutainment process will be ineffective. Educational games must focus on enjoyment to be effective in learning. While it is significant to automatically adapt the educational aspect of the game to the student's characteristics, it is equally important to adapt the game's elements automatically. If the emphasis is solely on the educational aspect, the game's appeal decreases significantly, defeating the purpose of using it for learning. Conversely, prioritizing the gaming aspect could lead to neglecting the educational component, resulting in undesired educational outcomes. Therefore, a game should adapt both aspects simultaneously, considering the student's educational needs. This approach can help maintain the student's interest to a high degree and provide an immersive environment to acquire the necessary knowledge.

However, no educational games consider both aspects [38, 80]. Adaptive educational games merely adapt the educational content to the student and do not attempt to modify game elements to enhance the player's interest [13, 55, 58]. Beyond the educational challenges they offer, at the game level, the content remains the same; thus, the time to keep the player engaged is limited. On the other hand, some games dynamically change their elements to keep the player's interest at high levels. However, they do not have educational material, so they are not educational, or the educational material is minimal and lacks adaptivity. Furthermore, in these games, the adaptivity of their elements has limited potential and may relate, for example, only to the game's difficulty level.

An educational game is a multifaceted entity that includes various elements, such as rules, characters, scenarios, and learning objectives. The scenario of an educational game is crucial and must be aligned with the educational content and objectives [2, 15, 39, 137]. Incorporating storytelling into a game allows for adapting the scenario based on the student's attributes, integrating additional game objectives, and increasing engagement. Modifying the game plot enhances the player's motivation and involvement in learning. Expanding the plot can introduce diverse educational goals, helping students overcome challenges while having fun. While there is potential for automatically tailoring the scenario to the player's needs, additional research is still needed. A significant challenge is developing a mechanism that considers the student's cognitive abilities and dynamically adjusts the learning content and game scenario. Such a mechanism could further boost the player's interest and prolong their engagement with the game by offering additional challenges and tasks, thereby improving the learning experience and enabling the player to acquire the expected knowledge.

Adventure games, characterized by their emphasis on storytelling, place players in the central character role and involve activities such as exploration, object gathering, conversations with non-playable characters (NPCs), and puzzle solving. Research has indicated that these games offer numerous educational advantages, aligning with learning objectives rather than purely focusing on action [84, 128, 136]. The narrative nature of adventure games promotes intrinsic motivation in players, encouraging them to explore, investigate, and solve problems [72]. When the game scenario effectively captures the player's empathy for the protagonist, players become emotionally invested in the virtual world, seamlessly engaging in learning activities within the game.

For instance, Papadimitriou and Virvou highlight the effectiveness of an online adventure game for teaching mathematics [71], while the same authors [122] discuss the educational benefits of a customizable "Guess Who" game for language learning [121, 122]. Additionally, Virvou and Papadimitriou demonstrate how an adaptive snake game facilitates simple arithmetic lessons, further emphasizing the potential of adventure games as educational tools [120]. These examples underscore the immersive nature of adventure games and their capacity to engage players in learning activities while enjoying gameplay.

It is essential to recognize that each user is unique, with different characteristics and preferences, not only in educational backgrounds but also in the type of games they enjoy. Tailoring a game to suit players' interests greatly enhances motivation and engagement. The entertainment factor is crucial in encouraging active participation, ultimately benefiting the player educationally. While an adventure game featuring puzzles, riddles, and a compelling storyline may be educational and enjoyable for many students, it might not appeal to others, making the learning process tedious and unengaging. It is, therefore, vital to consider each student's preferences when selecting games for educational purposes. An essential question in educational digital games is determining how much a student can benefit from being given the freedom to choose the games they play or the devices they use.

The bibliography contains various educational games designed to teach the same topic, and it's important for research to address this question. For instance, the integration of fuzzy reasoning into educational games has led to the development of cognitive-based adaptive scenarios, which can provide personalized learning experiences for students. Chrysafiadi, Papadimitriou, and Virvou explored the use of fuzzy states to dynamically adapt the plot of an educational game based on the learner's progress, ensuring tailored experiences that meet individual needs [22]. Similarly, Papadimitriou, Chrysafiadi, and Virvou introduced FuzzEG, a system integrating fuzzy logic for adaptive scenarios in an educational adventure game, showcasing its effectiveness in providing personalized learning experiences [75]. Furthermore, they underlined the importance of creating adaptable content for diverse educational contexts using web services, emphasizing the potential for technological and platform-independent learning materials [74]. These advancements underscore the potential of fuzzy reasoning in enhancing the adaptability and effectiveness of educational games, offering tailored learning experiences that cater to individual student needs.

It is important to use two separate games that provide similar personalized educational content to accurately evaluate the advantages students derive from selecting between two different educational games. This method will allow for a direct comparison of the elements of the two games, isolating their educational content. Both games should be able to access the user's educational profile and use their adaptive features to enhance the player's motivation.

Maintaining consistent data on players' cognitive abilities for student modeling is crucial. Student modeling is not only a technical consideration but is also essential for establishing the student model as it forms the basis for a system to deliver personalized content [94]. For example, an adaptive educational game develops a model to make informed educational decisions and adjusts its content based on the player's actions. However, if a player engages in a different game to learn about the same topic for the first time, that game must also create a student model. This could result in misguided directions and activities that do not align with the player's knowledge, thus impacting their learning.

Additionally, the game's progression may need to be enhanced, and players may need to be provided with extra incentives to maintain their engagement. If the system fails to retain the student's progress when they switch games, the student will have to begin anew and re-engage to reconstruct their profile. This could become a time-consuming and demoralizing endeavor for students, potentially leading them to give up on the game.

The book presents an innovative approach utilizing fuzzy logic to merge two rule-based systems to tailor learning content and game plots dynamically. This method assesses the difficulty of exercises from the student's perspective and enriches the game scenario by introducing new elements and tasks from the player's perspective. Both systems make decisions based on the user's knowledge level, using it to adjust educational objectives and game scenarios. This approach addresses the challenge of maintaining high player motivation in educational digital games by aligning with their educational progress and enhancing learning outcomes. It holds great promise for personalized learning in game-based education.

The book, also, discusses a system for integrating gaming and learning, in which the AI-based personalization process is generalized and abstracted for use across various platforms and game categories, including adventure games and others. This system utilizes web services, a standard technology in modern software development that facilitates data exchange between applications, to make the student model and personalized content accessible in any game. This innovative approach is essential because it allows any game to incorporate personalized educational content. Consequently, players can select their preferred game for educational purposes, accessing the same personalized educational material. Moreover, when multiple games feature the same educational content, the primary factor for selecting a game becomes its features. As a result, each student can choose their preferred game and acquire the desired knowledge level while enjoying an engaging gaming and educational experience.

The book presents research on developing, implementing, and evaluating an educational adventure game designed to teach HTML. This innovative game utilizes a rule-based system to meet students' individualized educational needs by adapting quizzes and altering scenarios based on their requirements. This approach allows students to advance their learning without sacrificing the element of surprise and motivation in the game. The study confirms that integrating adventure game elements, fuzzy logic intelligence, and adaptability in exercises and plots significantly enhances learning in educational games.

In addition, this work describes a VR exploration game that leverages the abstract mechanism to use the same educational content that the adventure game employs. We describe an evaluation that concerns the student's option to learn HTML by playing the game they prefer. The evaluation findings show that incorporating the same educational content in different games positively affects educational outcomes, as choosing the preferred game motivates students to participate in the learning process and increases their engagement in it.

This study involves ongoing work that requires evaluation, building on previous implementations and assessments. This evaluation's objective will be to address additional unresolved research questions. One such question concerns the potential for allowing the tutor to intervene in intelligent mechanisms for assessing the student's knowledge level through an authoring tool. Authoring tools are a distinct area of research due to the tutor's ability to monitor the student's progress and import educational material without requiring a programmer's involvement [32]. These tools are specialized software designed to streamline the tutor's educational work, making updating their course materials more convenient and efficient.

It is important to note that current authoring tools do not allow instructors to directly modify the operations of intelligent systems without the assistance of a programmer. This functionality could provide instructors with valuable insights into how the intelligent methodology works and how it assesses students' cognitive abilities. Instructors could adjust the methodology's parameters to generate more accurate educational insights and decisions. Accessibility of these capabilities to instructors without a programming background is a significant consideration. Additionally, it is crucial to explore whether giving tutors control over these systems' operations improves learning outcomes. To address these research questions, the study introduces a fully developed authoring tool with two customization approaches for evaluating student knowledge. The first approach is intended for AI experts, while the second approach does not require AI knowledge on the instructor's part. In both cases, the involvement of a programmer is not necessary. However, the effectiveness of the developed authoring tool has yet to be evaluated.

The ethical considerations surrounding computer games, especially those incorporating AI, are particularly crucial when the target audience includes children and adolescents. Furthermore, the impact of the new General Data Protection Regulation (GDPR) on adaptive educational games remains an open question. With the GDPR in effect since May 25, 2018, and its focus on users' rights in data collection and processing, adaptive educational games are directly influenced by its provisions. These games require collecting and processing user data to personalize the experience and provide appropriate support. Despite the extensive literature on the GDPR, more research is needed to analyze its specific impact on adaptive educational games. This study aims to elucidate the GDPR's effects on these games and outline design guidelines for GDPR-compliant games. Such analysis will serve as a springboard for further research to develop comprehensive design guidelines, a crucial step in ensuring that designers can create games that adhere to GDPR principles while also ensuring that players are fully informed about the data collected and the benefits they can expect in terms of both learning and entertainment. When students thoroughly understand the advantages of an educational game, they are more likely to engage with it.

Regarding this problem of GDPR protection, Mougiakou, Papadimitriou, and Virvou highlight the importance of transparency in ITSs, emphasizing the need for clear communication regarding the system's operation and data usage, particularly when catering to younger users [63]. Similarly, their work on automated decision-making and personal data protection in ITS emphasizes the necessity of designing guidelines that prioritize user privacy and data security [64]. Moreover, in light of GDPR, Mougiakou, Papadimitriou, and Virvou discuss the implications of synchronous and asynchronous learning methods, stressing the importance of compliance and ethical data handling practices in educational settings [65]. Papadimitriou, Mougiakou, and Virvou further examine the ethical dimensions of smart educational games, focusing on obtaining informed consent under the purview of data protection regulations [76]. These studies underscore the ethical imperative of ensuring transparency, privacy, and consent in AI-driven educational gaming environments, particularly when catering to vulnerable populations like children and adolescents.

In order to address the issues above, this project aimed at developing and implementing an adaptive educational game designed to maintain high levels of student engagement and satisfaction while enhancing their cognitive abilities. The game, Fuzzy EPATHLO, focuses on teaching HTML (HyperText Markup Language), the standard markup language for web browser-displayed documents [75]. Its versatile authoring tool allows it to accommodate content related to any learning object [73, 74]. Incorporating elements commonly found in adventure games, such as collecting objects, interacting with NPCs, and solving puzzles, the game integrates educational content through quizzes at various points in the story. After each quiz, a student modeling tool processes the answers to assess and categorize the student's knowledge level. This knowledge level serves as input for two rule-based systems. The first system adjusts the quiz question difficulty level, while the second expands the scenario by introducing new objectives, puzzles, NPCs, and objects. The student modeling tool and the rule-based system responsible for selecting appropriate educational content are interconnected but operate independently from the game, with the capacity to interact with external applications [43, 77]. The system's features include:

- It utilizes fuzzy logic theory, enabling the effective evaluation of ideas that involve ambiguity and uncertainty, like a student's grasp of knowledge.
- The system utilizes a set of rules to adapt to the complexity of questions based on the student's existing knowledge.
- The game uses a rule-based system to determine when to prolong the gaming experience, which provides students with further motivation to keep playing and take part in extra quizzes.
- Our user-friendly authoring tool, when used in conjunction with the tutor, provides a seamless experience for monitoring student progress and performance. The tutor can effortlessly add, modify, or remove concepts and questions and customize their difficulty level. This tool also allows for adjusting the fuzzy logic parameter using simple linguistic rules, eliminating the need for AI expert intervention. Moreover, AI experts can fine-tune the fundamental parameters of the fuzzy inference system (FIS) using mathematical functions or simple linguistic rules.
- The system uses RESTful web services technology, which allows the educational and game elements to operate independently. External applications can use these web services to retrieve the student model and access the exercises employed by Fuzzy EPATHLO.
- Fuzzy EPATHLO uses NPCs to assign the game's hero quests, which help students develop their soft skills.

1.2 21st Century Technical and Soft Skills

21st-century skills are essential competencies that equip individuals to navigate the complex demands of the modern world. These skills include critical thinking, collaboration, creativity, communication (often referred to as the "4Cs"), as well as digital literacy, adaptability, and problem-solving. Recent research has explored various methods and tools for developing these skills across different educational contexts, highlighting the potential of technology, innovative pedagogies, and experiential learning.

Educational Robotics and Gamification: Educational robotics is a promising approach for developing 21st-century skills in young learners, as it engages them in problem-solving and critical thinking through hands-on activities [7]. Similarly, gamification, as explored through the game DOTA 2, has been shown to enhance critical thinking, collaboration, and communication, offering insights into how game-based learning environments can foster these essential skills [85].

Digital Skills and E-Learning: The COVID-19 pandemic accelerated the shift to digital learning, highlighting both opportunities and challenges in enhancing students' digital skills. Carabregu-Vokshi et al. [14] discuss how e-learning platforms can improve higher education students' digital competencies, a critical aspect of 21st-century skills. Moreover, Rahimi [82] emphasizes the importance of digital competence for EFL teachers, suggesting that integrating digital skills with language teaching enhances both educators' and students' readiness for digital challenges.

Immersive Technologies and Interactive Learning: Using immersive technologies, such as Virtual (VR) and Augmented Reality (AR), can significantly support students' spatial abilities and 21st-century skills in K-12 education [87]. These technologies provide interactive and engaging learning experiences that can enhance creativity, collaboration, and problem-solving. Similarly, the development of interactive e-books for high school students has been shown to facilitate critical thinking and digital literacy, providing an innovative medium for skill acquisition [86].

Agile Methodologies and Game Development: Agile methodologies like Scrum have been effectively integrated into engineering education to develop the 4Cs, with students demonstrating improved collaboration, communication, and problem-solving skills [79]. Additionally, game development projects, such as escape room templates, offer experiential learning opportunities that require students to apply and assess their 21st-century skills [108].

Design and Project-Based Learning: Incorporating design projects into educational curricula can support the development of 21st-century skills while aligning with sustainable development goals [1]. Bhandari and Mathew [8] emphasize the importance of integrating design education in early schooling to foster creativity and innovation among young learners. Furthermore, using platforms like Minecraft Education Edition in college classrooms has been shown to enhance creativity and collaboration, demonstrating the potential of digital tools in higher education [53].

Professional Skills and Digital Economy: The growing demand for professional skills in the data-driven digital economy underscores the importance of 21st-century skills for career readiness. Demchenko et al. [27] highlight the need for educational programs to integrate these skills to prepare students for the evolving workforce. Similarly, Mononen [62] discusses how design sprints can enhance participants' digital skills and agency, promoting innovation and adaptability in professional settings.

Challenges and Future Directions: While there are numerous strategies for developing 21st-century skills, challenges remain, particularly in ensuring equitable access to technology and resources. The ongoing discussion about digital competence and emergent educational issues emphasizes the need for a "new normal" in education that prioritizes these skills [90]. As educators and policymakers continue exploring innovative teaching methods, integrating 21st-century skills into educational systems will be crucial for preparing students to thrive in an increasingly complex and interconnected world.

As part of 21st-century skills, the increasing importance of soft skills across various industries highlights their vital role in enhancing individual and organizational performance. A review of the literature regarding soft skills development and assessment across various contexts highlights the following issues:

- *Innovative Assessment Techniques*: The use of fuzzy logic for assessing soft skills, as discussed by Seixas de Novais et al. [67], highlights the effectiveness of innovative methods in evaluating competencies that are traditionally subjective and challenging to quantify. This approach can provide more nuanced insights into individual skill levels, supporting tailored educational interventions.
- *Integration of Technology and Traditional Methods*: Wiederhold [133] illustrates how modern technology can bridge traditional training methods, enhancing the development of soft skills, particularly in healthcare. This suggests a broader applicability of blended learning approaches to improve interpersonal skills across various sectors.
- *Impact on Well-being and Stress Reduction*: Molek-Winiarska and Kawka [61] demonstrate that soft skills training interventions can significantly reduce work-related stress, particularly in high-pressure environments like the mining industry. This emphasizes the importance of soft skills not only for job performance but also for enhancing employee well-being.
- *Relevance in Leadership and Management*: The study by Ferreira da Silva et al. [91] on using a fuzzy TOPSIS approach to evaluate project managers underscores the critical role soft skills play in leadership positions. Effective communication, decision-making, and interpersonal skills are pivotal in management roles, highlighting the need for their inclusion in leadership training programs.
- *Alignment of Educational Programs with Industry Needs*: Multiple studies, including those by Borges and de Souza [10] and Hanna [33], identify a gap between academic curricula and industry requirements, particularly regarding soft skills in software engineering and testing. This highlights the need for educational institutions to integrate soft skills training into their programs to better prepare students for the workforce.
- *Role of AI and Technology-Enhanced Learning*: The incorporation of AI in assessing and developing soft skills, as discussed by Gafni et al. [30] and Tomasi et al. [95], offers new opportunities for creating objective and personalized training programs. AI can enhance the precision of skill assessments and provide adaptive learning pathways tailored to individual needs.
- *Interactive and Engaging Learning Approaches*: The use of role-play, co-creation, and gamification, as highlighted by Torres de Macedo et al. [56] and Cerezo et al. [16], supports the development of soft skills through engaging and interactive methods. These approaches can make learning more enjoyable and effective, fostering better skill retention and application.
- *Global and Cross-Disciplinary Applicability*: The research works, such as those by Mahadevan et al. [57] on healthcare professionals and Calhau et al. [12] on enterprise architecture, demonstrate the universal importance of soft skills across

different sectors and cultures. The adaptability of soft skills training methods to various contexts underscores their broad relevance.

- *Soft Skills in Virtual and Remote Work Environments*: Kee et al. [46] highlight the importance of developing soft skills in virtual organizational cultures, emphasizing the need for specific metrics to assess these skills in remote work settings. As remote and hybrid work models become more prevalent, fostering soft skills in these environments is crucial for maintaining productivity and collaboration.
- *Empowering Underrepresented Groups*: Desidério et al. [28] focus on the development of soft skills in women involved in digital projects, pointing to the role these skills play in empowering underrepresented groups in technology fields. Training programs emphasizing soft skills can contribute to greater inclusivity and diversity in tech industries.

Overall, these research efforts highlight the critical importance of soft skills across various domains and propose innovative approaches to their assessment and development. Integrating technology, interactive learning methods, and aligning with industry needs are key factors in ensuring individuals are well-equipped with the skills necessary to thrive in dynamic and diverse work environments.

1.3 Related Fields and Open Research Questions

This study focuses on the user's central role within the educational process, functioning as both a student and a player. It explores various research questions related to enhancing satisfaction and enjoyment through gaming while achieving valuable learning outcomes. The following are the inquiries that stem from this study:

How can we develop educational games that address 21st century skills?
The open problem centers around the development of educational games that effectively cultivate and assess 21st-century skills. These skills encompass a broad spectrum of competencies, including technical proficiencies such as digital literacy and familiarity with AI technologies, as well as soft skills like problem-solving, collaboration, critical thinking, adaptability, and interdisciplinary competence. The challenge lies in creating educational games that not only teach academic content but also foster the development of these crucial 21st-century skills. This entails designing game mechanics, scenarios, and assessment methods that authentically simulate real-world challenges and encourage players to apply and refine these skills within the game environment. Additionally, ensuring that educational games adequately assess and provide feedback on 21st-century skills presents a significant challenge, as traditional assessment methods may not fully capture the complexity and nuance of these skills.

Can an educational game maintain a high level of student engagement until they achieve the desired level of knowledge?
Creating educational games that dynamically adjust content to maintain high levels of player engagement is a significant challenge in digital educational games. Despite the widespread popularity of educational games and the recognition of the importance of incorporating adaptive content, many existing games lack sufficient adaptive capabilities. In most cases, adaptivity is limited to the educational aspects of the game elements alone, without establishing a meaningful connection between the student's educational needs and the game elements. Therefore, there is a need for an approach that can identify the student's learning characteristics and customize the game's content to promote active participation in the learning process. Establishing a link between the student's knowledge level and the game's progression is important. This research introduces an innovative mechanism that identifies learning obstacles for students and enhances the game scenario with additional elements [75]. The utility of this mechanism lies in its ability to encourage continued play and facilitate the enhancement of the user's knowledge through an enjoyable process.

How can the game personalize its content to align with the individual student's current level of knowledge?
For a game to be successful, it must effectively motivate the player. When designing an educational game, it is essential to consider the user's dual roles as a student and a player. This study focuses on customizing the content for each role, considering the student's level of knowledge. The approach involves two rule-based strategies: the first assesses the difficulty of the exercises, while the second introduces additional game objectives to enhance the game environment. The primary goal of these adjustments is to maintain the player's motivation by providing exercises tailored to their learning level until they attain the desired knowledge [75].

Would the student's game choice improve the learning results?
It is crucial to consider each student's individual preferences and needs when it comes to gaming and learning. What may be engaging for one student could be dull for another, which can have a negative impact on their learning experience. Researchers recognize the diverse preferences of gamers and the importance of understanding each student's unique style. It would be valuable to conduct a study to evaluate how allowing students to choose their preferred educational games affects their performance in learning. Current research explores the

potential educational benefits for students when they can select their preferred game for learning purposes [77].

How can we ensure that the student model is accessible in every application?

When comparing two educational games, both must provide personalized educational content to the student. This ensures a fair comparison, allowing us to focus on their features and mechanics rather than just their educational aspects. If each game collects and maintains different data about the student, it may offer guidance, advice, and exercises unsuitable for the student's cognitive level. This could result in a meaningless learning process and hinder the student's progress. To address this issue, this study takes a unique approach by leveraging web services technology, a new method in educational games. Although web services are typically used in modern software development, they have not been previously utilized to deliver personalized educational content in multiple educational games or applications. Furthermore, each game can tailor its specific features to the player's educational profile, similar to how the main game in this study adjusts its scenario. The potential for leveraging unique attributes in an educational game to enhance player motivation is an essential discovery of this study.

The research includes evaluations to address the previously mentioned research queries. It also covers existing work that has been completed and is ready for assessment to offer accurate responses to the research above questions.

Can an authoring tool for creating content help a tutor modify the AI methods used in an Intelligent Tutoring System?

The question's significance lies in that authoring tools often have valuable but limited functionalities, offering options for importing educational material and tracking student progress. In contrast, the inner workings of student modeling and adaptivity algorithms are typically hidden or too complicated for non-specialists in AI to understand. The authoring tool examined in this study uses simple linguistic rules to configure intelligent system techniques for processing student data, drawing relevant conclusions, and making decisions. It also allows AI experts to adjust the AI parameters using mathematical functions without programmer intervention. Therefore, the authoring tool enables instructors to modify the AI mechanisms and assess the student's knowledge level, regardless of their AI expertise, eliminating the need for programmer involvement. This capability reduces the requirement for programmer intervention and facilitates continuous adjustment and experimentation with AI mechanisms, potentially leading to more reliable educational assessments and, consequently, more appropriate educational decision-making.

How does the GDPR impact the usage of intelligent educational games?
The GDPR, formally known as the EU General Data Protection Regulation (2016/679), is the most significant update to data protection laws in the past two decades. It has been in effect since May 25, 2018, and applies to all Member States without requiring additional state-level legislation. The GDPR impacts privacy issues in e-learning but also in educational games utilizing data collection and processing to offer personalized experiences and improve learning outcomes. However, there is a need for further in-depth research to comprehend the specific implications of the regulation of adaptive educational games. This research is crucial, as it leads to developing guidelines for game designers to ensure that intelligent educational games (IEG) fully adhere to the GDPR.

Implementing these guidelines will emphasize the significance of adaptive educational games and bring attention to the potential advantages of GDPR compliance for game developers. This project utilizes its adventure game as a prototype to demonstrate various aspects and features, such as the collection and utilization of student data, that may conflict with the GDPR. It then recommends that game developers design an intelligent educational game that aligns with GDPR regulations. These recommendations, combined with the necessity for further research in establishing a comprehensive set of guidelines, can instill optimism and hope regarding the potential benefits of GDPR compliance.

References

1. Abidin, A.F.B., Ahmad, A.: The element of integrated design project towards fulfilling the 21st century skills set and sustainable development goal (sdg). In: 2023 IEEE 12th International Conference on Engineering Education (ICEED), pp. 28–31. IEEE (2023)
2. Aguilar, J., Altamiranda, J., Díaz, F., de Mesa, J.G., Pinto, Á.: Adaptive plot system for serious emerging games based on the ant colony optimization algorithm. In: 2019 XLV Latin American Computing Conference (CLEI), pp. 1–10. IEEE (2019)
3. Alogogianni, E., Virvou, M.: Data mining for targeted inspections against undeclared work by applying the crisp-dm methodology. In: 2021 12th International Conference on Information, Intelligence, Systems & Applications (IISA), pp. 1–8. IEEE (2021)

4. Anastasiadis, T., Lampropoulos, G., Siakas, K.: Digital game-based learning and serious games in education. Int. J. Adv. Sci. Res. Eng. **4**(12), 139–144 (2018)
5. Andrews-Todd, J., Jiang, Y., Steinberg, J., Pugh, S.L., D'Mello, S.K.: Investigating collaborative problem solving skills and outcomes across computer-based tasks. Comput. Educ. **207**, 104928 (2023)
6. Ay, K., Dağhan, G.: The effect of the flipped learning approach designed with community of inquiry model to the development of students' critical thinking strategies and social, teaching and cognitive presences. Education and Information Technologies pp 1–33 (2023)
7. Bano, S., Atif, K., Mehdi, S.A.: Systematic review: Potential effectiveness of educational robotics for 21st century skills development in young learners. In: Education and Information Technologies, pp. 1–19 (2023)
8. Bhandari, U., Mathew, D.J.: Development of design education curriculum for grade 5 indian cbse schools to facilitate 21st century skills. In: 2023 IEEE 12th International Conference on Engineering Education (ICEED), pp. 165–170. IEEE (2023)
9. Bikanga Ada, M., Foster, M.E.: Enhancing postgraduate students' technical skills: perceptions of modified team-based learning in a six-week multi-subject bootcamp-style cs course. Comput. Sci. Educ. **33**(2), 186–210 (2023)
10. Borges, G.G., de Souza, R.C.G.: Skills development for software engineers: systematic literature review. Inf. Softw. Technol. 107395 (2024)
11. Burton, E., Dueber, D., Goldsmith, J., Goldstein, B., Sampson, S., Toland, M.D.: Assessment of cs students' ethical reasoning skills. In: Proceedings of the 51st ACM Technical Symposium on Computer Science Education, pp. 1406–1406 (2020)
12. Calhau, R.F., Almeida, J.P.A., Kokkula, S., Guizzardi, G.: Modeling competences in enterprise architecture: from knowledge, skills, and attitudes to organizational capabilities. Softw. Syst. Model. 1–40 (2024)
13. Capuano, N., King, R.: Adaptive serious games for emergency evacuation training. In: 2015 International Conference on Intelligent Networking and Collaborative Systems, pp. 308–313. IEEE (2015)
14. Carabregu-Vokshi, M., Ogruk-Maz, G., Yildirim, S., Dedaj, B., Zeqiri, A.: 21st century digital skills of higher education students during covid-19-is it possible to enhance digital skills of higher education students through e-learning? Educ. Inf. Technol. **29**(1), 103–137 (2024)
15. Carro, R.M., Breda, A.M., Castillo, G., Bajuelos, A.L.: A methodology for developing adaptive educational-game environments. In: Adaptive Hypermedia and Adaptive Web-Based Systems: Second International Conference, AH 2002 Málaga, Spain, 29–31 May 2002 Proceedings, vol. 2, pp. 90–99. Springer (2002)
16. Cerezo, E., González-González, C.S., Bonillo, C.: Empowering soft skills in children with adhd through the co-creation of tangible tabletop games. Univers. Access Inf. Soc. **23**(1), 3–21 (2024)
17. Chaudy, Y., Connolly, T.: Specification and evaluation of an assessment engine for educational games: integrating learning analytics and providing an assessment authoring tool. Entertain. Comput. **30**, 100294 (2019)
18. Chrysafiadi, K., Tsichrintzi, E.A.: An intelligent fuzzy-based emergency alert generation for persons with episodic memory decline problems. Intell. Dec. Technol. **15**(4), 681–690 (2021)
19. Chrysafiadi, K., Virvou, M.: Student modeling approaches: a literature review for the last decade. Expert Syst. Appl. **40**(11), 4715–4729 (2013)
20. Chrysafiadi, K., Virvou, M.: Fuzzy logic for adaptive instruction in an e-learning environment for computer programming. IEEE Trans. Fuzzy Syst. **23**(1), 164–177 (2014)
21. Chrysafiadi, K., Papadimitriou, S., Virvou, M.: Which is better for learning: a web-based educational application or an educational game? In: 2019 International Symposium on Performance Evaluation of Computer and Telecommunication Systems (SPECTS), IEEE, pp 1–6 (2019)
22. Chrysafiadi, K., Papadimitriou, S., Virvou, M.: Fuzzy states for dynamic adaptation of the plot of an educational game in relation to the learner's progress. In: 2020 11th International Conference on Information. Intelligence, Systems and Applications (IISA), pp. 1–7. IEEE (2020)

23. Chrysafiadi, K., Virvou, M., Tsihrintzis, G.A.: A fuzzy-based evaluation of e-learning acceptance and effectiveness by computer science students in greece in the period of covid-19. Electronics **12**(2), 428 (2023)
24. Chrysafiadi, K., Virvou, M., Tsihrintzis, G.A., Hatzilygeroudis, I.: Evaluating the user's experience, adaptivity and learning outcomes of a fuzzy-based intelligent tutoring system for computer programming for academic students in greece. Educ. Inf. Technol. **28**(6), 6453–6483 (2023)
25. Daly, T., Hancock, A., Phillips, J., Mainz, M., Perry, B.: Development of the assessment toolkit for leader adaptability skills (atlas). In: International Conference on Human-Computer Interaction, pp. 271–285. Springer (2022)
26. Dele-Ajayi, O., Strachan, R., Sanderson, J., Pickard, A.: A modified tam for predicting acceptance of digital educational games by teachers. In: 2017 IEEE Global Engineering Education Conference (EDUCON), pp. 961–968. IEEE (2017)
27. Demchenko, Y., Degeler, V., Opresu, A., Brewer, S.: Professional and 21st century skills for data driven digital economy. In: 2023 IEEE Global Engineering Education Conference (EDUCON), pp. 1–8. IEEE (2023)
28. Desidério, S.B., Lelis, M.R.L., Rodrigues, M.E., Marques, A.B.: How ready for hci? a qualitative analysis of the practice of soft skills related to hci by women involved in the digital girls program partners projects. J. Interact. Syst. **15**(1), 504–516 (2024)
29. Falcão, T.P.: e Peres FMdA, de Morais DCS, da Silva Oliveira G,: Participatory methodologies to promote student engagement in the development of educational digital games. Comput. Educ. **116**, 161–175 (2018)
30. Gafni, R., Aviv, I., Kantsepolsky, B., Sherman, S., Rika, H., Itzkovich, Y., Barger, A.: Objectivity by design: The impact of ai-driven approach on employees' soft skills evaluation. Inf. Softw. Technol. **170**, 107430 (2024)
31. Galuret, S., Vallée, N., Tronchot, A., Thomazeau, H., Jannin, P., Huaulmé, A.: Gaze behavior is related to objective technical skills assessment during virtual reality simulator-based surgical training: a proof of concept. Int. J. Comput. Assist. Radiol. Surg. 1–9 (2023)
32. Grammatikopoulou, A., Laraba, S., Sahbenderoglu, O., Dimitropoulos, K., Douka, S., Grammalidis, N.: An adaptive framework for the creation of exergames for intangible cultural heritage (ich) education. J. Comput. Educ. **6**(3), 417–450 (2019)
33. Hanna, S.: Analysis of the gap between software testing courses at universities and the needed skills by industry. Int. J. Learn. Technol. **19**(2), 203–242 (2024)
34. Hatzilygeroudis, I., Tsihrintzis, G., Virvou, M., Perikos, I.: Special issue on information, intelligence, systems and applications. Neural Comput. Appl. **35**(1), 1–2 (2023)
35. Hawlitschek, A., Joeckel, S.: Increasing the effectiveness of digital educational games: the effects of a learning instruction on students' learning, motivation and cognitive load. Comput, Hum. Behav. **72**, 79–86 (2017)
36. Hendrix, M., Bellamy-Wood, T., McKay, S., Bloom, V., Dunwell, I.: Implementing adaptive game difficulty balancing in serious games. IEEE Transactions on Games **11**(4), 320–327 (2018)
37. Hocine, N., Gouaïch, A., Cerri, S.A., Mottet, D., Froger, J., Laffont, I.: Adaptation in serious games for upper-limb rehabilitation: an approach to improve training outcomes. User Model. User Adapt. Interact. **25**, 65–98 (2015)
38. Hooshyar, D., Malva, L., Yang, Y., Pedaste, M., Wang, M., Lim, H.: An adaptive educational computer game: effects on students' knowledge and learning attitude in computational thinking. Comput. Hum. Behav. **114**, 106575 (2021)
39. Hussaan, A.M., Sehaba, K., Mille, A.: Tailoring serious games with adaptive pedagogical scenarios: a serious game for persons with cognitive disabilities. In: 2011 IEEE 11th International Conference on Advanced Learning Technologies, pp. 486–490. IEEE (2011)
40. Jou, M., Chen, P.C., Wang, J.: The developmental characteristics of computational thinking and its relationship with technical skills: taking the department of engineering as an example. Interact. Learn. Environ. **31**(6), 3380–3395 (2023)

41. Jumaa, Y.M., Moussa, S.M., Khalifa, M.E.: The main aspects of adaptive educational games for normal and disabled/disordered learners: a comprehensive study. In: 2017 Eighth International Conference on Intelligent Computing and Information Systems (ICICIS), pp. 348–355. IEEE (2017)
42. Kabassi, K., Virvou, M.: Combination of a cognitive theory with the multi-attribute utility theory. In: International Conference on Knowledge-Based and Intelligent Information and Engineering Systems, pp. 944–950. Springer (2003)
43. Kamitsios, M., Papadimitriou, S., Chrysafiadi, K., Virvou, M.: Incorporating quizzes into a 3d educational game calling web services. In: EdMedia+ Innovate Learning, Association for the Advancement of Computing in Education (AACE), pp. 1170–1179 (2018)
44. Katsionis, G., Virvou, M.: A cognitive theory for affective user modelling in a virtual reality educational game. In: 2004 IEEE International Conference on Systems, Man and Cybernetics (IEEE Cat. No. 04CH37583), vol. 2, pp. 1209–1213. IEEE (2004)
45. Katsionis, G., Virvou, M.: Personalised e-learning through an educational virtual reality game using web services. Multimed. Tools Appl. **39**, 47–71 (2008)
46. Kee, K., Jankowski, H., Knepper, R., Snapp-Childs, W., Brady, R.: Fostering a thriving virtual organizational culture: the development of metrics for ten soft skills in the case of xsede. SN Comput. Sci. **5**(5), 615 (2024)
47. Khine, M.S., Liu, Y., Afari, E.: Mediating effects of social and emotional skills on cognitive ability and academic achievement: Suzhou study. Int. J. Knowl. Learn. **17**(1), 16–27 (2024)
48. Kickmeier-Rust, M.D., Göbel, S., Albert, D.: 80days: Melding adaptive educational technology and adaptive and interactive storytelling in digital educational games. In: Proceedings of the First International Workshop on Story-Telling and Educational Games (STEG'08), Citeseer, vol. 8 (2008)
49. Kickmeier-Rust, M.D., Mattheiss, E., Steiner, C., Albert, D.: A psycho-pedagogical framework for multi-adaptive educational games. In: Developments in Current Game-Based Learning Design and Deployment, pp. 103–117. IGI Global (2013)
50. Kim, Y.R., Yang, J., Lee, Y., Earwood, B.: Assessing cybersecurity problem-solving skills and creativity of engineering students through model-eliciting activities using an analytic rubric. IEEE Access (2024)
51. Kontogianni, A., Alepis, E., Virvou, M., Patsakis, C.: Smart Tourism–The Impact of Artificial Intelligence and Blockchain. Springer Nature Switzerland (2024)
52. Koppikar, U., Kandakatla, R., Mallibhat, K., Joshi, G.: Exploration of skills required by engineering faculty to mentor freshmen undergraduate students for interdisciplinary design projects. IEEE Trans, Educ (2023)
53. Lee, S., Jang, W.W., Rollins, M.: Using minecraft education edition to enhance 21st century skills in the college classroom: a mixed methods study (2024)
54. Liagkouras, K., Metaxiotis, K., Tsihrintzis, G.: Incorporating environmental and social considerations into the portfolio optimization process. Ann. Oper. Res. 1–26 (2020)
55. Lindberg, R.S., Laine, T.H.: Formative evaluation of an adaptive game for engaging learners of programming concepts in k-12. Int. J. Ser. Games **5**(2), 3–24 (2018)
56. de Macedo, G.T., Fontão, A., Gadelha, B.: Building soft skills through a role-play based approach for requirements engineering remote education. J. Braz. Comput. Soc. **30**(1), 1–16 (2024)
57. Mahadevan, A., Rivera, R., Najhawan, M., Saadat, S., Strehlow, M., Rao, G.R., Youm, J.: Assessing the efficacy of a novel massive open online soft skills course for south asian healthcare professionals. J. Med. Syst. **48**(1), 32 (2024)
58. Mees, M., Jay, T., Habgood, J.: Designing an Adaptive Learner Model for a Mathematics Game. Academic Conferences and Publishing International Limited (2018)
59. Micarelli, A., Gasparetti, F., Sciarrone, F., Gauch, S.: Personalized search on the world wide web. In: Brusilovsky P, Kobsa A, Nejdl W (eds.) The Adaptive Web (2007)
60. Miljanovic, M.A., Bradbury, J.S.: Making serious programming games adaptive. In: Joint International Conference on Serious Games, pp. 253–259. Springer (2018)

61. Molek-Winiarska, D., Kawka, T.: Reducing work-related stress through soft-skills training intervention in the mining industry. Hum. Factors **66**(5), 1633–1649 (2024)
62. Mononen, A.: Perceptions of 21st-century digital skills and agency among design sprint participants in Laurea UAS. Finland. J, Data Mining Digital Humanit (2024)
63. Mougiakou, E., Papadimitriou, S., Virvou, M.: Intelligent tutoring systems and transparency: the case of children and adolescents. In: 2018 9th International Conference on Information, Intelligence, Systems and Applications (IISA), pp. 1–8. IEEE (2018)
64. Mougiakou, E., Papadimitriou, S., Virvou, M.: Automated decision making and personal data protection in intelligent tutoring systems: design guidelines. In: Knowledge-Based Software Engineering: 2018: Proceedings of the 12th Joint Conference on Knowledge-Based Software Engineering (JCKBSE 2018) Corfu, Greece, vol. 12, pp. 231–241. Springer (2019)
65. Mougiakou, E., Papadimitriou, S., Virvou, M.: Synchronous and asynchronous learning methods under the light of general data protection regulation. In: 2020 11th International Conference on Information, Intelligence, Systems and Applications (IISA), pp. 1–7. IEEE (2020)
66. Moundridou, M., Virvou, M.: Authoring intelligent tutoring systems over the world wide web. In: Proceedings First International IEEE Symposium Intelligent Systems, vol. 1, pp. 160–165. IEEE (2002)
67. Novais, A.S.D., Matelli, J.A., Silva, M.B.: Fuzzy soft skills assessment through active learning sessions. Int. J. Artif. Intell. Educ. **34**(2), 416–451 (2024)
68. Olsen, J.K., Belenky, D.M., Aleven, V., Rummel, N.: Intelligent tutoring systems for collaborative learning: enhancements to authoring tools. In: Artificial Intelligence in Education: 16th International Conference, AIED 2013, Memphis, TN, USA, 9–13 July 2013. Proceedings, vol. 16, pp. 900–903. Springer (2013)
69. Panagoulias, D.P., Palamidas, F.A., Virvou, M., Tsihrintzis, G.A.: Evaluating the potential of llms and chatgpt on medical diagnosis and treatment. In: 2023 14th International Conference on Information, Intelligence, Systems & Applications (IISA), pp. 1–9. IEEE (2023)
70. Panagoulias, D.P., Virvou, M., Tsihrintzis, G.A.: A novel framework for artificial intelligence explainability via the technology acceptance model and rapid estimate of adult literacy in medicine using machine learning. Expert Syst. Appl. **248**, 123375 (2024)
71. Papadimitriou, S., Virvou, M.: An online adventure game for teaching math. In: 2016 7th International Conference on Information, Intelligence, Systems & Applications (IISA), pp. 1–5. IEEE (2016)
72. Papadimitriou, S., Virvou, M.: Adaptivity in scenarios in an educational adventure game. In: 2017 8th International Conference on Information, Intelligence, Systems & Applications (IISA), pp. 1–6. IEEE (2017)
73. Papadimitriou, S., Chrysafiadi, K., Kamitsios, M., Virvou, M.: Fsp creator: a novel web service api creator of fuzzy students progress profile. In: 2018 9th International Conference on Information, Intelligence, Systems and Applications (IISA), pp. 1–7. IEEE (2018)
74. Papadimitriou, S., Chrysafiadi, K., Virvou, M.: Authoring technological and platform independent learning material and student's progress profile using web services. In: Knowledge-Based Software Engineering: 2018: Proceedings of the 12th Joint Conference on Knowledge-Based Software Engineering (JCKBSE 2018) Corfu, Greece, vol. 12, pp. 242–251. Springer (2019)
75. Papadimitriou, S., Chrysafiadi, K., Virvou, M.: Fuzzeg: Fuzzy logic for adaptive scenarios in an educational adventure game. Multimed. Tools Appl. **78**(22), 32023–32053 (2019). https://doi.org/10.1007/s11042-019-07955-w
76. Papadimitriou, S., Mougiakou, E., Virvou, M.: Smart educational games and consent under the scope of general data protection regulation. In: 2019 10th International Conference on Information, Intelligence, Systems and Applications (IISA), pp. 1–8. IEEE (2019)
77. Papadimitriou, S., Kamitsios, M., Chrysafiadi, K., Virvou, M.: Learn-and-play personalised reasoning from point-and-click to virtual reality mobile educational games. Intell. Dec. Technol. **15**(2), 321–332 (2021)
78. Papadimitriou, S., Chrysafiadi, K., Virvou, M.: Adaptive quizzes using fuzzy genetic algorithm. In: 2023 14th International Conference on Information, Intelligence, Systems & Applications (IISA), pp. 1–8. IEEE (2023)

79. Paraskeva, F., Choustoulakis, E., Karampa, V., Lepenitsa, S., Nikoloudakis, D.: Investigating the impact of agile scrum processes on student engineers' 21st century (4cs) skills. In: International Workshop on Learning Technology for Education Challenges, pp. 96–109. Springer (2024)
80. Park, K., Mott, B.W., Min, W., Boyer, K.E., Wiebe, E.N., Lester, J.C.: Generating educational game levels with multistep deep convolutional generative adversarial networks. In: 2019 IEEE Conference on Games (CoG), pp. 1–8. IEEE (2019)
81. Peirce, N., Conlan, O., Wade, V.: Adaptive educational games: providing non-invasive personalised learning experiences. In: 2008 Second IEEE International Conference on Digital Game and Intelligent Toy Enhanced Learning, pp. 28–35. IEEE (2008)
82. Rahimi, A.R.: Beyond digital competence and language teaching skills: the bi-level factors associated with efl teachers' 21st-century digital competence to cultivate 21st-century digital skills. Educ. Inf. Technol. **29**(8), 9061–9089 (2024)
83. Rahimi, S., Shute, V., Kuba, R., Dai, C.P., Yang, X., Smith, G., Fernández, C.A.: The use and effects of incentive systems on learning and performance in educational games. Comput. Educ. **165**, 104135 (2021)
84. Riemer, V., Schrader, C.: Learning with quizzes, simulations, and adventures: students' attitudes, perceptions and intentions to learn with different types of serious games. Comput. Educ. **88**, 160–168 (2015)
85. Samala, A.D., Bojic, L., Vergara-Rodríguez, D., Klimova, B., Ranuharja, F.: Exploring the impact of gamification on 21st-century skills: Insights from dota 2. Int. J. Interact. Mob. Technol. **17**(18) (2023)
86. dos Santos Silva, D.E., Sobrinho, M.C., Valentim, N.M.C.: Development of 21st-century skills and competencies in high school students through the interactive e-books creation. Revista Brasileira de Informática na Educação **31**, 971–1004 (2023)
87. Serrano-Ausejo, E., Mårell-Olsson, E.: Opportunities and challenges of using immersive technologies to support students' spatial ability and 21st-century skills in k-12 education. Educ. Inf. Technol. **29**(5), 5571–5597 (2024)
88. Shneiderman, B.: Responsible ai: Bridging from ethics to practice. Commun. ACM **64**(8), 32–35 (2021)
89. Shute, V., Rahimi, S., Lu, X.: Supporting learning in educational games: promises and challenges. In: Learning in a Digital World: Perspective on Interactive Technologies for Formal and Informal Education, pp. 59–81 (2019)
90. Siddiq, F., Olofsson, A.D., Lindberg, J.O., Tomczyk, L.: What will be the new normal? digital competence and 21st-century skills: critical and emergent issues in education. Educ. Inf. Technol. **29**(6), 7697–7705 (2024)
91. Silva, L.F.D., Oliveira, P.S.G.D., Grander, G., Penha, R., Bizarrias, F.S.: Soft skills fuzzy topsis ranked multi-criteria to select project manager. Int. J. Inf. Decis. Sci. **16**(1), 19–45 (2024)
92. Sotiropoulos, D.N., Alepis, E., Kabassi, K., Virvou, M.K., Tsihrintzis, G.A., Sakkopoulos, E.: Artificial immune system-based learning style stereotypes. Int. J. Artif. Intell. Tools **28**(04), 1940008 (2019)
93. Supriana, I., Agustin, R.D., Bakar, M.A., Zin, N.A.M.: Serious games for effective learning. In: 2017 6th International Conference on Electrical Engineering and Informatics (ICEEI), pp 1–6. IEEE (2017)
94. Thomson, D.: Towards a negotiable student model for constraint-based itss (2008)
95. Tomasi, S., Liu, J., Sun, L.: Ai helps, doesn't it: The impact of ai on soft skills (2024)
96. Torrente, J., Moreno-Ger, P., Fernandez-Manjon, B.: Learning models for the integration of adaptive educational games in virtual learning environments. In: Technologies for E-Learning and Digital Entertainment: Third International Conference, Edutainment 2008 Nanjing, China, 25–27 June 2008 Proceedings, vol. 3, pp. 463–474. Springer (2008)
97. Tourtoglou, K., Virvou, M.: Simulated annealing in finding optimum groups of learners of uml. In: Intelligent Interactive Multimedia Systems and Services, pp. 147–156. Springer (2010)

98. Tsihrintzis, G.A., Virvou, M., Jain, L.C.: Multimedia Services in Intelligent Environments: Recommendation Services. Springer (2013)
99. Tsihrintzis, G.A., Virvou, M., Phillips-Wren, G.: Surveys in artificial intelligence-based technologies. Intell. Decis. Technol. **13**(4), 393–394 (2019)
100. Tsihrintzis, G.A., Virvou, M., Hatzilygeroudis, I.: Special collection of extended selected papers on "novel research results presented in the 12th international conference on information, intelligence, systems and applications (iisa2021), 12–14 july 2021, chania, crete, greece. https://easyconferences.eu/iisa2021/". Intell. Decis. Technol. **15**(4):641–643 (2021)
101. Tsihrintzis, G.A., Phillips-Wren, G., Jain, L.C., Watada, J.: Intelligent decision technologies: an international journal (idt) enters its 16th year: a short note from the editors. Intell. Decis. Technol. **16**(1), 1–4 (2022)
102. Tsihrintzis, G.A., Virvou, M., Jain, L.C.: Advances in Machine Learning/Deep Learning-based Technologies. Springer (2022)
103. Tsihrintzis, G.A., Virvou, M., Saruwatari, T.: Special Collection of Extended Selected Papers on "Novel Research Results Presented at the 14th International Joint Conference on Knowledge-Based Software Engineering (jckbse2022), 22–24 Aug 2022, Larnaca, Cyprus https:://easyconferences.eu/jckbse2022/" (2022)
104. Tsihrintzis, G.A., Virvou, M., Hatzilygeroudis, I.: Special Issue on Selected Papers from the 33rd Annual IEEE International Conference on Tools with Artificial Intelligence (ictai-2021) (2023)
105. Tsihrintzis, G.A., Virvou, M., Doukas, H., Jain, L.C.: Introduction to advances in artificial intelligence-empowered decision support systems. In: Advances in Artificial Intelligence-Empowered Decision Support Systems: Papers in Honour of Professor John Psarras, pp. 1–11. Springer (2024)
106. Vanbecelaere, S., Van den Berghe, K., Cornillie, F., Sasanguie, D., Reynvoet, B., Depaepe, F.: The effectiveness of adaptive versus non-adaptive learning with digital educational games. J. Comput. Assist. Learn. **36**(4), 502–513 (2020)
107. Vanbecelaere, S., Van den Berghe, K., Cornillie, F., Sasanguie, D., Reynvoet, B., Depaepe, F.: The effects of two digital educational games on cognitive and non-cognitive math and reading outcomes. Comput. Educ. **143**, 103680 (2020)
108. Vigneau, C., Mader, S., Rolland, C., Buendia, A.: Acquiring and assessing 21st century skills through game development: a case study with an escape game template. In: International Conference on Entertainment Computing, pp. 150–158. Springer (2023)
109. Virvou, M.: Automatic reasoning and help about human errors in using an operating system. Interact. Comput. **11**(5), 545–573 (1999)
110. Virvou, M.: Intelligence and adaptivity in human-computer interaction concerning biotechnological software users. In: Scattering And Biomedical Engineering: Modeling and Applications, pp. 448–458. World Scientific (2002)
111. Virvou, M.: The emerging era of human-ai interaction: Keynote address. In: 2022 13th International Conference on Information, Intelligence, Systems & Applications (IISA), pp. 1–10. IEEE (2022)
112. Virvou, M.: Artificial intelligence and user experience in reciprocity: contributions and state of the art. Intell. Decis. Technol. **17**(Preprint), 1–53 (2023)
113. Virvou, M.: The emerging role of artificial intelligence in teaching 21st century skills in software engineering. In: Keynote 1 in Proceedings of the 35th IEEE Conference on Software Engineering Education and Training, CSEE&T 2023 (2023)
114. Virvou, M., Kabassi, K.: Reasoning about users' actions in a graphical user interface. Hum. Comput. Interact. **17**(4), 369–398 (2002)
115. Virvou, M., Katsionis, G.: Virge: tutoring english over the web through a game. In: Proceedings 3rd IEEE International Conference on Advanced Technologies, p. 469. IEEE (2003)
116. Virvou, M., Katsionis, G.: On the usability and likeability of virtual reality games for education: The case of vr-engage. Comput. Educ. **50**(1), 154–178 (2008)
117. Virvou, M., Manos, K.: Individualizing a cognitive model of students' memory in intelligent tutoring systems. In: Knowledge-Based Intelligent Information and Engineering Systems: 7th

INternational Conference, KES 2003, Oxford, UK, Sept 2003. Proceedings, Part I, vol. 7, pp. 893–897. Springer (2003)
118. Virvou, M., Maras, D.: Error diagnosis in an english tutor. In: Proceedings of the HCI International'99 (The 8th International Conference on Human-Computer Interaction) on Human-Computer Interaction: Communication, Cooperation, and Application Design, vol. 2, pp. 657–661 (1999)
119. Virvou, M., Moundridou, M.: A web-based authoring tool for algebra-related intelligent tutoring systems. J. Educ. Technol. Soc. **3**(2), 61–70 (2000)
120. Virvou, M., Papadimitriou, S.: Simple arithmetic lessons through an adaptive snake game. In: 2013 International Conference on Computer, Information and Telecommunication Systems (CITS), pp. 1–4. IEEE (2013)
121. Virvou, M., Papadimitriou, S.: Use of digital 'guess who board game' in teaching english. In: IISA 2014, The 5th International Conference on Information, Intelligence, Systems and Applications, pp. 439–443. IEEE (2014)
122. Virvou, M., Papadimitriou, S.: Playing, learning, extending: educational "guess who" game that is renewable by a teacher. In: 2015 6th International Conference on Information, Intelligence, Systems and Applications (IISA), pp. 1–5. IEEE (2015)
123. Virvou, M., Sidiropoulos, S.C.: An intelligent tutoring system over a social network for mathematics learning. In: IISA 2013, pp 1–4. IEEE (2013)
124. Virvou, M., Tsihrintzis, G.A.: Is chatgpt beneficial to education? A holistic evaluation framework based on intelligent tutoring systems. In: 2023 14th International Conference on Information, Intelligence, Systems & Applications (IISA), pp. 1–8. IEEE (2023)
125. Virvou, M., Tsihrintzis, G.A.: A novel trust state-chart model for requirements engineering of trustful ai-empowered software. In: 2023 14th International Conference on Information, Intelligence, Systems & Applications (IISA), pp. 1–6. IEEE (2023)
126. Virvou, M., Tsihrintzis, G.A.: Pre-made empowering artificial intelligence and chatgpt: the growing importance of human ai-experts. In: 2023 14th International Conference on Information, Intelligence, Systems & Applications (IISA), pp. 1–8. IEEE (2023)
127. Virvou, M., Manos, C., Katsionis, G., Tourtoglou, K.: Vr-engage: a virtual reality educational game that incorporates intelligence. In: Proceedings of IEEE International Conference on Advanced Learning Technologies, pp. 16–19 (2002)
128. Virvou, M., Katsionis, G., Manos, K.: Combining software games with education: Evaluation of its educational effectiveness. J. Educ. Technol. Soc. **8**(2), 54–65 (2005)
129. Virvou, M., Lampropoulos, A.S., Tsihrintzis, G.A.: Inma: a knowledge-based authoring tool for music education. In: International Conference on Knowledge-Based and Intelligent Information and Engineering Systems, pp. 376–383. Springer (2006)
130. Virvou, M., Tsihrintzis, G.A., Bourbakis, N.G., Jain, L.C.: Handbook on Artificial Intelligence-Empowered Applied Software Engineering. Smart Software Applications in Cyber-Physical Systems, vols. 2 and 3. Springer (2022)
131. Virvou, M., Tsihrintzis, G.A., Bourbakis, N.G., Jain, L.C.: Introduction to handbook on artificial intelligence-empowered applied software engineering—Novel methodologies to engineering smart software systems. In: Handbook on Artificial Intelligence-Empowered Applied Software Engineering, Novel Methodologies to Engineering Smart Software Systems, vol. 1, pp. 1–8. Springer (2022)
132. Virvou, M., Tsihrintzis, G.A., Sotiropouloss, D.N., Chrysafiadi, K., Sakkopoulos, E., Tsichrintzi, E.A.: Chatgpt in artificial intelligence-empowered e-learning for cultural heritage: the case of lyrics and poems. In: 2023 14th International Conference on Information, Intelligence, Systems & Applications (IISA), pp. 1–9. IEEE (2023)
133. Wiederhold, B.K.: Bridging tradition and technology: soft skills training for the modern day "house call" (2024)
134. Yang, J.C., Quadir, B., Chen, N.S.: Effects of children's trait emotional intelligence on digital game-based learning. Int. J. Hum. Comput. Interact. **35**(4–5), 374–383 (2019)
135. Yuce, A., Abubakar, A.M., Ilkan, M.: Intelligent tutoring systems and learning performance: Applying task-technology fit and is success model. Online Inf. Rev. **43**(4), 600–616 (2019)

136. Zahari, A.S., Rahim, L.A., Mehat, M.: A review of modelling languages for adventure educational games. In: 2016 3rd International Conference on Computer and Information Sciences (ICCOINS), pp. 495–500. IEEE (2016)
137. Zarraonandía, T., Díaz, P., Aedo, I.: Modeling games for adaptive and personalized learning: the a-grem. In: Learning Designs for Emerging Pedagogies, The Future of Ubiquitous Learning, pp. 217–239 (2016)
138. Zeng, J., Parks, S., Shang, J.: To learn scientifically, effectively, and enjoyably: A review of educational games. Hum. Behav. Emerg. Technol. **2**(2), 186–195 (2020)
139. Zhao, F., Yan, F., Jin, H., Yang, L.T., Yu, C.: Personalized mobile searching approach based on combining content-based filtering and collaborative filtering. IEEE Syst. J. **11**(1), 324–332 (2015)

Chapter 2
Computer Games for Entertainment and Education: A Literature Review and Exploration on Artificial Intelligence Integration

Abstract This chapter includes extensive literature and papers related to games, educational tools, and student modeling techniques. We describe the various categories of games and their characteristics and provide typical examples of popular games. In education, we present Intelligent Tutoring Systems (ITS) that identify students' learning needs and adapt their content accordingly. Educational games that function as ITSs are charged with the additional role of entertaining the player. Adventure games, in particular, a category whose main features are the development of a story through exploration and conversations with non-playable characters (NPCs), are ideal for use as educational tools. They help the player engage in trial-and-error exercises at their own pace and provide motivation to solve problems and puzzles to unfold the story. The chapter also describes student modeling techniques, particularly fuzzy logic. Fuzzy logic simulates human thinking and can handle concepts that involve ambiguity, such as whether the student knows a lesson. Finally, the chapter contains an extensive list of authoring tools that have been implemented both for educational applications and for educational games.

Keywords Educational software · Educational games · Artificial intelligence in educational games · Fuzzy logic in education · User experience · AI-empowered software games · User modeling · Human-AI interaction · e-learning games · 21st-century skills · Human-centred artificial intelligence

2.1 Introduction to Computer Games

Computer games have evolved from simple entertainment mediums into complex systems that offer significant educational, cognitive, and technological benefits. Having performed a review of recent relevant literature on computer games, we have focused on their applications in education, their impact on cognitive skills, and the latest advancements in game development.

S. Papadimitriou and M. Virvou, *Artificial Intelligence—Based Games as Novel Holistic Educational Environments to Teach 21st Century Skills*, Intelligent Systems Reference Library 93, https://doi.org/10.1007/978-3-031-77464-5_2

Educational Applications of Computer Games

Computer games are increasingly used as educational tools due to their interactive and engaging nature. Hooshyar et al. [71] explore the use of educational computer games to model learners' performance and predict learning outcomes. This approach enables educators to personalize learning experiences, providing targeted interventions that can enhance student achievement.

López-Fernández et al. [99] compare the effectiveness of different genres of educational video games in computer science education. Their research highlights how serious games can improve students' understanding of complex subjects by providing an immersive learning environment that encourages exploration and experimentation.

Melchor-Ferrer and Davia-Rodriguez [105] examine the role of computer games in teaching terminology and concepts related to national accounts. By integrating game-based learning with traditional educational methods, educators can make abstract topics more accessible and engaging for students.

Computer games are increasingly recognized for their educational potential, particularly in teaching programming and other technical skills. Quintero and Álvarez [139] highlight the use of serious games in developing computer programming competencies. By integrating game mechanics with educational content, these games provide an engaging and effective way to learn complex programming concepts.

Jaimez-González et al. [76] introduce BlockCode, a web application designed to teach programming logic through game-based learning. BlockCode exemplifies how games can simplify complex subjects, making them accessible and enjoyable for learners and fostering a deeper understanding of programming principles.

Rubio-Manzano et al. [143] investigate imitative learning in computer games, using linguistic descriptions and decision trees to teach complex phenomena. Their study illustrates how games can facilitate learning through imitation, enabling players to develop new skills and knowledge through interactive and engaging experiences.

Cognitive and Social Benefits

Computer games also offer cognitive and social benefits that extend beyond education. Gao and Izadpanah [57] investigate the relationship between computer games, self-efficacy, and academic engagement. Their research suggests that games can enhance creativity and problem-solving skills, leading to improved academic performance.

Bouchillon and Stewart [24] explore the social impact of role-playing games that simulate immigrant experiences in the South. These games can foster empathy and understanding by immersing players in diverse perspectives, promoting trust and immediacy among participants.

Uludagli and Oguz [169] focus on non-player character (NPC) decision-making in games, highlighting the use of Artificial Intelligence (AI) to create dynamic and realistic interactions. This advancement enhances the overall gaming experience, providing players with more engaging and unpredictable gameplay.

Computer games also offer cognitive and social benefits, contributing to the development of critical skills and behaviors. Basu et al. [20] assess the impact of computer games on attention and working memory among young adults, finding that games can enhance cognitive functions critical for academic and professional success.

Aminoleslami et al. [11] use eye-tracking data from computer games to classify autistic and normal children, highlighting the potential of games in behavioral research and therapy. This application underscores the importance of games in understanding and supporting diverse cognitive needs.

Kahraman and Kazancoglu [82] explore young people's motivations for playing online multiplayer games, providing insights into the social dynamics and cultural influences that drive gaming behavior. Their research reveals the complex interplay between individual motivations and the social environment in shaping gaming experiences.

Technological Advancements and Innovations

The development of computer games is closely linked to advancements in technology and AI. Chen et al. [29] report on the latest developments in computer Chinese chess, showcasing how AI has transformed traditional games into sophisticated platforms for strategic thinking and skill development.

Cohen-Solal and Cazenave [40] highlight the achievements of Athénan, an AI program that won sixteen gold medals at the Computer Olympiad. These accomplishments demonstrate the potential of AI in elevating the complexity and appeal of computer games, pushing the boundaries of what is possible in game design.

Hartisch et al. [66] provide insights from the Advances in Computer Games (ACG 2023) conference, which presented cutting-edge research and innovations in gaming technology. The conference emphasized the growing importance of AI and Machine Learning (ML) in shaping the future of computer games, offering new opportunities for creativity and exploration.

Evolution and Impact of Computer Chess

Computer chess has played a significant role in the development of AI and computer games. Hartmann [67] provides an in-depth look at the life of Hans Berliner, a pioneer in computer chess whose contributions have shaped the field. Berliner's work not only advanced the capabilities of computer chess programs but also laid the groundwork for modern AI applications in gaming.

Iida et al. [74] discuss the 2023 Computer Olympiad, highlighting the advancements in AI-driven games. The Olympiad showcases the latest developments in game AI, emphasizing the competitive and collaborative nature of computer games as a field of research. These competitions have become a platform for testing new algorithms and techniques, pushing the boundaries of what AI can achieve in gaming.

Pijl [134] explores the future of computer chess tournaments, emphasizing the shift towards more complex and dynamic AI systems. As these systems continue to evolve, they offer new challenges and opportunities for both researchers and players, reflecting the ongoing innovation in the field.

Schaeffer [149] marks the end of an era with the 2024 World Computer Chess Championships, reflecting on the event's historical significance. The championships have been instrumental in advancing AI research, showcasing progress over the past five decades, and setting the stage for future developments in AI-driven games.

Our current research is centered on enhancing students' cognitive abilities and enriching their academic profile through an educational adventure game. This game is designed to dynamically adjust its content and offer the player additional motivation, incentives, and satisfaction. Initially, we provide an overview of existing game genres and their characteristics [32, 62]. We then explore various educational games documented in the literature. Finally, we examine the specific advantages of educational adventure games compared to other game genres when used in an educational context.

The field overview also encompasses Intelligent Tutoring Systems (ITSs), which can automatically tailor educational content to provide a personalized learning experience. Additionally, we discuss authoring tools that empower non-programmers to create educational materials for their students without requiring programming knowledge. Moreover, tutors can use these authoring tools to track their students' progress and adapt the curriculum.

In summary, we highlight the importance of student modeling as the foundation for creating personalized content that caters to students' needs. This concept is a crucial component of ITSs. When student modeling effectively identifies each student's unique characteristics, it dramatically enhances the delivery of tailored educational content. The user interfaces of software must incorporate adaptivity and intelligence to automatically adapt to each user's specific needs, interests, and knowledge [176]. It is worth noting that adaptivity can also be integrated into educational games to maintain the player's engagement and motivation to continue learning while playing.

We focus on fuzzy logic theory and associated methodologies. Our discussion encompasses research demonstrating the advantageous application of fuzzy logic in modeling student behavior and assessing cognitive levels. Additionally, we feature educational implementations employing fuzzy techniques while noting their underrepresentation in educational games despite their benefits.

After the field review, we encapsulate the findings and obstacles from the earlier discussions. The subsequent part of the book delves into addressing these obstacles and providing solutions.

2.2 Traditional Games

This category includes subcategories such as card, board, and puzzle games. Traditional games, whether in video game formats like classic arcade games or non-video game formats like card or dice games, quizzes, and classic board games, have been around for a long time. Puzzle games use puzzles or enigmas, challenging players to manipulate game objects and utilize reasoning and problem-solving skills [90, 107]. The main objective is to solve dilemmas and puzzles, with some games requiring extended time and careful thought, while others test the player's alertness. Classic examples of this genre include Tetris, Minesweeper, Sokoban, and Lemmings [60].

More contemporary puzzle games include Flow Free and Bejeweled. In Flow Free, players must match dots of the same color by moving and dragging them, while in Bejeweled, players manipulate jewels on an 8x8 grid to create rows or columns containing at least three identical jewels, causing them to disappear and new ones to appear to fill the gaps in the grid.

Digital card games are a modern interpretation of traditional card games, allowing players to collect cards, create decks, and compete against other players or the computer online [79, 91]. This genre has a long history, dating back to the 10th century when playing cards were first invented. It falls within the broader category of card games. It has been found to benefit students, helping them grasp and apply essential aspects of developmental psychology, such as Piaget's stages of cognitive development theory [205]. Card games involve managing cards with specific attributes, forming strategic combinations, and exploring probability, classification, and grouping concepts. While most card games use the standard four-suit deck, some feature specialized cards. Examples of card games include 1000 Miles, Blackjack, and Eric's Ultimate Solitaire [209].

Technological advancements have enabled traditional board games to transition onto various digital platforms such as websites, smartphones, tablets, and digital tabletop devices. Popular games like Scrabble, Carcassonne, Apples to Apples, and Stone Age have seamlessly connected players through a wide digital network [211]. Traditionally played using physical components like boards, cards, and tokens, these games have now been adapted into digital versions for smartphones, computers, and gaming consoles. However, there is an ongoing discussion about balancing preserving the original interactive experience of the board games with utilizing the unique features offered by the digital medium [141]. Board games are known for their structured gameplay and clear core mechanics, making them suitable for analysis [16].

2.3 Simulation Games

Simulation games encompass video games that replicate sports, flying, driving, or the complex dynamics of urban and rural settings [16, 25]. As noted by Qaffas, simulation games provide a virtual platform for engaging with real-world events [138]. These games aim to recreate various real-life activities within a gaming context for education, analysis, and prediction. Typically, simulation games do not impose strict objectives, allowing players to manipulate characters and environments freely. For instance, The Sims simulates life and everyday activities, enabling players to construct homes and care for in-game families [117, 131]. Another example is the Farming Simulator series by Giants Software, where players can engage in activities such as farming, breeding livestock, and cultivating crops and vegetables [215].

There are a plethora of video games that emulate various athletic and team sports. Some aim to faithfully reproduce the intensity of the sport, such as the Madden NFL series, while others prioritize strategy and planning, exemplified in games like Football Manager [44, 78, 204]. For instance, the Football Manager series is renowned for its exceptional sports simulation. It allows players to manage some of the world's most prominent soccer teams and partake in realistic soccer management experiences [88]. Additionally, there are vehicle simulation games covering subgenres like racing games—Gran Turismo, train simulator games—Train Simulator series or Train Sim World, truck simulator games—Euro Truck Simulator series, flight simulators, and submarine simulator games [75, 94].

Business simulation games are designed to mimic economic processes in a business setting [53, 157]. These games are popular because they mirror real-world situations, connecting players' choices to their potential outcomes. For instance, Railroad Tycoon allows players to run a railway company, facilitating the construction of rail lines, establishing stations, and acquiring and scheduling trains [113]. Theme Park requires players to attract customers and maximize profits [47]. In Capitalism, players compete globally to make their business the most profitable, vying in various markets against other corporations [146]. The game tasks players with acting as the CEO of a company, ensuring its financial stability and protecting it from takeovers. Lastly, in Detroit, players manage an automobile company, overseeing finance, research, design, testing, production, and marketing of automobile product lines.

2.4 Strategy Games

Strategy games demand skillful reasoning and precise planning in order to achieve victory. They present players with strategic, tactical, and sometimes logistical challenges [16, 30, 56, 64, 80]. Players must carefully consider their moves as they engage in actions to overcome opposing forces. In these games, players typically wield godlike control over the game world and its units, often incorporating elements of warfare alongside tactical and strategic decision-making. Beyond combat, strategy games frequently test players' abilities to manage economies and explore their environments, with many incorporating economic challenges and exploration.

Strategy games stand out from other genres, such as action games, due to their expansive scope. The primary focus of strategy games is on the player's skill in outmaneuvering their opponent rather than on physical challenges, making them particularly appealing to players with a strategic mindset. Unlike other genres, strategy games typically involve maintaining a balance between opposing sides, providing players with similar resources and actions. There is an equilibrium between the strengths and weaknesses of each side. Additionally, strategy games differentiate themselves from construction and management simulations by emphasizing immediate action against opponents rather than purely economic challenges.

In Sid Meier's Civilization, players can build vast empires across different periods and continents [28, 203]. As the leader of a specific ancient civilization, such as the Greeks, Indians, or Chinese, the player navigates through history, managing technological progress, diplomacy, and warfare. StarCraft, a popular real-time strategy game developed by Blizzard Entertainment in 1998, is set in a futuristic world where players control one of three balanced races [120]. The game involves resource collection, base building, unit creation, technological research, and military strategy. Similarly, Warcraft is a real-time strategy game focused on multiplayer competition, where players strive to overcome their opponents' factions [26]. It also features a single-player campaign mode where players progress through missions like establishing a settlement, gathering resources, amassing troops, and engaging in battles.

2.5 Action Games

The action games genre comprises games that put physical skills, hand-eye coordination, and quick reflexes to the test [30]. These games involve swift, high-speed movements, necessitate careful observation of the game environment, and often require simultaneous tracking of multiple targets [48]. Various video game genres, such as fighting, platform, and shooter games, can fall into the action category. Typically, the player guides a character through levels, collecting items, avoiding obstacles, and using innate abilities and weapons to combat foes [2]. After certain levels, players often face a more challenging boss enemy, usually integral to the game's storyline. Enemy attacks and obstacles deplete the player's energy and health until the game concludes. In action games, players progress through several levels to achieve a final objective [172].

Fighting games, a subgenre of action games, involve two or more characters engaging in combat [17]. These games are known for featuring special moves triggered by quick button presses and joystick movements. An example of a classic fighting game is Street Fighter II: The World Warrior, which offers a two-player mode that encourages direct player versus player competition [58]. On the other hand, platform games center around a protagonist who navigates through levels, collecting bonuses, and overcoming enemies and bosses in various ways [140]. Super Mario Bros., a renowned platform game, follows the adventures of the heroic plumber Mario in his quest to navigate the Mushroom Kingdom, defeat the main antagonist, Bowser, and rescue Princess Toadstool. The game is a two-dimensional side-scrolling platformer in which the player continuously moves to the right, tackling various challenges before reaching the end of each level. Mario can move left and right, jump, and throw fireballs [132, 159].

The first-person shooter (FPS) is a subcategory of action shooter games that involve weapon-based battles from a first-person perspective, allowing players to experience the action through the eyes of the main character in a three-dimensional space [210]. Wolfenstein 3D, one of the earliest FPS games, used simple 3D graphics and featured levels constructed with areas and rooms forming a grid-based pattern of walls and doors [147]. Players navigate these levels to find elevators and face a challenging enemy in the final level. Doom, developed by id Software, surpassed Wolfenstein 3D and placed players as a space marine combating hordes of invading demons and beasts [63]. Considered one of the most influential games in video game history, Doom led to the emergence of a new game subcategory known as Doom clones and inspired the creation of numerous similar games.

2.6 Fantasy Games

Fantasy games consist of two main sub-genres: role-playing games and adventure games. Role-playing games (RPGs) involve the player controlling a game character or a group of characters, often referred to as a party, in order to complete a series of quests [19, 22, 135, 138, 208]. Players explore a game world, solve puzzles, and engage in combat. As the game progresses, the characters become stronger and gain new skills [95]. The character development, storytelling, and narrative elements in video role-playing games are influenced by early tabletop, leading to the adoption of similar terminology, settings, and gameplay mechanics.

The Dungeon Master game incorporated real-time battles and immersive features such as dynamic lighting and sound effects to signal the presence of nearby monsters [21, 170]. These immersive features not only enhance the gaming experience but also keep the players engaged. In Eye of the Beholder, players experience a first-person perspective in a three-dimensional dungeon and lead a party of four members to combat monsters using a point-and-click interface [136, 155]. Players can increase their party size to six by resurrecting skeletons from deceased encountering NPCs during their dungeon exploration. In Lands of Lore: The Throne of Chaos by Westwood Studios, players seek to rescue the kingdom from a sorceress by traversing diverse environments, collecting items, and battling creatures [19, 150]. New World Computed developed the Might and Magic series, featuring gameplay set in medieval fantasy worlds, with later sections of the games introducing science fiction settings [22]. The expansive game worlds in the Might and Magic series and the intense combat scenes against large groups of enemy creatures offer hours of immersive gameplay experience.

An adventure game is a type of video game in which the player assumes the role of the protagonist in an interactive story driven by exploration and puzzle-solving, with minimal action elements [156, 172]. This genre does not align with the traditional definition of 'adventure' and is distinct from role-playing games (RPGs), which typically involve combat, party management, and resource allocation. Early adventure games, which emerged in the 1970s and early 1980s, relied on text-based gameplay, interpreting players' text commands to progress. As technology advanced, these games transitioned to graphical formats, replacing text with visual cues and evolving into point-and-click interfaces. Modern adventure games commonly feature immersive, real-time 3D visuals and pre-rendered or full-motion video sequences.

The advent of the adventure game genre can be traced back to the release of Colossal Cave Adventure (also known as ADVENT or Adventure) in 1976 [49, 100, 144]. This text-based game allowed players to navigate a cave, collect treasures, and ultimately escape, earning points. Similarly, the King's Quest series follows the adventures of the royal family of the Kingdom of Daventry, facing challenges like battling witches and wizards [43, 65, 116, 145]. Another influential game, The Secret of Monkey Island, places players in the shoes of Guybrush Threepwood, a young aspiring pirate who solves puzzles and unravels mysteries on fantasy islands [42, 77]. Known for its humour, visual and audio elements, and engaging gameplay, it is widely regarded as one of history's most impactful video games.

2.7 Educational Games and Targeted Skill Acquisition and Cultivation

The educational games genre encompasses games that provide entertainment and serve as a means of teaching across diverse fields [68]. These games aim to educate students about specific subjects, expand their knowledge, foster growth, and help them grasp concepts while playing. They offer an effective alternative to traditional teaching methods by making learning enjoyable and motivating students to learn. Additionally, they promote active learning, motivation, and teamwork, facilitate classroom participation, and foster positive attitudes toward learning, leading to successful learning outcomes. Moreover, they contribute to increased self-esteem, drive learning through challenges, encourage exploratory behavior and a desire to learn, and enable students to learn at their own pace without facing negative consequences [98]. Students more engaged in learning can better comprehend theoretical or complex concepts [89].

In the realm of gaming, a single game often encompasses multiple genres. According to various researchers, adventure games show promise as a beneficial tool for learning, yielding positive results. These educational games combine elements from both educational software and adventure games [104]. Adventure games engage players in reasoning, deduction, inference, memory, visualization, and problem-solving. They require players to resolve puzzles and mysteries by manipulating in-game objects or interacting with fictional characters. Importantly, they do not have action-packed or time-bound objectives, allowing students to enhance their skills at their own pace, free from pressure. Therefore, they are valuable resources for enhancing students' cognitive abilities [12, 214]. Moreover, these games facilitate the smooth integration of educational content by focusing on narrative, character interactions, and puzzle-solving. This combination creates various conducive learning conditions: educational material is presented through captivating storytelling, seamlessly woven into the game's world and puzzles, enabling students to absorb the presented knowledge at their own pace.

Given the information provided, we will examine different educational game studies. Specifically, we will emphasize educational games in the adventure game genre and illustrate that their integration within the academic community still requires improvement despite researchers confirming the advantages of educational adventure games.

2.7.1 General Educational Games

The reference list includes various educational games designed to achieve multiple learning goals. One such game, Decimal Point: The Fantastically Fabulous World of Fractional Fun, is a single-player game that makes learning about decimal places an exciting adventure [102]. Players explore different themed settings, such as the Haunted House and the Wild West, participating in mini-games to strengthen their grasp of decimal places. The game includes six virtual characters who mentor and motivate students to play, praising their achievements. An assessment of the game revealed that it delivers an enjoyable experience to students while enhancing learning outcomes compared to traditional teaching methods.

Es-Sajjade et al. developed MATHERIAL, a two-player educational game for elementary school students to help them learn mathematical equations [52]. In this game, players can either challenge others or accept challenges and see a list of incomplete games they are involved in. Each exercise consists of a series of slots, some of which are empty while others contain numbers and operators. Students use different elements, such as numbers and operators, to solve the equation and place them in the empty slots. Points are earned based on the number of items used. The project evaluation demonstrated that students who played the game achieved higher math scores than those who did not. The authors also emphasized that educators can achieve significant results in teaching mathematics by employing simple games that integrate learning theory and principles without resorting to expensive commercial games.

Lin et al. developed a role-playing game to help students grasp specific concepts of English grammar [96]. In the game, players engage in a story where the main character needs to deliver a task report to her teacher. However, a sudden gust of wind scatters the report cards, prompting players to assist the main character in collecting the pieces while participating in various learning activities that drive the plot forward. The game progressively increases the difficulty of the exercises to maintain the students' interest. The study's evaluation demonstrated that students who used the game showed a reduction in English grammar errors compared to those who received traditional instructional methods.

Vanbecelaere et al. investigated the efficacy of two digital educational games focusing on mathematics and reading skills [171]. The Number Sense Game (NSG) comprises two types of exercises: a comparison game and a number line estimation game (NLE). The game gradually increases the difficulty level of the exercises, and players must achieve a minimum of 80% correct answers to progress to the next level. The Reading Game aims to develop students' phonological awareness by helping them recognize and manipulate oral parts of words and sentences while fostering letter knowledge acquisition. Both games offer comprehensive educational support to students. The study revealed that while the games did not significantly impact students' attitudes toward math and reading, they did enhance their understanding of the near/far transfer mathematics model and contributed to improved reading skills.

The bibliography contains references to various educational games that employ adaptivity to deliver a customized learning experience to students. One example is AutoThinking, an adaptive educational game developed by Hooshyar D et al., which focuses on promoting computational thinking [70]. Players must devise different strategies in this game and find multiple solutions to complete the three levels. The game gathers data from the player's actions, which is input into a Bayesian network. This network assesses the player's use of computational thinking, influencing the movement of the intelligent cat in the game (game-level adaptation) and determining whether to provide feedback or hints to the player (education-level adaptation). This adaptivity ensures that each student receives a personalized learning experience, enhancing their understanding and retention of the material.

Research has shown that spatial imaging skills play a crucial role in achieving success in science, technology, engineering, and mathematics (STEM). In light of the increasing significance of STEM in education, Xiao et al. developed Cubicle [212] to foster these skills. Cubicle consists of eight modules that cover a wide range of topics, such as converting orthographic drawings into 3D models, transitioning from 2D to 3D, and manipulating 3D objects. The game also monitors the player's actions, including mouse clicks, to incorporate adaptive elements in future versions.

Euclid: The Game is an engaging online interactive game that aims to instruct players on various aspects of Euclidean Geometry. These aspects include angles, lines, creating straight lines between points, and calculating the circumference of a circle [148]. The game comprises 25 puzzles that prompt players to utilize various tools to connect points, create line segments, and form complete circles. Players can also construct geometric shapes such as isosceles, triangles, and parallel or perpendicular lines and find the intersecting point of two lines. Upon completing a level, players are awarded a silver medal. Achieving the minimum number of moves awards them a gold medal. A study revealed that the game significantly improved the performance of 49 civil engineering students who used it before and after evaluation.

2.7.2 *Educational Adventure Games*

Tesla's Revenge is an engaging 2D educational adventure game designed to teach various information literacy concepts [93]. The game aims to help university students familiarize themselves with essential aspects of information literacy, including keyword searching, source location, and proper referencing. It incorporates an exciting storyline that unfolds as players progress, enhancing student engagement. Additionally, the game features avatar self-presentation, storytelling, narrative context, feedback mechanisms, reputation management, grading, leveling, rule-based competition, and time constraints. It also includes a variety of mini-games to facilitate the learning of fundamental information literacy concepts. Feedback from students who evaluated the game indicates that it significantly improved their understanding of information literacy. Moreover, 29% of students enjoyed the game's scenario and storytelling, while 25% appreciated the gameplay.

Chiang and colleagues designed an educational adventure game to teach children about safety concerns [31]. The game features realistic dangerous situations that children may encounter, providing them with essential knowledge on handling these scenarios. For example, children are taught to cross the road safely and check food items' expiration dates before consumption. The game integrates design principles with child development strategies to offer a balanced and engaging learning experience. Additionally, gamification elements have been incorporated, such as awarding badges and unlocking new characters upon mission completion. The evaluation of the game revealed that it was both enjoyable and engaging, and it effectively improved children's retention of information related to electrical safety, external safety, and road safety.

The Em Busca do Santo Grau is a 2D adventure game crafted as a versatile educational tool that caters to educators and students across all levels [115]. It is also adjustable to suit players with visual impairments. The game revolves around the journey of Ariel, a young student who embarks on a quest to solve a series of puzzles within a fictional college in order to obtain her undergraduate diploma. Through interactions with various items and characters, players are required to solve puzzles relevant to the course material, tailored by the professor for each stage. The game incorporates educational principles such as clear objectives, immediate feedback, progressively challenging problems, balanced tasks and abilities, and readily available information and assistance. The authors employed the eGameFlow framework to assess students' enjoyment and determined that the students had an enjoyable gaming experience [55].

The Siret Security game is an immersive adventure game where players take on the role of a corporate employee [51]. The main objective is safeguarding the company's confidential data from threats like spies and hackers. The game aims to educate players on security concepts, virus defense, fraud protection, and cryptography principles. It is designed for IT specialists and company employees who want to learn how to implement security protocols and understand the importance of protecting sensitive data. The game mechanics follow the elements of classic adventure games, with the protagonist managing an inventory of objects, interacting with NPCs, and solving puzzles. Alongside traditional puzzles, there are challenges related to computer security principles to enhance the educational aspect of the game. If the system detects that a player is struggling with puzzles or making multiple incorrect attempts, a mentor character, a senior security officer, will reach out to provide guidance and advice. Game assessments have shown that the educational objectives were achieved, the educational content was seamlessly integrated, the game controls were user-friendly, and the game's storyline was easily comprehensible.

Moreover, researchers suggest that problem-solving is important in facilitating cognitive learning processes by allowing students to apply their prior knowledge and cognitive strategies [50, 73]. In line with this idea, Hou and Chou developed an interactive game called Boom Room to improve students' understanding of computer assembly through problem-solving activities [72]. In the game, players assist the main character in collecting hardware components and assembling a PC desktop to deactivate a time bomb and escape from a room within ten minutes. The game provides helpful hints and tips as players interact with various objects apart from computer parts. Once players gather the required computer components, they can assemble the computer. An initial assessment demonstrated that the game was an effective educational tool with well-rounded game elements. A considerable number of students expressed that the game enhanced their learning experience. They felt that the game aided them in grasping concepts related to computer hardware structure, and many believed that they could understand this structure better through the game than through traditional textbooks.

2.8 Educational Games with Artificial Intelligence

Recent advancements in AI have profoundly influenced the multimedia and gaming industries by introducing cutting-edge tools and methodologies that boost interactivity and user engagement. AI's transformative impact is evident through advanced decision support systems [168] that enhance game design and user experience by analyzing player behavior and preferences in real-time using Machine Learning (ML) and Deep Learning (DL) techniques [165, 178]. This enables the creation of adaptive, personalized gaming experiences where intelligent NPCs dynamically respond to player actions, enhancing immersion and engagement. Additionally, integrating AI in multimedia services allows for the development of intelligent systems that can manage and deliver content more efficiently, resulting in a more immersive and engaging user experience [163].

The application of AI in software engineering, particularly in teaching and implementing 21st-century skills, further underscores its importance in creating sophisticated multimedia and gaming applications [192]. AI's involvement in engineering smart software systems is highlighted in various works, where novel methodologies are developed to enhance interactive environments and cyber-physical systems [199, 200]. These advancements not only enhance the technical capabilities of gaming and multimedia platforms but also pave the way for more interactive and intelligent environments, ultimately transforming how users interact with digital content and games. The evolution of AI in these areas illustrates its critical role in developing next-generation multimedia and gaming experiences, as explored in various scholarly works [166, 167].

Recent advancements in AI technologies have the potential to significantly impact the multimedia and gaming industries by enhancing interactivity, personalization, and user engagement. Large Language Models (LLMs) like ChatGPT (Open AI) are transforming how multimedia content is created and consumed by enabling natural language interactions and personalized experiences tailored to individual users [190]. In gaming, AI can create adaptive environments and intelligent NPCs that respond dynamically to player actions, thus improving engagement and immersion. The development of trust models, such as the novel trust state-chart model, seeks to enhance AI systems' reliability and capability of effective collaboration with users, fostering trust in AI-driven games and multimedia applications [191]. LLMs can be integrated into ITSs to provide personalized, interactive experiences that adapt to student needs, promoting both hard and soft skills. Virvou and Tsihrintzis highlight the growing importance of human AI experts in designing AI-empowered educational systems [192], who emphasize the need for expertise in creating effective learning environments. A trust state-chart model proposed by Virvou and Tsihrintzis [191] aims to ensure that AI software meets user trust expectations, which is important for successful implementation. In e-learning for cultural heritage, LLMs like ChatGPT enhance engagement and creativity, as shown by Virvou et al. [201], offering new ways to explore subjects like literature and history. The VIRTSI model introduced by Virvou, Tsihrintzis, and Tsichrintzi [202] emphasizes trust dynamics in AI-human collaboration, highlighting the importance of transparency in educational technology. Furthermore, Panagoulias et al. [124] discuss the adaptability of LLMs in fields such as medical diagnosis, underscoring their versatility and potential to enrich educational experiences across disciplines. However, these technologies are yet to be explored further and refined to maximize their effectiveness and reliability in educational applications.

Moreover, AI's integration into multimedia platforms allows for improved content management and delivery, offering personalized recommendations and enhancing the creative process [202]. The ability of AI to understand and generate human language, as demonstrated by ChatGPT, provides opportunities for interactive storytelling and dynamic content creation, leading to more engaging multimedia experiences [137]. Additionally, AI technologies are increasingly seen as tools that empower human creators and experts, highlighting the growing importance of collaboration between AI systems and human professionals in developing innovative multimedia and gaming content [192]. While these recent developments are yet to be fully consolidated, the incorporation of AI into ITSs, which can be combined with games, offers a more mature technology that can be employed for the benefit of education.

Educational games are increasingly recognized for their transformative impact on learning by integrating interactive gameplay with educational content, focusing on targeted skill acquisition and cultivation. Game-based learning systems like VIRGE, developed by Virvou and Katsionis [181], are designed to enhance student motivation and engagement through immersive environments. These environments enable students to actively practice and develop essential skills such as critical thinking, problem-solving, and collaboration. By providing a fun and engaging platform, educational games encourage learners to apply concepts in real-world scenarios, fostering both technical and interpersonal skills.

ITSs complement educational games by personalizing education through AI-driven, tailored instruction and feedback. Virvou and Sidiropoulos [189] demonstrated how ITS integrated with social networks can enhance mathematics learning, leveraging social interactivity to build communication and analytical skills. As explored by Virvou and Manos [183], cognitive modeling customizes learning experiences to match students' cognitive capabilities, which is crucial for effective skill development. Kabassi and Virvou [81] advanced this approach by integrating cognitive modeling with decision theory to refine decision-making skills in learners.

Affective user modeling, discussed by Katsionis and Virvou [85], enhances ITS by enabling them to adapt to students' emotional states, ensuring a responsive and supportive learning experience that fosters emotional intelligence alongside academic skills. Error diagnosis systems, such as those developed by Virvou and Maras [184], play a critical role in identifying and correcting misconceptions, providing timely feedback that facilitates continuous improvement in technical and analytical skills. During the COVID-19 pandemic, Chrysafiadi et al. [38] used fuzzy logic to evaluate e-learning effectiveness, showcasing ITS's adaptability in various learning contexts and their ability to sustain skill development under different conditions.

Moreover, ITSs enhance collaborative learning by fostering peer interaction and knowledge sharing, as studied by Olsen et al. [119]. Research by Tourtoglou and Virvou [160] applied simulated annealing to optimize learner groups in UML education, promoting teamwork, communication, and collaborative problem-solving skills alongside technical learning. These systems not only enhance individual learning but also cultivate essential soft skills required for effective collaboration and leadership in group settings.

As AI, ML, and DL continue to evolve, they are fundamentally reshaping how skills are developed and applied across various fields [165]. AI enhances user experiences across critical sectors, including healthcare, emergency response, the environment, education, and tourism, underscoring its profound impact on modern education and skill development. The integration of AI technologies into educational games and ITS exemplifies the potential of these tools to prepare learners for the complexities of the modern world by equipping them with the necessary skills to succeed.

2.8.1 Intelligent Tutoring Systems

Bloom's 1984 study revealed that students who received personalized one-on-one tutoring from a human teacher performed better than those who underwent traditional group teaching methods [9, 23]. Subsequent research further corroborated the effectiveness of individualized human tutoring [13, 14]. Skilled human tutors play a crucial role in assisting students struggling with the material, aiding them in overcoming their challenges. Recognizing the potential benefits of personalized tutoring, numerous researchers in AI have focused on creating systems capable of identifying students' unique characteristics and providing customized instruction and guidance to enhance learning outcomes. This initiative led to the development of intelligent computer-assisted instruction (ICAI) systems and computerized instructional programs enhanced with intelligence, offering course content through simulations, games, tutorials, and practice sessions. Over time, the term Intelligent Tutoring Systems (ITSs) emerged as the preferred designation, becoming synonymous with ICAI [54].

ITSs refer to computer programs utilizing AI techniques to serve as virtual trainers, delivering personalized instruction to students [118]. These digital learning environments integrate insights from cognitive science, learning sciences, computational linguistics, AI, and mathematics. Their primary aim is to emulate the effectiveness of individualized human tutoring through digital platforms. For instance, an ITS can assess a student's unique characteristics and needs, such as their knowledge level, emotional state, motivations, and skills, and adjust the educational content accordingly. This capability enables the ITS to provide similar educational advantages as a human tutor [15, 61].

In their study, Conde et al. stated that the ITS initially includes the following characteristics [41].

- It enables teaching tasks to individuals with weaknesses, thus enhancing their independence in work environments.
- It includes an intuitive, robust, versatile, and ergonomically well-adapted multimedia interface.
- It enables integration with a mobile device, such as a mobile phone or personal digital assistants (PDAs).
- Non-technical stakeholders can easily customize it, allowing simple and flexible access.
- It features a multimodal Task Management System for integrating data from various sources (speech, images, video, and text) linked to each personalized profile.
- It includes human emotional prediction to prevent potential risks, emergencies, and obstacles that may negatively impact individuals and hinder their ability to participate in work and social activities.
- It can be exported to other groups, such as the older people, to benefit them.

2.8.2 Architecture of Intelligent Tutoring Systems and Functions

The architecture of an ITS consists of four components: the domain model, the student model, the tutoring model, and the user interface model [10]. Figure 2.1 illustrates the generic architecture of an ITS.

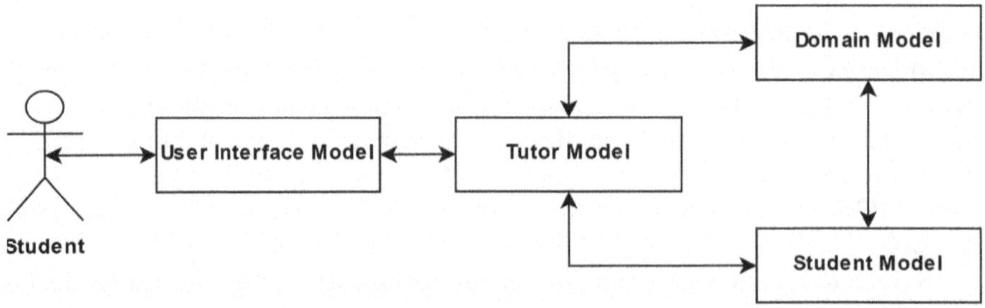

Fig. 2.1 Generic architecture of ITS

Domain model

The domain model, also known as a cognitive or expert knowledge model, encompasses the domain knowledge and the procedures employed by experts in their respective fields. It draws on learning theories to examine all potential problem-solving steps. Furthermore, it incorporates knowledge of specific subjects, including computer science, physics, history, and mathematics. To delve deeper, this model covers the concepts, principles, and problem-solving techniques within the studied domain. It also assesses student performance and identifies any errors or deficiencies.

Student model

The student model observes the student's behavior and delivers an in-depth assessment of the student's cognitive abilities and emotional state. Its objective is to produce valuable insights to tailor personalized feedback effectively. Furthermore, it exchanges information with other instructional components. The model stores details about the learner, including skills, goals, performance, the number of accurately answered questions per session, and learning styles and preferences.

Three common methods for representing student misconceptions are:

- *The overlay model*: This model compares the student's behavior with an expert's. The difference between the student and the expert lies in the skills and knowledge the student has not yet mastered.
- *The perturbation model*: This model integrates a bug library into the overlay model to capture the student's correct knowledge acquired and the known errors and misconceptions made in the domain.
- *Student-based modeling*: is an alternative approach to student modeling that focuses on gathering knowledge by encouraging students to disclose their misconceptions about the subject matter. Using ML techniques, we can develop rules for problem-solving to understand the sequence of steps that lead to a student's misconception.

Tutor model

The tutor model, also known as the pedagogical, teaching, or expert model, takes input from the domain and student models and then chooses teaching strategies and actions. In problem-solving scenarios, the student may seek guidance on their next steps based on their current position in the model. The student model contains the knowledge base for selecting and developing instructional components that align with the student model. It determines the appropriate course of action to address the student's interaction with the system, such as providing feedback or a hint.

Some of the tasks the tutor model is responsible for are the following:

- It selects a tutoring strategy depending on the learning process.
- It selects and produces questions to verify the learning progress.
- It selects and generates productive feedback.
- It provides assistance and additional information to address student knowledge gaps.
- It takes action to enhance the student's motivation throughout the learning process.

User Interface model

The user interface model, also known as the communication model, is designed to facilitate interaction between the user and the system. Its role is to present educational material to the student and to capture the student's responses to the questions posed by the ITS. This communication can take various forms, including text, voice, simulations, and games. The user interface can also be accessed via different devices such as computers, smartphones, and tablets. It is responsible for recording the user's interactions and generating the corresponding student model whenever the user interacts with the system.

The literature encompasses various works on ITSs. Almasri et al. conducted a systematic review covering various works from 2000 to 2018 [10]. These works span diverse learning objectives. For instance, [106] discusses the implementation of a tool designed to enhance literacy skills for deaf high school students. The authors propose a system that can analyze any text written by a deaf student, identify grammatical errors, and engage the student in a conversational exchange to facilitate the necessary corrections.

KERMIT is a valuable resource for Computer Science students who wish to grasp the entity-relationship (ER) model employed in database design [154]. This system monitors students' interactions and customizes its content to suit their skill levels and learning capabilities. Additionally, it employs constraint-based modeling (CBM) to generate domain knowledge and create student models, which is crucial for facilitating various learning design tasks across diverse domains.

Al Rekhawi introduced a web-based ITS aimed at helping students tackle the obstacles they face when creating Android applications [4]. The system is designed to acquaint students with different aspects of Android programming, proactively address their challenges, and adapt to their knowledge level and learning pace in real time. Initial assessments have shown that the system has positively impacted students' academic performance.

Kusairi et al. have introduced iTutor, a computer-based learning system aimed at helping students understand challenging physics principles, particularly in fluid statics [92]. This ITS can identify students' difficulties and provide customized tutorials to address their needs. It includes multiple-choice problems covering critical concepts related to fluid statics, such as hydrostatic pressure, Pascal's Principle, and Archimedes' Principle. If students encounter difficulties with a particular problem, the system offers remedial instructions to help them overcome obstacles. The evaluation of the system revealed that the provision of adaptive instructions significantly enhanced students' average scores.

Abu-Naser et al. have developed a web-based ITS to assist students in understanding Java programming concepts [1]. The system covers Java object topics and automatically generates problems for students to solve while adjusting its content dynamically to match the student's progress. Moreover, it can accommodate customized presentation structures. JavaTutor is another ITS focused on teaching essential Java topics, offering cognitive (skills and knowledge) and emotional feedback to aid students throughout their learning process [207]. Weragama and Reye have implemented a programming learning system using PHP, with an evaluation confirming that students using the PHP ITS demonstrated significant improvement in their examination scores [206].

Mathtutor is a tool specifically developed to create an intelligent multi-agent teaching system. It aims to support various formalisms, making it more convenient for teachers to develop course content. The system simplifies content creation and offers adaptability and flexibility in presenting the material [27]. It uses a Petri net to manage the interaction between students and agents, which is automatically converted to a rule-based expert system. Within the Petri net, the tokens contain data objects with pointers to the learner's model and domain knowledge stored in a database of texts, examples, and exercises. The transitions of the Petri net objects are controlled by logical conditions linked to the learner's model. Upon activation of these transitions, corresponding actions are generated to update the learner's model accordingly.

Karaci et al. designed an ITS that operates on a web-based platform and examined its impact on academic performance and student retention [84]. The ITS incorporates several features, including providing feedback through scores or language, directing students to areas they need to enhance their learning, and allowing access to specific sections based on their current knowledge level. Moreover, the ITS utilizes a neural network to analyze students' page-viewing behaviors, such as the frequency and duration of page visits and their responses to activity questions. The study involved 80 undergraduate students from different departments within the Faculty of Education at Kastamonu University, focusing on teaching electronic spreadsheet software. The evaluation findings revealed that the ITS is an effective tool for enhancing academic achievement.

The Chinese dialogue-based Mathematical ITS is an online learning platform created by [122]. The system's domain model presents educational content centered on multiplying and dividing time expressions. It encompasses ten core problems addressing seven time-related concepts. Each problem includes an ideal solution, expected answers, common misconceptions, and instructional approaches. The interconnections of these concepts signify that students need to comprehend specific concepts before advancing to others.

In order to enhance learning, the authors have integrated a five-step tutoring framework and expectation and misconception-adapted dialogues within the tutoring model. In this approach, students utilize a computer-based tutor to master the lessons by solving equations or selecting correct answers. The system evaluates the correctness of these responses, identifies any misconceptions, and provides immediate and suitable feedback. Furthermore, it leverages an automatic analysis technique to recognize error patterns and construct a pertinent student model [213].

Evaluation of the system's effectiveness showed that students using the ITS significantly improved their mathematics performance. The system's instructional strategies also helped students understand problem-solving steps and mathematical concepts.

Droid-Tutor is a sophisticated tutoring system developed to instruct Android application development [5]. Its architecture comprises several essential modules: a pedagogical module, a problem generation module, an expert module, a student module, and a tutoring process module. The pedagogical module introduces Android application development topics to students and automatically generates problems for them to solve. The problem generation module utilizes a BNF-like grammar technique to create problems and generates code templates covering various Android concepts, such as user interface, application components, activity lifecycle, layout manager, views, dialogues, fragments, and advanced Android material design. A question and possible solutions accompany each template. The expert module analyzes and resolves the generated problems, compares its solution with the student's answer, and provides corresponding feedback. The student module stores information about the student's profile, including their name, login dates, session scores, and progress throughout each session. Lastly, the tutoring process module oversees the overall functionality of the system. The evaluation of the system yielded promising results and included recommendations for expanding the use of ITSs to other teaching domains.

I-Prolog represents a web-based ITS crafted to facilitate acquiring the Prolog programming language [152]. The system strongly emphasizes the individual learner, aiming to help them recognize their preferred learning style to foster a positive attitude towards the learning process. Consequently, the ITS holds the potential to enhance traditional educational methodologies, with its remote capabilities potentially serving as a substitute for in-person teaching. The principal objective is to ignite the interest of both students and educators and to prompt further research and refinement. According to the feedback, students found the system engaging and highly beneficial. Additionally, they provided suggestions for improving real-time communication between students and teachers during practice sessions and enhancing the feedback delivered after identifying errors in the Prolog code.

2.8.3 Evaluation of ITSs

Evaluating educational games that serve as ITSs involves a comprehensive assessment to ensure their efficacy in facilitating learning. Within the broader context of intelligent interactive systems, Virvou highlights three critical dimensions for evaluating edutainment systems: educational effectiveness, attractiveness, and usability [34]. Formative evaluation, as discussed by, plays an important role in refining the design of an intelligent system through empirical studies [162] or other methods. ITS evaluation encompasses various dimensions. Examples of research include research by Moundridou and Virvou [112] on evaluating the persona effect of interface agents in tutoring systems, Chrysafiadi et al. on assessing which is better for learning: a web-based educational application or an educational game [36], and on the evaluation of intelligent interactive systems through comparison with human experts [174, 180]. Moreover, insights from Virvou, Manos, and Katsionis on evaluating an evaluation agent simulating students' behavior in ITS and the comparison of computer educational games with non-gaming environments provide valuable perspectives on effective evaluation methodologies [196], while the Technology Acceptance Model and Rapid Estimate of Adult Literacy in Medicine using ML is used by Panagoulias and colleagues [125]. These diverse approaches ensure that educational games functioning as ITS meet educational objectives while offering an engaging and effective learning experience for students [7, 33–35, 39, 162–164, 193, 199].

Integrating LLMs such as ChatGPT into educational settings represents a cutting-edge advancement in AI technology, offering new opportunities for enhancing learning experiences. Virvou and Tsihrintzis emphasize the significance of ChatGPT in education, highlighting its potential to empower e-learning initiatives in cultural heritage domains, specifically in the analysis of lyrics and poems [192]. Furthermore, their work underscores the importance of evaluating ChatGPT's impact on education through a comprehensive framework, shedding light on its benefits within ITSs [190]. Panagoulias et al. [125] extend this evaluation to the realm of healthcare AI literacy among stakeholders, demonstrating the utility of LLM-generated multimodal diagnosis from medical images and symptom analysis. These studies collectively underscore the growing recognition of ChatGPT's role in transforming educational practices and the need for rigorous assessment methodologies to elucidate its efficacy and potential in educational contexts.

2.8.4 Combination of ITS with Educational Games

A. Types of Game

- *Virtual Reality (VR) Game*: Virvou, Manos, Katsionis, and Tourtoglou introduced VR-ENGAGE, a VR educational game integrating intelligence [194]. They explored the potential of VR environments to enhance learning experiences, revealing high educational effectiveness and motivation, especially for students with prior poor performance in the domain.
- *Board Game for Teaching English*: Virvou and Papadimitriou developed a digital version of the 'Guess Who Board Game' for teaching English, offering additional features for language learning and progress monitoring [188].
- *Board Game for Teaching Arithmetic*: Virvou and Papadimitriou They developed a digital variant of the classic game 'Snakes and Ladders,' where the student practices their knowledge of arithmetic by placing the pawn at the right place on the board, according to the value of the dice [187]. The game contains adaptive elements by selecting primarily arithmetic operations that are challenging for the student. The same authors developed a digital board game where the student has to manage the coins at their disposal to buy products. The game adapts the numerical operations to the student's knowledge level and provides adaptive hints [186].
- *Adventure Game for Teaching Math*: Papadimitriou and Virvou described an educational adventure game for learning arithmetic, leveraging adaptive features and error diagnosis to enhance the learning process [126]. Papadimitriou and Virvou

presented an educational adventure game for learning computer programming, featuring adaptive scenarios based on fuzzy logic to provide individualized help and extend scenarios based on player weaknesses [127].
- *Various Types of Game with the Same Inner Reasoning*: Papadimitriou, Kamitsios, Chrysafiadi, and Virvou discussed personalized reasoning across different game categories, emphasizing the importance of incorporating educational content into various game types for enhanced engagement and learning outcomes [83, 129].

B. Beyond Educational Aims Referring to Entertainment and Usability

- *Motivation and Attractiveness*: Virvou, Katsionis, and Manos investigated the motivation and attractiveness of educational VR games, highlighting the potential for enjoyable learning experiences [197]. Additionally, Virvou and Katsionis addressed usability and likeability issues in VR educational games, emphasizing the need for enhancing educational benefits while maintaining appeal [182].
- *Usability of the Educational Game*: Dynamic Difficulty Adjustment (DDA) of an Educational 3D-Game: Chrysafiadi, Kamitsios, and Virvou presented a fuzzy-based DDA mechanism for educational 3D games, ensuring optimal challenge levels for players' skills and enhancing motivation without compromising learning [33].

C. Student Modeling Approaches and Reasoning

- *Cognitive-Psychology Ebbinghaus' forgetting curve*: Virvou and Manos discussed individualizing cognitive models of students' memory in ITSs, adapting teaching strategies based on cognitive psychology principles [183].
- *Affective reasoning*: Emotion recognition, an important aspect of affective computing, has significantly increased by integrating various modalities. Studies have explored the fusion of visual-facial indications with supplementary information, such as keyboard-stroke patterns and audio-lingual signals, to enhance accuracy and robustness in emotion identification processes [7, 161]. These efforts aim to create bi-modal interaction systems capable of deciphering emotions more effectively by employing multiple data sources [6, 162]. Additionally, research has investigated the utility of incorporating keyboard-stroke pattern information to assist visual-facial affect recognition systems, further refining their capabilities [153]. Such interdisciplinary approaches emphasize the dynamic nature of emotion recognition research, continually aiming to improve understanding and application in diverse domains. Regarding educational games, Katsionis and Virvou [86] describe how the Ortony, Clore & Collins cognitive theory of emotions [121] can be adapted in an educational game in order to model students' emotions while they learn. The system's inferences about students' emotions adapt interaction to each student's needs in particular educational and playing situations. In [87], affective reasoning is employed in a VR game that operates through web services so that the reasoning reaches each individual student-player at a distance through the Web.
- *Fuzzy logic-based quiz generation*: Papadimitriou, Chrysafiadi, and Virvou introduced FuzzEG, an adaptive educational game using fuzzy logic to tailor quizzes

and scenarios to learners' progress and needs, providing a personalized learning experience [128]. Chrysafiadi, Papadimitriou, and Virvou also presented a genetic algorithm that embeds a fuzzy rule-based mechanism in the fitness calculation phase to select the most appropriate questions for each learner and create personalized quizzes more realistically. Also, the presented algorithm aims to describe the appropriateness of each question with linguistic terms rather than obscure mathematics forms. In such a way, estimating the suitability of a quiz question is closer to the human way of thinking and becomes more explainable to tutors [130].

- *Cognitive-based adaptive scenarios*: Chrysafiadi, Papadimitriou, and Virvou explored cognitive-based adaptive scenarios in educational games using fuzzy reasoning, dynamically adjusting difficulty levels and plot scenarios based on individual student needs [37].
- *Pivacy issues on students' data*: The fact that many actions of the students are recorded so that data analytics may be performed to infer the needs and preferences of students leads to the requirement of applying privacy controls which conform with regulations such as GDPR [108, 109].

2.9 Authoring Tools

When developing computer-based instruction, there are typically three primary approaches: programming languages, authoring languages, and authoring tools [158]. While programming languages provide flexibility, they can be complex due to their varying syntax and commands. Authoring languages also require coding skills but are less intricate than programming languages, offering tailored commands and syntax for instructional purposes.

On the other hand, authoring tools are user-friendly, providing on-screen tools such as menus and icons to assist educators in creating educational material and interactive instruction without programming. As a result, authoring tools are an excellent choice for individuals without programming skills or access to programmers and those who are not willing to invest the time and effort to learn programming.

The research by Da Silva, Teichrieb, and Smith investigates the world of authoring tools for Augmented Reality (AR) applications in education. They advocate for a specialized authoring tool that caters to the unique skills and requirements of teachers [151]. AR technology has gained significant popularity due to its ability to enhance real-world environments by integrating virtual elements. This allows users to have an interactive experience, such as using a smartphone camera to view augmented content. For instance, visitors at a museum can use their mobile phones to access additional information about a painting or utilize GPS to gather details about their surroundings. Despite the widespread use of AR in other areas, its adoption by teachers in educational settings has been limited. A significant contributing factor to this underutilization is the absence of authoring tools with specific features tailored for educational use. This points to the crucial role such tools can play in enhancing the teaching and learning experience. The research findings underscore the necessity for teachers to have access to more specialized authoring tools to facilitate practical classroom exploration. Additionally, the study highlights key considerations for developers when creating such tools, including prioritizing user-friendliness, creating an offline version, ensuring mobile compatibility, offering personalization options, and enabling collaboration among users.

TaleBlazer, on the other hand, is a free, open-source platform that supports the development and play of location-based experiences on Android or iOS devices [103]. It includes agents, actions that govern their behavior, the game areas, and scenarios, allowing game designers to define the game's logic using block scripting.

ARIS is a cloud-based, open-source platform that enables the creation and enjoyment of location-based augmented reality experiences on iOS devices [69]. It employs various elements such as interactive objects, triggers for linking real-world actions to ARIS actions, and scenes that facilitate the design of stories and game flow. Additionally, ARIS offers tools for constructing dialogues between players and in-game characters.

Vert and Andone have recognized the positive impact of VR games on education and the challenges instructors encounter in creating personalized VR experiences. As a result, they have put forth a set of criteria for selecting authoring tools suitable for educational purposes [173]. They specifically opted for authoring tools meeting the following requirements: no need for programming skills, continuous maintenance and development, web-based platform, free of charge, and accommodating many students across various platforms such as web, Android, iOS, and VR headsets. They have chosen CoSpaces to demonstrate how it manages and disseminates VR experiences. Additionally, they have highlighted that this approach is more immersive and educational than exploring a location via Google Maps. The authors have also emphasized the importance of VR authoring tools in supporting collaboration among educators, providing a multiplayer experience, enabling assessment of students' knowledge (e.g., through the integration of quizzes), assigning students to classes, importing VR items, and having an offline version for usage by students with limited or no internet access.

Diaz-Escudero, Torrado, Gomez, and Montoro have developed a mobile authoring tool that includes functionality allowing authors, such as caregivers, teachers, and relatives, to create emotional self-regulation strategies for individuals with autism spectrum disorder and cognitive disabilities [46]. These strategies include steps or cards with multimedia content, such as pictograms, photos, text, or animations. Authors can also create regulation methods comprising multiple strategies or maintain multiple profiles with different regulation methods and strategies. Individuals with autism spectrum disorder wear a smartwatch that syncs with their mobile phones to receive the authors' profiles. For example, when the smartwatch detects high heart rates, self-regulation strategies and clear instructions appear on the screen, and the individual follows them until the heart rate drops. Subsequently, the watch sends the reduced pulse rate to the mobile phone for recording, allowing the authors to monitor the individual's pulse evolution, the strategies followed, and their effectiveness in promoting emotional self-regulation. The system evaluation shows that teachers using it can organize interventions more efficiently since the system records a complete history of the interventions and the individuals' responses. Additionally, the ease of creating strategies enables them to offer interventions to larger groups of students without sacrificing the ability to personalize them according to each student's needs.

The team of Lytridis and Tsinakos has ventured into the realm of augmented reality with their creation. ARTutor is an educational software that seamlessly integrates augmented reality into the academic process for both teachers and students [101]. Its main goal is to augment digital content alongside traditional academic textbooks and other materials to facilitate independent learning and enhance students' comprehension of the subject matter. The platform also provides a web authoring tool that enables educators to develop AR books and categorize them based on subject and language. Teachers can upload PDF files and multimedia components such as videos, images, and audio. The authoring tool boasts a user-friendly interface, making it simple for educators to enrich their teaching materials, even with limited computer proficiency. The authors evaluated the use, effectiveness, and suitability of the AR-based learning tool. Both educators and students reacted positively to integrating digital content into books and using the mobile application to access interactive digital material. However, the educational benefits of using the authoring tool are still being explored and are part of the authors' plans.

DEDOS-editor is a versatile authoring tool designed to enable educators to create interactive multimedia activities suitable for a wide range of devices, including personal computers, smartphones, tablets, digital blackboards, and multitouch tabletops [142]. The tool offers a user-friendly, card-based interface, where each card represents a virtual object for importing learning content and aligning it with educational goals. Users can easily create activities by dragging and dropping cards to create multiple-choice exercises, pairing-match activities, connecting-the-dots tasks, or a combination of these formats. Additionally, the tool can adapt to the technological proficiency of educators by selectively locking or unlocking advanced features. In a comparative study, the DEDOS-editor was significantly more user-friendly than another authoring tool, JClic-Author, based on feedback from survey participants.

Enook (Evolutionary notebook) is a specialized authoring tool used to create activity-based learning objects [3]. These activities provide students with hands-on learning experiences and access to educational resources, including real e-books that serve as textbooks and notebooks. Additionally, educators can collect feedback from students on the quality of the lessons and activities, enabling them to monitor student progress and identify areas where additional support may be necessary. Compared to other authoring tools like eXelearning and SmartBuilder, Enook offers additional features, including support for XML or JSON format, team-based learning, learning data, and learning analytics. Furthermore, Enook offers various student evaluation methods, including peer and self-assessment, question-and-answer sessions, and survey questionnaires.

Game analytics involves the collection and analysis of user interactions within a game. For example, studying a player's movements can offer crucial insights into their behavior, including their challenges in completing specific levels. Developers can then use this information to enhance the game and create a more enjoyable experience for the player. Perez-Colado et al. introduced SG-xAPI, an authoring tool designed to integrate game analytics into uAdventure, a platform for developing educational adventure games [133]. SG-xAPI records events in uAdventure and categorizes them based on type while also capturing additional details about the item the player interacted with and the event's outcome. This enables the author to track the player's progress and behavior by accessing information such as when the player accessed the NPC or item action menu, initiated/skipped a cutscene and responded to a dialogue question.

The authors Paiva, Leal, and Queirós highlight the potential of gamification, which involves integrating game elements into a process to enhance student motivation and engagement in learning [123]. Game elements such as leaderboards, badges, tournaments, and achievements can captivate participants. However, creating these elements can be challenging and require programming skills. To address this, the authors introduce the authoring component of Asura, a game-based programming assessment environment that enables educators to create programming challenges without requiring technical expertise. In Asura, students are tasked with coding a software agent capable of solving various game-form programming tests. The platform includes a component for evaluating the code of the software agent and assigning grades accordingly. After the agent development phase is completed, educators can organize tournaments, specifying the point system and tournament structure. The authors compared the process of creating Asura challenges to creating problems for the International Collegiate Programming Contest (ICPC) and found that Asura enables faster challenge creation. They assert that the authoring component of Asura is user-friendly and valuable for educators.

Baldeón, Puig, Rodriguez, Muriel, and Zardain acknowledged educators' challenges in customizing educational games to meet specific learning objectives and students' learning styles. In response, they proposed a conceptual domain model to facilitate the development of authoring tools for educational games [18]. This model has two main components: an educational section for incorporating and customizing learning activities and a game domain section for adding and modifying game elements. It incorporates three fundamental models: the Outcome-Based Education (OBE) Educational Model for educational design, the Game Design Model for game design considerations, and the Digital Game-Based Learning Model, which integrates elements from the previous models. The authors applied this model to create authoring tools for two math games. Fracsland is an adventure game designed to help students understand the concept of fractions. It allows teachers to create exercises, define solutions, provide specific fractions, and choose their visual representation (pie or bar). In GeoPieces, players use 2D shapes to construct complex 3D objects. Each game's curriculum aligns with the intended learning outcomes specified by the teacher using the authoring tool.

As mentioned above, ITSs utilize AI techniques to offer personalized instruction to students based on their comprehension of the subject matter and their current emotional state [45]. A well-crafted ITS has the potential to effectively complement or even replace traditional teaching models across various academic disciplines. However, the design and development of ITSs is a complex and resource-intensive undertaking. Furthermore, despite their proven efficacy, many educators are hesitant to embrace ITSs due to challenges in implementation, increased workload for instructors, and sometimes a lack of clearly defined learning objectives [114].

Hence, developing an authoring tool poses several challenges for ITSs, including ensuring the flexibility of the integrated techniques, achieving domain independence, assessing the tool's effectiveness for prospective authors (instructors), and evaluating its impact on students who will engage with the resulting ITSs. Over time, researchers have developed authoring tools to streamline the creation of ITSs, lower production efforts, and diversify the range and quantity of available tutors [3, 8, 18, 45, 46, 59, 69, 97, 101, 103, 114, 123, 133, 142, 151, 173, 177, 179]. For example, WEAR has been a tool for authoring and delivering adaptive Web-based textbooks using WEAR [110]. Although there are commercially available authoring tools for traditional computer-assisted instruction (CAI) and multimedia tutoring, there remains a need for more user-friendly systems to create intelligent tutors. Commercial authoring tools tend to prioritize creating visually appealing outputs, often neglecting the integration of techniques essential for building intelligent tutors that offer tailored and meaningful learning experiences. Consequently, shortly after the start of ITS research, researchers began delving into ITS authoring tools, resulting in the creation of numerous specialized authoring tools. These tools, tailored for ITS, aim to facilitate the cost-effective development of ITSs that cater to the unique needs of individual learners. Ultimately, the primary goal of an ITS is to provide highly personalized guidance to students [8, 179].

Authoring tools may be suitable for achieving reusability in Intelligent Tutoring as well as gaming environments. They enable various forms of user modeling, such as student modeling for educational processes, user experience modeling for enhancing user-friendliness, and player modeling for users interacting with educational games. For example, Virvou et al. [195] describe VR-INTEGATE, which creates VR educational games and allows for player modeling by tracking user interactions. Virvou and Moundridou [185] discuss an algebra authoring tool that facilitates personalized student modeling, enabling ITSs to adapt to learners' needs. Moundridou and Virvou [111] highlight the benefits of web-based ITS authoring for user experience modeling, offering broader educational access and customization. Virvou and Alepis [179] emphasize mobile learning features that improve user experience by supporting flexible, on-the-go education. Virvou [177] examines authoring tools that tailor content to users' cognitive processes, enhancing system responsiveness, while Virvou [175] introduces a domain-independent reasoning mechanism to boost versatility across subjects. Additionally, the INMA tool, designed for music education, combines education and entertainment to enhance music perception and understanding using knowledge-based techniques [198].

The following sections provide an overview of the current book's authoring tool, highlighting the author's expertise and critical role in shaping the delivery of personalized educational content and exercises using AI techniques tailored to students' knowledge levels. The present authoring tool not only allows for the import of learning content but also enables authors to customize the features of the intelligent component that powers the integrated ITS, thereby enriching the personalized learning experience.

References

1. Abu-Naser, S.S., Ahmed, A., Almasri, N., Abu Sultan, Y.S.: Human computer interaction design of the lp-its: linear programming intelligent tutoring systems. Int. J. Artif. Intell, Appl (2011)
2. Adams, E., Rollings, A.: Fundamentals of game design (game design and development series). Prentice-Hall, Inc. (2006)
3. Ahn, J.Y., Mun, G.S., Han, K.S., Choi, S.H.: An online authoring tool for creating activity-based learning objects. Educ. Inf. Technol. **22**, 3005–3015 (2017)
4. Al Rekhawi, H.A.: Android applications development intelligent tutoring system (2020)
5. Al Rekhawi, H.A., Abu Naser, S.S.: An intelligent tutoring system for learning android applications ui development. Int. J. Eng. Inf. Syst. (IJEAIS) **2**(1), 1–14 (2018)
6. Alepis, E., Virvou, M.: Emotional intelligence: constructing user stereotypes for affective bi-modal interaction. In: International Conference on Knowledge-Based and Intelligent Information and Engineering Systems, Springer, pp. 435–442 (2006)
7. Alepis, E., Stathopoulou, I.O., Virvou, M., Tsihrintzis, G.A., Kabassi, K.: Audio-lingual and visual-facial emotion recognition: towards a bi-modal interaction system. In: 2010 22nd IEEE International Conference on Tools with Artificial Intelligence, IEEE, vol. 2, pp. 274–281. IEEE (2010)
8. Aleven, V., McLaren, B.M., Sewall, J., Koedinger, K.R.: The cognitive tutor authoring tools (ctat): preliminary evaluation of efficiency gains. In: Intelligent Tutoring Systems: 8th International Conference, ITS 2006, Jhongli, Taiwan, 26–30 June 2006. Proceedings, vol. 8, pp. 61–70. Springer (2006)
9. Alkhatlan, A., Kalita, J.: Intelligent tutoring systems: A comprehensive historical survey with recent developments (2018). arXiv:1812.09628
10. Almasri, A., Ahmed, A., Almasri, N., Abu Sultan, Y.S., Mahmoud, A.Y., Zaqout, I.S., Akkila, A.N., Abu-Naser, S.S.: Intelligent tutoring systems survey for the period 2000–2018 (2019)
11. Aminoleslami, S., Maghooli, K., Sammaknejad, N., Haghipour, S., Sadeghi-Firoozabadi, V.: Classification of autistic and normal children using analysis of eye-tracking data from computer games. Signal Image Video Process. **17**(8), 4357–4365 (2023)
12. Amory, A.: Building an educational adventure game: Theory, design, and lessons. J. Interact. Learn. Res. **12**(2), 249–263 (2001)
13. Anania, J.: The Effects of Quality of Instruction on the Cognitive and Affective Learning of Students. The University of Chicago (1981)
14. Anania, J.: The influence of instructional conditions on student learning and achievement. Eval. Educ. **7**(1), 1–92 (1983)
15. Anderson, J.R., Boyle, C.F., Reiser, B.J.: Intelligent tutoring systems. Science **228**(4698), 456–462 (1985)
16. Apperley, T.H.: Genre and game studies: toward a critical approach to video game genres. Simul. Gaming **37**(1), 6–23 (2006)
17. Ashcraft, B., Snow, J., Williams, K., Crecente, B.: Arcade mania!: the turbo-changed world of japan's game centers. (No Title) (2008)
18. Baldeón, J., Puig, A., Rodríguez, I., Muriel, C., Zardain, L.: A conceptual model for educational game authoring: A showcase in math games. In: Design, User Experience, and Usability: Designing Pleasurable Experiences: 6th International Conference, DUXU 2017, Held as Part of HCI International 2017, Vancouver, BC, Canada, 9–14 July 2017, Proceedings, Part II, vol. 6, pp. 347–361. Springer (2017)
19. Barton, M., Stacks, S.: Dungeons and Desktops: The History of Computer Role-Playing Games 2e. CRC Press (2019)
20. Basu, S., Saha, S., Das, S., Guha, R., Mukherjee, J., Mahadevappa, M.: Assessment of attention and working memory among young adults using computer games. J. Ambient Intell. Hum. Comput. **14**(3), 2413–2428 (2023)
21. Bateman, C., Zagal, J.P.: Game design lineages. Transactions **3**(3), 13–46 (2018)

22. Björk, S., Zagal, J.P.: Game design and role-playing games. In: Role-Playing Game Studies, Routledge, pp. 323–336 (2018)
23. Bloom, B.S.: The 2 sigma problem: The search for methods of group instruction as effective as one-to-one tutoring. Educ. Res. **13**(6), 4–16 (1984)
24. Bouchillon, B.C., Stewart, P.A.: Computer games, trust, and immediacy: Role-playing as immigrants in the south. Comput. Hum. Behav. **140**, 107571 (2023)
25. Braun, B., Stopfer, J.M., Müller, K.W., Beutel, M.E., Egloff, B.: Personality and video gaming: Comparing regular gamers, non-gamers, and gaming addicts and differentiating between game genres. Comput. Hum. Behav. **55**, 406–412 (2016)
26. Buchanan-Oliver, M., Seo, Y.: Play as co-created narrative in computer game consumption: the hero's journey in warcraft iii. J. Consum. Behav. **11**(6), 423–431 (2012)
27. Cardoso, J., Bittencourt, G., Frigo, L.B., Pozzebon, E., Postal, A.: Mathtutor: a multi-agent intelligent tutoring system. In: Artificial Intelligence Applications and Innovations: IFIP 18th World Computer Congress TC12 First International Conference on Artificial Intelligence Applications and Innovations (AIAI-2004) 22–27 Aug 2004 Toulouse, France, vol. 1, pp. 231–242. Springer (2004)
28. Chapman, A.: Is sid meier's civilization history? Rethink. Hist. **17**(3), 312–332 (2013)
29. Chen, C.H., Chen, J.C., Wu, I., et al.: Special issue on advances in computer chinese chess. ICGA J. (Preprint), 1–2 (2024)
30. Chesham, A., Wyss, P., Müri, R.M., Mosimann, U.P., Nef, T., et al.: What older people like to play: genre preferences and acceptance of casual games. JMIR Ser. Games **5**(2), e7025 (2017)
31. Chiang, F.K., Chun-Hao, C., Hu, D., Zhang, G., Liu, Y.: Design and development of a safety educational adventure game. Int. J. Emerg. Technol. Learn. (Online) **14**(3), 201 (2019)
32. Choi, E., Shin, S.H., Ryu, J.K., Jung, K.I., Kim, S.Y., Park, M.H.: Commercial video games and cognitive functions: video game genres and modulating factors of cognitive enhancement. Behav. Brain Funct. **16**, 1–14 (2020)
33. Chrysafiadi, K.: Fuzzy logic-based software systems. In: Fuzzy Logic-Based Software Systems, pp. 31–129. Springer (2023)
34. Chrysafiadi, K., Virvou, M.: Evaluating the integration of fuzzy logic into the student model of a web-based learning environment. Expert Syst. Appl. **39**(18), 13127–13134 (2012)
35. Chrysafiadi, K., Virvou, M.: Persiva: An empirical evaluation method of a student model of an intelligent e-learning environment for computer programming. Comput. Educ. **68**, 322–333 (2013)
36. Chrysafiadi, K., Papadimitriou, S., Virvou, M.: Which is better for learning: a web-based educational application or an educational game? In: 2019 International Symposium on Performance Evaluation of Computer and Telecommunication Systems (SPECTS), pp. 1–6. IEEE (2019)
37. Chrysafiadi, K., Papadimitriou, S., Virvou, M.: Cognitive-based adaptive scenarios in educational games using fuzzy reasoning. Knowl. Based Syst. **250**, 109111 (2022)
38. Chrysafiadi, K., Virvou, M., Tsihrintzis, G.A.: A fuzzy-based evaluation of e-learning acceptance and effectiveness by computer science students in greece in the period of covid-19. Electronics **12**(2), 428 (2023)
39. Chrysafiadi, K., Virvou, M., Tsihrintzis, G.A., Hatzilygeroudis, I.: Evaluating the user's experience, adaptivity and learning outcomes of a fuzzy-based intelligent tutoring system for computer programming for academic students in greece. Educ. Inf. Technol. **28**(6), 6453–6483 (2023)
40. Cohen-Solal, Q., Cazenave, T.: Athénan wins sixteen gold medals at the computer olympiad. ICGA J. (Preprint), 1–2 (2023)
41. Conde, A., de Ipiña, K.L., Larrañaga, M., Garay-Vitoria, N., Irigoyen, E., Ezeiza, A., Rubio, J.: LAGUNTXO: A rule-based intelligent tutoring system oriented to people with intellectual disabilities. Springer (2009)
42. Courtois, C.: Jouer dans l'espace urbain capitaliste de disneyland avec the secret of monkey island: Monkey island 1 & 2. Géographie et cultures, pp. 99–118 (2019)

43. Cox, A.R.: Women by women: a gender analysis of sierra titles by women designers. Feminism in play, pp. 21–35 (2018)
44. Crawford, G.: The cult of champ man: the culture and pleasures of championship manager/football manager gamers. Inf. Commun. Soc. **9**(4), 496–514 (2006)
45. Dermeval, D., Paiva, R., Bittencourt, I.I., Vassileva, J., Borges, D.: Authoring tools for designing intelligent tutoring systems: a systematic review of the literature. Int. J. Artif. Intell. Educ. **28**, 336–384 (2018)
46. Díaz-Escudero, A., Torrado, J.C., Gomez, J., Montoro, G.: Technology: aided authoring tools for asd coaching support. In: Proceedings of the XIX International Conference on Human Computer Interaction, pp. 1–6 (2018)
47. Dille, F., Platten, J.Z.: The Ultimate Guide to Video Game Writing and Design. Lone Eagle (2008)
48. Dobrowolski, P., Hanusz, K., Sobczyk, B., Skorko, M., Wiatrow, A.: Cognitive enhancement in video game players: the role of video game genre. Comput. Hum. Behav. **44**, 59–63 (2015)
49. Dubinsky, A., Zhykharieva, Y.: The project of text adventure game for studying the paramedical algorithms. In: eLearning & Software for Education, vol. 1 (2020)
50. Duffield, J.A.: Designing computer software for problem-solving instruction. Educ. Technol. Res. Dev. **39**(1), 50–62 (1991)
51. D'Apice, C., Grieco, C., Piscopo, R., Liscio, L.: Dms2015short-2: advanced learning technologies for elearning in the enterprise: design of an educational adventure game to teach computer security. J. Vis. Lang. Comput. **31**, 260–266 (2015)
52. Es-Sajjade, A., Paas, F.: Educational theories and computer game design: lessons from an experiment in elementary mathematics education. Educ. Technol. Res. Dev. **68**(5), 2685–2703 (2020)
53. Faria, A.J.: Business simulation games: current usage levels-an update. Simul. Gaming **29**(3), 295–308 (1998)
54. Fletcher-Flinn, C.M., Gravatt, B.: The efficacy of computer assisted instruction (cai): a meta-analysis. J. Educ. Comput. Res. **12**(3), 219–241 (1995)
55. Fu, F.L., Su, R.C., Yu, S.C.: Egameflow: a scale to measure learners' enjoyment of e-learning games. Comput. Educ. **52**(1), 101–112 (2009)
56. Gabbiadini, A., Greitemeyer, T.: Uncovering the association between strategy video games and self-regulation: a correlational study. Personal. Individ. Differen. **104**, 129–136 (2017)
57. Gao, F., Izadpanah, S.: The relationship between computer games and computer self-efficacy with academic engagement: the mediating role of students' creativity. Educ. Inf. Technol. **28**(11), 14229–14248 (2023)
58. Gingold, Y.I.: From rock, paper, scissors to street fighter ii: proof by construction. In: Proceedings of the 2006 ACM SIGGRAPH symposium on Videogames, pp. 155–158 (2006)
59. Gordillo, A., Barra, E., Quemada, J.: Sgame: an authoring tool to easily create educational video games by integrating scorm-compliant learning objects. IEEE Access **9**, 126414–126430 (2021)
60. Grace, L.: Game type and game genre (2005). Accessed 22 Feb 2009. (8)
61. Graesser, A., Conley, M., Olney, A.: Intelligent tutorıng systems, apa educational psychology handbook (2012)
62. Greenberg, B.S., Sherry, J., Lachlan, K., Lucas, K., Holmstrom, A.: Orientations to video games among gender and age groups. Simul. Gaming **41**(2), 238–259 (2010)
63. Grossman, L., Ressner, J.: The age of doom. Time **164**(6), 82–84 (2004)
64. Hahn, B., Nithyanand, R., Gill, P., Johnson, R.: Games without frontiers: investigating video games as a covert channel. In: 2016 IEEE European Symposium on Security and Privacy (EuroS&P), pp. 63–77. IEEE (2016)
65. Hanson, C.: Daventry and the worlds of king's quest. In: Exploring Imaginary Worlds, pp. 137–155. Routledge (2020)
66. Hartisch, M., Hsueh, C.H., Schaeffer, J.: Advances in computer games (acg 2023) conference report. ICGA J. (Preprint), 1–4 (2024)

67. Hartmann, D.: Hans berliner: a life in chess, computer chess and much more. ICGA J. (Preprint), 1–10 (2023)
68. Hawlitschek, A., Joeckel, S.: Increasing the effectiveness of digital educational games: The effects of a learning instruction on students' learning, motivation and cognitive load. Comput. Hum. Behav. **72**, 79–86 (2017)
69. Holden, C.L., Gagnon, D.J., Litts, B.K., Smith, G.: Aris: an open-source platform for widespread mobile augmented reality experimentation. In: Technology Platform Innovations and Forthcoming Trends in Ubiquitous Learning, pp. 19–34. IGI Global (2014)
70. Hooshyar, D., Malva, L., Yang, Y., Pedaste, M., Wang, M., Lim, H.: An adaptive educational computer game: Effects on students' knowledge and learning attitude in computational thinking. Comput. Hum. Behav. **114**, 106575 (2021)
71. Hooshyar, D., El Mawas, N., Milrad, M., Yang, Y.: Modeling learners to early predict their performance in educational computer games. IEEE Access **11**, 20399–20417 (2023)
72. Hou, H.T., Li, M.C.: Evaluating multiple aspects of a digital educational problem-solving-based adventure game. Comput. Hum. Behav. **30**, 29–38 (2014)
73. Howland, J., Jonassen, D., Marra, R., Moore, J.: Learning to Solve Problems with Technology: A Constructivist Perspective. Upper Saddle River, New Jersey, Merrill Prentice Hall (2003)
74. Iida, H., Schaeffer, J., Wu, I., et al.: The 2023 computer olympiad. ICGA J. (Preprint), 1–18 (2023)
75. Ip, B., Jacobs, G.: Quantifying game design. Design Stud. **25**(6), 607–624 (2004)
76. Jaimez-González, C.R., Erazo-Palacios, J., García-Mendoza, B.: Blockcode: a web application to create games that support the learning of computer programming logic. Int. J. Emerg. Technol. Learn. **18**(15) (2023)
77. Janebi Enayat, M., Haghighatpasand, M.: Exploiting adventure video games for second language vocabulary recall: A mixed-methods study. Innov. Lang. Learn. Teach. **13**(1), 61–75 (2019)
78. Jenny, S.E., Schary, D.: Exploring the effectiveness of learning american football through playing the video game "madden nfl.âŁž Int. J. Technol. Teach. Learn. **10**(1), 72–87 (2014)
79. Johansson, S.J.: What makes online collectible card games fun to play? In: DiGRA Conference, p. 34 (2009)
80. Johnson, D., Wiles, J.: Computer games with intelligence. In: 10th IEEE International Conference on Fuzzy Systems. (Cat. No. 01CH37297), vol. 3, pp. 1355–1358. IEEE (2001)
81. Kabassi, K., Virvou, M.: Combination of a cognitive theory with the multi-attribute utility theory. In: International Conference on Knowledge-Based and Intelligent Information and Engineering Systems, pp. 944–950. Springer (2003)
82. Kahraman, A., Kazançoğlu, I.: A qualitative research of young people's motivation to start, continue, reduce and quit playing online multiplayer games on computer. Int. J. Hum. Comput. Interact. **39**(17), 3289–3311 (2023)
83. Kamitsios, M., Papadimitriou, S., Chrysafiadi, K., Virvou, M.: Incorporating quizzes into a 3d educational game calling web services. In: EdMedia+ Innovate Learning, Association for the Advancement of Computing in Education (AACE), pp. 1170–1179 (2018)
84. Karaci, A., Akyüz, H.I., Bilgici, G., Arici, N.: Effects of web-based intelligent tutoring systems on academic achievement and retention. Online Submiss. **181**(16), 35–41 (2018)
85. Katsionis, G., Virvou, M.: A cognitive theory for affective user modelling in a virtual reality educational game. In: 2004 IEEE International Conference on Systems, Man and Cybernetics (IEEE Cat. No. 04CH37583), vol. 2, pp. 1209–1213. IEEE (2004)
86. Katsionis, G., Virvou, M.: Adapting occ theory for affect perception in educational software. In: Proceedings of the International Conference on Human-Computer Interaction (2005)
87. Katsionis, G., Virvou, M.: Personalised e-learning through an educational virtual reality game using web services. Multimed. Tools Appl. **39**, 47–71 (2008)
88. Kaye, L.K.: Football manager as a persuasive game for social identity formation. In: Cases on the Societal Effects of Persuasive Games, pp. 1–17. IGI Global (2014)
89. Kim, S., Chang, M., Deater-Deckard, K., Evans, M.A., Norton, A., Samur, Y.: Educational games and students' game engagement in elementary school classrooms. J. Comput. Educ. **4**, 395–418 (2017)

90. Kordaki, M., Gousiou, A.: Computer card games in computer science education: a 10-year review. J. Educ. Technol. Soc. **19**(4), 11–21 (2016)
91. Kordaki, M., Gousiou, A.: Digital card games in education: a ten year systematic review. Comput. Educ. **109**, 122–161 (2017)
92. Kusairi, S., Alfad, H., Zulaikah, S.: Development of web-based intelligent tutoring (itutor) to help students learn fluid statics. J. Turk. Sci. Educ. **14**(2), 1–11 (2017)
93. Kwak, M., Koohang, A., Floyd, K., Choi, A.: An educational adventure game for teaching information literacy and student engagement (2018)
94. Leino, O.T.: Chaos at the europoort: performance, empathy and materiality in euro truck simulator 2. In: Proceedings of Philosophy of Computer Games Conference 2017 (2017)
95. Lewis, M.L., Weber, R., Bowman, N.D.: "They may be pixels, but they're my pixels:" developing a metric of character attachment in role-playing video games. Cyberpsychol. Behav. **11**(4), 515–518 (2008)
96. Lin, C.J., Hwang, G.J., Fu, Q.K., Cao, Y.H.: Facilitating efl students' english grammar learning performance and behaviors: a contextual gaming approach. Comput. Educ. **152**, 103876 (2020)
97. Locatis, C., Ullmer, E., Carr, V., Banvard, R., Le, Q., Lo, R., Williamson, M.: Authoring systems: an introduction and assessment. J. Comput. Higher Educ. **3**, 23–35 (1991)
98. de Lope, R.P., Arcos, J.R.L., Medina-Medina, N., Paderewski, P., Gutiérrez-Vela, F.: Design methodology for educational games based on graphical notations: designing urano. Entertain. Comput. **18**, 1–14 (2017)
99. López-Fernández, D., Mayor, J., García-Pérez, M., Gordillo, A.: Are virtual reality serious video games more effective than web video games? IEEE Comput. Graph. Appl. **43**(2), 32–42 (2023)
100. Lyömiö, S.: Level and story design in choose your own adventure games (2017)
101. Lytridis, C., Tsinakos, A.: Evaluation of the artutor augmented reality educational platform in tertiary education. Smart Learn. Environ. **5**(1), 1–15 (2018)
102. McLaren, B.M., Adams, D.M., Mayer, R.E., Forlizzi, J.: A computer-based game that promotes mathematics learning more than a conventional approach. Int. J. Game Based Learn. (IJGBL) **7**(1), 36–56 (2017)
103. Medlock-Walton, M.P.: Taleblazer: a platform for creating multiplayer location based games. Ph.D. thesis, Massachusetts Institute of Technology (2012)
104. Mehm, F., Göbel, S., Steinmetz, R.: An authoring tool for educational adventure games: concept, game models and authoring processes. Int. J. Game Based Learn. (IJGBL) **3**(1), 63–79 (2013)
105. Melchor-Ferrer, E., Davia-Rodriguez, M.A.: Computer games and the study of terminology: an application to national accounts. Educ. Inf. Technol. **28**(1), 135–153 (2023)
106. Michaud, L.N., McCoy, K.F., Pennington, C.A.: An intelligent tutoring system for deaf learners of written english. In: Proceedings of the Fourth International ACM Conference on Assistive Technologies, pp. 92–100 (2000)
107. Mora-Cantallops, M.: Transhistorical perspective of the puzzle video game genre. In: Proceedings of the 13th International Conference on the Foundations of Digital Games, pp. 1–9 (2018)
108. Mougiakou, E., Virvou, M.: Based on gdpr privacy in uml: case of e-learning program. In: 2017 8th International Conference on Information, Intelligence, Systems & Applications (IISA), pp. 1–8. IEEE (2017)
109. Mougiakou, E., Papadimitriou, S., Virvou, M.: Automated decision making and personal data protection in intelligent tutoring systems: design guidelines. In: Knowledge-Based Software Engineering: 2018: Proceedings of the 12th Joint Conference on Knowledge-Based Software Engineering (JCKBSE 2018) Corfu, Greece, vol. 12, pp. 231–241. Springer (2019)
110. Moundridou, M., Virvou, M.: Authoring and delivering adaptive web-based textbooks using wear. In: Proceedings IEEE International Conference on Advanced Learning Technologies, pp. 185–188. IEEE (2001)

111. Moundridou, M., Virvou, M.: Authoring intelligent tutoring systems over the world wide web. Proc. First Int. IEEE Sympos. Intell. Syst. IEEE **1**, 160–165 (2002)
112. Moundridou, M., Virvou, M.: Evaluating the persona effect of an interface agent in a tutoring system. J. Comput. Assist. Learn. **18**(3), 253–261 (2002)
113. Mullgardt, B.: Gaming in the gilded age. Councilor A J. Soc. Stud. **75**(1), 2 (2014)
114. Murray, T.: An overview of intelligent tutoring system authoring tools: Updated analysis of the state of the art. In: Toward Cost-Effective Adaptive, Interactive and Intelligent Educational Software, Authoring Tools for Advanced Technology Learning Environments, pp. 491–544 (2003)
115. Neto, L.V., Fontoura Junior, P.H., Bordini, R.A., Otsuka, J.L., Beder, D.M.: Design and implementation of an educational game considering issues for visually impaired people inclusion. Smart Learn. Environ. **7**(1), 4 (2020)
116. Nooney, L.: Let's begin again: Sierra on-line and the origins of the graphical adventure game. Am. J. Play **10**(1), 71–98 (2017)
117. Nutt, D., Railton, D.: The sims: Real life as genre. Inf. Commun. Soc. **6**(4), 577–592 (2003)
118. Nwana, H.S.: Intelligent tutoring systems: an overview. Artif. Intell. Rev. **4**(4), 251–277 (1990)
119. Olsen, J.K., Belenky, D.M., Aleven, V., Rummel, N.: Intelligent tutoring systems for collaborative learning: enhancements to authoring tools. In: Artificial Intelligence in Education: 16th International Conference, AIED 2013, Memphis, TN, USA, 9–13 July 2013. Proceedings, vol. 16, pp. 900–903. Springer (2013)
120. Ontanón, S., Synnaeve, G., Uriarte, A., Richoux, F., Churchill, D., Preuss, M.: A survey of real-time strategy game ai research and competition in starcraft. IEEE Trans. Comput. Intell. AI Games **5**(4), 293–311 (2013)
121. Ortony, A., Clore, G.L., Collins, A.: The Cognitive Structure of Emotions Cambridge, vol. 9. Cambridge University Press, UK (1988)
122. Pai, K.C., Kuo, B.C., Liao, C.H., Liu, Y.M.: An application of chinese dialogue-based intelligent tutoring system in remedial instruction for mathematics learning. Educ. Psychol. **41**(2), 137–152 (2021)
123. Paiva, J.C., Leal, J.P., Queirós, R.: Authoring game-based programming challenges to improve students' motivation. In: The Challenges of the Digital Transformation in Education: Proceedings of the 21st International Conference on Interactive Collaborative Learning (ICL2018), vol. 1, pp. 602–613. Springer (2020)
124. Panagoulias, D.P., Palamidas, F.A., Virvou, M., Tsihrintzis, G.A.: Evaluating the potential of llms and chatgpt on medical diagnosis and treatment. In: 2023 14th International Conference on Information, Intelligence, Systems & Applications (IISA), pp. 1–9. IEEE (2023)
125. Panagoulias, D.P., Virvou, M., Tsihrintzis, G.A.: A novel framework for artificial intelligence explainability via the technology acceptance model and rapid estimate of adult literacy in medicine using machine learning. Expert Syst. Appl. **248**, 123375 (2024)
126. Papadimitriou, S., Virvou, M.: An online adventure game for teaching math. In: 2016 7th International Conference on Information, Intelligence, Systems & Applications (IISA), pp. 1–5. IEEE (2016)
127. Papadimitriou, S., Virvou, M.: Adaptivity in scenarios in an educational adventure game. In: 2017 8th International Conference on Information, Intelligence, Systems & Applications (IISA), pp. 1–6. IEEE (2017)
128. Papadimitriou, S., Chrysafiadi, K., Virvou, M.: Fuzzeg: Fuzzy logic for adaptive scenarios in an educational adventure game. Multimed. Tools Appl. **78**(22), 32023–32053 (2019). https://doi.org/10.1007/s11042-019-07955-w
129. Papadimitriou, S., Kamitsios, M., Chrysafiadi, K., Virvou, M.: Learn-and-play personalised reasoning from point-and-click to virtual reality mobile educational games. Intell. Decis. Technol. **15**(2), 321–332 (2021)
130. Papadimitriou, S., Chrysafiadi, K., Virvou, M.: Adaptive quizzes using fuzzy genetic algorithm. In: 2023 14th International Conference on Information, Intelligence, Systems & Applications (IISA), pp. 1–8. IEEE (2023)

131. Pearce, C.: Towards a game theory of game. In: First Person: New Media as Story, Performance, and Game (2004)
132. Pedersen, C., Togelius, J., Yannakakis, G.N.: Modeling player experience in super mario bros. In: 2009 IEEE Symposium on Computational Intelligence and Games, pp. 132–139. IEEE (2009)
133. Perez-Colado, I.J., Perez-Colado, V.M., Freire-Moran, M., Martínez-Ortiz, I., Fernández-Manjón, B.: Integrating learning analytics into a game authoring tool. In: Advances in Web-Based Learning–ICWL 2017: 16th International Conference, Cape Town, South Africa, 20–22 Sept 2017, Proceedings, vol. 16, pp. 51–61. Springer (2017)
134. Pijl, R.: The future of computer chess tournaments. ICGA J. (Preprint), 1–6 (2023)
135. Pinelle, D., Wong, N., Stach, T.: Using genres to customize usability evaluations of video games. In: Proceedings of the 2008 Conference on Future Play: Research, Play, Share, pp. 129–136 (2008)
136. Grande-de Prado, M., Baelo, R., García-Martín, S., Abella-García, V.: Mapping role-playing games in Ibero-America: an educational review. Sustainability **12**(16), 6298 (2020)
137. Psarologou, A., Bourbakis, N., Virvou, M.: A mapping mechanism of nl sentences onto an spn state machine for understanding purposes. In: IISA 2014, The 5th International Conference on Information. Intelligence, Systems and Applications, pp. 321–324. IEEE (2014)
138. Qaffas, A.: An operational study of video games' genres (2020)
139. Quintero, C.A., Álvarez, G.A.C.: Serious games and computer programming competencies development in educational contexts. IEEE Revista Iberoamericana de Tecnologias del Aprendizaje **18**(1), 48–53 (2023)
140. Reyno, E.M., Cubel, JÁ.C.: Model driven game development: 2d platform game prototyping. In: GAMEON, pp. 5–7 (2008)
141. Rogerson, M.J., Gibbs, M.R., Smith, W.: Digitising boardgames: issues and tensions. In: DiGRA Conference (2015)
142. Roldán-Álvarez, D., Martín, E., Haya, P.A., García-Herranz, M., Rodríguez-González, M.: Dedos: An authoring toolkit to create educational multimedia activities for multiple devices. IEEE Trans. Learn. Technol. **11**(4), 493–505 (2018)
143. Rubio-Manzano, C., Lermanda, T., Martínez-Araneda, C., Vidal, C., Segura, A.: Teach me to play, gamer! imitative learning in computer games via linguistic description of complex phenomena and decision trees. Soft Comput. **27**(6), 3023–3035 (2023)
144. Sadler, R.W.: The continuing evolution of virtual worlds for language learning. The handbook of technology and second language teaching and learning, pp. 184–201 (2017)
145. Salter, A.: King's quest. How to play video games, p. 29 (2019)
146. Saltzman, M.: Game Design: Secrets of the Sages. Macmillan Digital Publishing (1999)
147. Sanglard, F.: Game Engine Black Book: Wolfenstein 3D. Software Wizards (2017)
148. Sarría, Í., González, R., Magreñán, ÁA., Narváez, S.P., Orcos, L.: Games math. Adaptive video game to evaluate basic mathematic concepts. In: Learning Technology for Education Challenges: 6th International Workshop, LTEC 2017, Beijing, China, August 21–24, 2017, Proceedings, vol. 6, pp. 27–35. Springer (2017)
149. Schaeffer, J.: 2024 world computer chess championships: The end of an era 1974–2024. ICGA J. (Preprint), 1–3 (2024)
150. Schules, D., Peterson, J., Picard, M.: Single-player computer role-playing games, pp. 107–129. Transmedia Foundations, Role-Playing Game Studies (2018)
151. da Silva, M.M.O., Teichrieb, V., Smith, P.: What are teachers needs concerning augmented reality digital authoring tools? In: Anais dos Workshops do Congresso Brasileiro de Informática na Educação, vol. 8, p. 1452 (2019)
152. Stathaki, A., Kondylakis, H., Marakakis, E., Kalogerakis, M.: i-prolog: a web-based intelligent tutoring system for learning prolog. In: Interactivity, Game Creation, Design, Learning, and Innovation: 6th International Conference, ArtsIT 2017, and Second International Conference, DLI 2017, Heraklion, Crete, Greece, 30–31 Oct 2017, Proceedings, vol. 6, pp. 337–346. Springer (2018)

153. Stathopoulou, I.O., Alepis, E., Tsihrintzis, G.A., Virvou, M.: On assisting a visual-facial affect recognition system with keyboard-stroke pattern information. Knowl. Based Syst. **23**(4), 350–356 (2010)
154. Suraweera, P., Mitrovic, A.: An intelligent tutoring system for entity relationship modelling. Int. J. Artif. Intell. Educ. **14**(3–4), 375–417 (2004)
155. Švelch, J.: Encoding monsters:"ontology of the enemy" and containment of the unknown in role-playing games. In: Philosophy of Computer Games (2018)
156. Sweetser, P., Johnson, D., Wyeth, P., Anwar, A., Meng, Y., Ozdowska, A.: Gameflow in different game genres and platforms. Comput. Entertain. (CIE) **15**(3), 1–24 (2017)
157. Tao, Y.H., Cheng, C.J., Sun, S.Y.: What influences college students to continue using business simulation games? The taiwan experience. Comput. Educ. **53**(3), 929–939 (2009)
158. Thomson, D.: Towards a negotiable student model for constraint-based itss (2008)
159. Togelius, J., Karakovskiy, S., Koutník, J., Schmidhuber, J.: Super mario evolution. In: 2009 IEEE Symposium on Computational Intelligence and Games, pp. 156–161. IEEE (2009)
160. Tourtoglou, K., Virvou, M.: Simulated annealing in finding optimum groups of learners of uml. In: Intelligent Interactive Multimedia Systems and Services, pp. 147–156. Springer (2010)
161. Tsihrintzis, G.A., Virvou, M., Alepis, E., Stathopoulou, I.O.: Towards improving visual-facial emotion recognition through use of complementary keyboard-stroke pattern information. In: Fifth International Conference on Information Technology: New Generations (itng 2008), pp. 32–37. IEEE (2008)
162. Tsihrintzis, G.A., Virvou, M., Stathopoulou, I.O., Alepis, E.: On improving visual-facial emotion recognition with audio-lingual and keyboard stroke pattern information. In: 2008 IEEE/WIC/ACM International Conference on Web Intelligence and Intelligent Agent Technology, vol. 1, pp. 810–816. IEEE (2008)
163. Tsihrintzis, G.A., Virvou, M., Jain, L.C.: Multimedia Services in Intelligent Environments: Recommendation Services. Springer (2013)
164. Tsihrintzis, G.A., Virvou, M., Phillips-Wren, G.: Surveys in artificial intelligence-based technologies. Intell. Decis. Technol. **13**(4), 393–394 (2019)
165. Tsihrintzis, G.A., Virvou, M., Jain, L.C.: Advances in Machine Learning/Deep Learning-based Technologies. Springer (2022)
166. Tsihrintzis, G.A., Virvou, M., Saruwatari, T.: Special collection of extended selected papers on "novel research results presented at the 14th International Joint Conference on Knowledge-Based Software Engineering (jckbse2022), 22–24 Aug 2022, Larnaca, Cyprus" (2022). https://easyconferences.eu/jckbse2022/
167. Tsihrintzis, G.A., Virvou, M., Hatzilygeroudis, I.: Special Issue on Selected Papers from the 33rd Annual IEEE International Conference on Tools with Artificial Intelligence (ictai-2021) (2023)
168. Tsihrintzis, G.A., Virvou, M., Doukas, H., Jain, L.C.: Introduction to advances in artificial intelligence-empowered decision support systems. In: Advances in Artificial Intelligence-Empowered Decision Support Systems: Papers in Honour of Professor John Psarras, pp. 1–11. Springer (2024)
169. Uludağlı, M.Ç., Oğuz, K.: Non-player character decision-making in computer games. Artif. Intell. Rev. **56**(12), 14159–14191 (2023)
170. Vahlo, J.: An enactive account of the autonomy of videogame gameplay. Game Stud. **17**(1) (2017)
171. Vanbecelaere, S., Van den Berghe, K., Cornillie, F., Sasanguie, D., Reynvoet, B., Depaepe, F.: The effects of two digital educational games on cognitive and non-cognitive math and reading outcomes. Comput. Educ. **143**, 103680 (2020)
172. Vargas-Iglesias, J.J.: Making sense of genre: the logic of video game genre organization. Games Cult. **15**(2), 158–178 (2020)
173. Vert, S., Andone, D.: Virtual reality authoring tools for educators. In: ITM Web of Conferences, EDP Sciences, vol. 29, p. 03008 (2019)
174. Virvou, M.: Automatic reasoning and help about human errors in using an operating system. Interact. Comput. **11**(5), 545–573 (1999)

175. Virvou, M.: A domain-independent reasoning mechanism for an its authoring tool. In: EdMedia+ Innovate Learning, Association for the Advancement of Computing in Education (AACE), pp. 1762–1764 (2000)
176. Virvou, M.: Intelligence and adaptivity in human-computer interaction concerning biotechnological software users. In: Scattering And Biomedical Engineering: Modeling and Applications, pp. 448–458. World Scientific (2002)
177. Virvou, M.: Modelling the knowledge and reasoning of users in a knowledge-based authoring tool. Int. J. Bus. Perform. Manag. **5**(2–3), 399–412 (2003)
178. Virvou, M.: Artificial intelligence and user experience in reciprocity: contributions and state of the art. Intell. Dec. Technol. **17**(Preprint), 1–53 (2023)
179. Virvou, M., Alepis, E.: Mobile educational features in authoring tools for personalised tutoring. Comput. Educ. **44**(1), 53–68 (2005)
180. Virvou, M., Kabassi, K.: Evaluating an intelligent graphical user interface by comparison with human experts. Knowl. Based Syst. **17**(1), 31–37 (2004)
181. Virvou, M., Katsionis, G.: Virge: tutoring english over the web through a game. In: Proceedings 3rd IEEE International Conference on Advanced Technologies, p. 469. IEEE (2003)
182. Virvou, M., Katsionis, G.: On the usability and likeability of virtual reality games for education: The case of vr-engage. Comput. Educ. **50**(1), 154–178 (2008)
183. Virvou, M., Manos, K.: Individualizing a cognitive model of students' memory in intelligent tutoring systems. In: Knowledge-Based Intelligent Information and Engineering Systems: 7th International Conference, KES 2003, Oxford, UK, Sept 2003. Proceedings, Part I, vol 7, Springer, pp 893–897 (2003)
184. Virvou, M., Maras, D.: Error diagnosis in an english tutor. In: Proceedings of the HCI International'99 (the 8th International Conference on Human-Computer Interaction) on Human-Computer Interaction: Communication, Cooperation, and Application Design, vol. 2, pp. 657–661 (1999)
185. Virvou, M., Moundridou, M.: An authoring tool for algebra-related domains. In: HCI (2), Citeseer, pp. 647–651 (1999)
186. Virvou, M., Papadimitriou, S.: Learning addition by playing an adaptive game with coins. In: IISA 2013, pp. 1–4. IEEE (2013)
187. Virvou, M., Papadimitriou, S.: Simple arithmetic lessons through an adaptive snake game. In: 2013 International Conference on Computer. Information and Telecommunication Systems (CITS), pp. 1–4. IEEE (2013)
188. Virvou, M., Papadimitriou, S.: Use of digital 'guess who board game' in teaching english. In: IISA 2014, The 5th International Conference on Information. Intelligence, Systems and Applications, pp. 439–443. IEEE (2014)
189. Virvou, M., Sidiropoulos, S.C.: An intelligent tutoring system over a social network for mathematics learning. In: IISA 2013, pp. 1–4. IEEE (2013)
190. Virvou, M., Tsihrintzis, G.A.: Is chatgpt beneficial to education? A holistic evaluation framework based on intelligent tutoring systems. In: 2023 14th International Conference on Information, Intelligence, Systems & Applications (IISA), pp. 1–8. IEEE (2023)
191. Virvou, M., Tsihrintzis, G.A.: A novel trust state-chart model for requirements engineering of trustful ai-empowered software. In: 2023 14th International Conference on Information, Intelligence, Systems & Applications (IISA), pp. 1–6. IEEE (2023)
192. Virvou, M., Tsihrintzis, G.A.: Pre-made empowering artificial intelligence and chatgpt: the growing importance of human ai-experts. In: 2023 14th International Conference on Information, Intelligence, Systems & Applications (IISA), pp. 1–8. IEEE (2023)
193. Virvou, M., Tsiriga, V., Maras, D.: Evaluation of an its for the passive voice of the english language using the ciao! framework. In: EdMedia+ Innovate Learning, Association for the Advancement of Computing in Education (AACE), pp. 1757–1758 (2000)
194. Virvou, M., Manos, C., Katsionis, G., Tourtoglou, K.: Vr-engage: a virtual reality educational game that incorporates intelligence. In: Proceedings of IEEE International Conference on Advanced Learning Technologies, pp. 16–19 (2002)

195. Virvou, M., Manos, K., Katsionis, G., Tourtoglou, K.: Vr-integate: a knowledge-based authoring tool for virtual reality educational games. In: Proceedings of ED-MEDIA 2002 World Conferences on Educational Multimedia, Hypermedia and Telecommunications (2002)
196. Virvou, M., Manos, K., Katsionis, G.: An evaluation agent that simulates students' behaviour in intelligent tutoring systems. In: SMC'03 Conference Proceedings. 2003 IEEE International Conference on Systems, Man and Cybernetics. Conference theme-System security and assurance (Cat. No. 03CH37483), vol. 5, pp. 4872–4877. IEEE (2003)
197. Virvou, M., Katsionis, G., Manos, K.: On the motivation and attractiveness scope of the virtual reality user interface of an educational game. In: Computational Science-ICCS 2004: 4th International Conference, Kraków, Poland, 6–9 June 2004, Proceedings, Part III, vol 4, pp. 962–969. Springer (2004)
198. Virvou, M., Lampropoulos, A.S., Tsihrintzis, G.A.: Inma: a knowledge-based authoring tool for music education. In: International Conference on Knowledge-Based and Intelligent Information and Engineering Systems, pp. 376–383. Springer (2006)
199. Virvou, M., Tsihrintzis, G.A., Bourbakis, N.G., Jain, L.C.: Handbook on Artificial Intelligence-Empowered Applied Software Engineering: Vol. 2: Smart Software Applications in Cyber-Physical Systems, vol. 3. Springer (2022)
200. Virvou, M., Tsihrintzis, G.A., Bourbakis, N.G., Jain, L.C.: Introduction to handbook on artificial intelligence-empowered applied software engineering—vol. 1: Novel methodologies to engineering smart software systems. In: Handbook on Artificial Intelligence-Empowered Applied Software Engineering: Novel Methodologies to Engineering Smart Software Systems, vol. 1, pp. 1–8. Springer (2022)
201. Virvou, M., Tsihrintzis, G.A., Sotiropouloss, D.N., Chrysafiadi, K., Sakkopoulos, E., Tsichrintzi, E.A.: Chatgpt in artificial intelligence-empowered e-learning for cultural heritage: the case of lyrics and poems. In: 2023 14th International Conference on Information, Intelligence, Systems & Applications (IISA), pp. 1–9. IEEE (2023)
202. Virvou, M., Tsihrintzis, G.A., Tsichrintzi, E.A.: Virtsi: A novel trust dynamics model enhancing artificial intelligence collaboration with human users-insights from a chatgpt evaluation study. Inf. Sci. **675**, 120759 (2024)
203. Voorhees, G.A.: I play therefore i am: Sid meier's civilization, turn-based strategy games and the cogito. Games Cult. **4**(3), 254–275 (2009)
204. Weber, B., John, M., Mateas, M., Jhala, A.: Modeling player retention in madden nfl 11. Proc. AAAI Conf. Artif. Intell. **25**, 1701–1706 (2011)
205. Weisskirch, R.S.: Dealing with piaget: Analyzing card games for understanding concepts (2003)
206. Weragama, D., Reye, J.: The php intelligent tutoring system. In: Artificial Intelligence in Education: 16th International Conference, AIED 2013, Memphis, TN, USA, 9–13 July 2013. Proceedings, vol. 16, pp. 583–586. Springer (2013)
207. Wiggins, J.B., Boyer, K.E., Baikadi, A., Ezen-Can, A., Grafsgaard, J.F., Ha, E.Y., Lester, J.C., Mitchell, C.M., Wiebe, E.N.: Javatutor: an intelligent tutoring system that adapts to cognitive and affective states during computer programming. In: Proceedings of the 46th ACM Technical Symposium on Computer Science Education, pp. 599–599 (2015)
208. Wirth, W., Ryffel, F., Von Pape, T., Karnowski, V.: The development of video game enjoyment in a role playing game. Cyberpsychol. Behav. Soc. Netw. **16**(4), 260–264 (2013)
209. Wolf, M.J.: The Medium of the Video Game. University of Texas Press (2002)
210. Wolf, M.J., Perron, B.: The Routledge Companion to Video Game Studies. Taylor & Francis (2023)
211. Wu, C.J., Chen, G.D., Huang, C.W.: Using digital board games for genuine communication in efl classrooms. Educ. Technol. Res. Dev. **62**, 209–226 (2014)
212. Xiao, Z., Wauck, H., Peng, Z., Ren, H., Zhang, L., Zuo, S., Yao, Y., Fu, W.T.: Cubicle: an adaptive educational gaming platform for training spatial visualization skills. In: 23rd International Conference on Intelligent User Interfaces, pp. 91–101 (2018)
213. Yang, C.W., Kuo, B.C., Liao, C.H.: A ho-irt based diagnostic assessment system with constructed response items. Turkish Online J. Educ. Technol. TOJET **10**(4), 46–51 (2011)

214. Zahari, A.S., Rahim, L.A., Mehat, M.: A review of modelling languages for adventure educational games. In: 2016 3rd International Conference on Computer and Information Sciences (ICCOINS), pp. 495–500. IEEE (2016)
215. Zendle, D., Meyer, R., Ballou, N.: The changing face of desktop video game monetisation: An exploration of exposure to loot boxes, pay to win, and cosmetic microtransactions in the most-played steam games of 2010–2019. PloS one **15**(5), e0232780 (2020)

Chapter 3
User-Player and Student Modeling in Personalized Educational Games: A Literature Review

Abstract Personalizing educational games involves sophisticated research challenges due to users' roles as both students and players. Unlike traditional educational settings, where learning outcomes are the primary focus, educational games blur the lines between education and entertainment, necessitating student models that accommodate diverse user behaviors and motivations. Users interact with educational game software not only to acquire knowledge but also for enjoyment and immersion in the gaming experience. Consequently, student models must align instructional objectives with the interactive and dynamic nature of the gameplay, ensuring seamless integration of personalized learning experiences within the entertainment framework. This review synthesizes the literature on both user-player and student modeling in educational gaming research, emphasizing their relevance and the need for integration in addressing the complexities of personalization in educational games. By examining advances and challenges in this field, the review provides insights into future directions for research and development, guiding efforts to enhance the effectiveness and engagement of personalized educational games.

Keywords Educational software · Personalized educational games · Artificial Intelligence in educational games · Fuzzy logic in education · User experience · AI-empowered software games · User modeling · Human-AI interaction · e-learning games · 21st-century skills · Human-centred Artificial Intelligence

3.1 Introduction to User, Player and Student Modeling

User modeling is the process of creating representations or profiles of users based on their behavior, preferences, goals, and other relevant characteristics [110]. User modeling employs the technology of Artificial Intelligence (AI) [99] and data analytics or data mining [2]. Within the domain of personalized educational systems, the user primarily refers to a student, thereby giving rise to the concept of a student model, which serves as the foundational element for personalization within educational software applications. Similarly, in the context of computer gaming, the user model evolves into what is commonly termed a player model.

S. Papadimitriou and M. Virvou, *Artificial Intelligence—Based Games as Novel Holistic Educational Environments to Teach 21st Century Skills*, Intelligent Systems Reference Library 93, https://doi.org/10.1007/978-3-031-77464-5_3

These representations, whether of users, students, or players, are deeply intertwined with the advancements in AI and Intelligent Applications [100]. Intelligent applications aim to provide personalized experiences to users by tailoring content, recommendations, and interactions based on individual preferences and behaviors. User modeling helps these applications understand users' preferences, interests, and past interactions to deliver personalized experiences effectively. Moreover, user modeling frequently encompasses the evolution of intelligent multimedia [132]. Intelligent multimedia technologies generate rich, diverse data types, including audio, video, and interactive content. By integrating these data sources into user modeling processes, researchers can capture a more comprehensive understanding of user behavior, preferences, and reactions, leading to more accurate and insightful user models. For example, integrating various modalities for student modeling, such as visual-facial affect recognition and keyboard-stroke patterns [87, 94, 96], further refines the personalization process. Facial micro-expression recognition evaluates students' emotional states in online learning environments, providing insights into their learning rates [54].

User modeling also benefits from advances in AI-empowered applied software engineering [126]. AI-empowered applied software engineering aims to automate and optimize various aspects of software development, including testing, debugging, and deployment. User modeling can benefit from these advancements by streamlining the process of building and maintaining user models, reducing development time and costs, and improving the scalability and reliability of personalized systems as well as innovative methodologies in smart software systems [128].

For the aims of user modeling, a diverse array of AI techniques have been employed. AI encompasses a broad range of techniques and approaches to simulate human-like intelligence in machines, allowing them to perceive, reason, learn, and make decisions.

One of the key challenges in student modeling within educational games lies in the context of edutainment, where education is seamlessly integrated with entertainment [109]. As such, the users of educational games have a dual role as students and players. This duality complicates the process of understanding and predicting user behaviors, as individuals may exhibit distinct preferences and tendencies depending on whether they are in learning or gaming modes. Developing effective student models in edutainment necessitates a sophisticated approach that accounts for the multifaceted nature of user interactions and aligns educational content with gaming mechanics to foster engagement and learning.

Authoring tools may be used for achieving reusability of user modeling, whether this extends to student modeling concerning educational processes of users, user experience modeling focusing on user-friendliness and assistance, [123] describe VR-INTEGATE, which creates immersive Virtual Reality (VR) educational games, enabling player modeling by addressing how users interact with game elements. Virvou and Moundridou [115] highlight an authoring tool for algebra that provides personalized student modeling, allowing Intelligent Tutoring Systems (ITSs) to adapt to individual learner needs. Moundridou and Virvou [63] discuss web-based ITS authoring, facilitating user experience modeling by expanding educational access and customization. Virvou and Alepis [112] emphasize mobile features that enhance user experience and educational effects by enabling on-the-go learning and improving convenience and engagement. Virvou [108] explores authoring tools that align content with users' cognitive processes, enhancing responsiveness in educational systems, while Virvou [106] presents a domain-independent reasoning mechanism that increases versatility across various subjects.

3.2 Student Modeling

Student modeling is a fundamental aspect of personalized education. Various methodologies and technologies have made significant advancements in this area. These advances embody a shift towards sophisticated educational systems that prioritize adaptivity and personalized learning. Student modeling encompasses several works in the literature regarding the evolution of ITSs.

The literature emphasizes that student modeling is crucial in addressing domain knowledge issues, including analyzing student achievement, identifying underlying misunderstandings, understanding students' goals and plans, tracking prior learning, and assessing personality traits [82]. As a critical component of adaptive and intelligent teaching tools, it encompasses various student characteristics, such as proficiency [9]. Consequently, by maintaining individual profiles for each student, the system can effectively tailor its learning content and make optimal use of the available learning resources [47].

It is essential to think of the student profile as a virtual representation of an actual student, where the dimensions of the profile reflect different aspects of the student and the profile's properties mirror the student's traits [134]. This emphasizes the critical role of student modeling in enabling ITSs to make informed learning judgments and provide solutions [49]. By understanding the student's needs, a student model plays a crucial role in identifying areas that require attention [88]. Student modeling involves gathering the correct information to calculate the student's cognitive state and deliver adaptive content [91].

Constructing an effective student model is crucial for developing an adaptive education system. Creating a student model requires careful consideration of various factors, including the collection of information and data necessary to gain insights about the student, strategies to ensure the model remains current and accurate, and methods to utilize the model for offering customized content and experiences within the system [110].

When creating a student model, a crucial initial step is identifying the student's characteristics during their first interaction with the educational system and ITS. Selecting the student attributes is a primary concern for researchers developing ITSs [31]. These attributes, which can be domain-specific or independent, play a key role in enabling the ITS to personalize the student [134] effectively. Additionally, static characteristics of the students, such as their age and prior knowledge, are essential inputs for the ITS and are determined before the learning process begins.

These static features are meant to remain unchanged throughout the training session, although in some cases, students may have the option to modify them through an accessible menu. In contrast, dynamic features are derived from the student's ongoing interactions with the ITS. The ITS continuously updates these dynamic features based on the data collected during learning sessions.

Identifying students' dynamic characteristics is essential for adapting educational content to their needs. These characteristics encompass the students' knowledge level, abilities, mistakes, learning styles and preferences, and emotional and cognitive aspects. Knowledge level pertains to the students' prior and current understanding of the subject matter. One approach to assessing their existing knowledge is to administer pre-learning tests. By analyzing the test results and monitoring students' interactions, the system can pinpoint areas of misunderstanding and where additional support is needed. Learning styles refer to the individual skills and preferences that shape how students perceive, organize, and process educational content [40].

Moreover, teachers can detect and respond to students' emotional states in traditional classrooms to support and enhance their learning experiences. Similarly, an ITS should be able to identify students' emotions and adapt its approach to customizing individual needs by responding appropriately to those emotions [48]. As a result, emotional factors are essential student characteristics for an ITS to consider when creating a student model. These emotional states can include happiness, sadness, anger, worry, disappointment, boredom, distraction, concentration, and confusion. Rodrigo et al. found that specific emotional states, such as boredom or disappointment, can lead students to exhibit off-task behavior [81]. Off-task behavior indicates that students have lost focus and engage in activities irrelevant to the educational system or the intended learning objectives [14]. Examples of off-task behavior include surfing the internet, spending time on non-subject content, and chatting with peers without any learning-related outcomes [4]. These behaviors are associated with motivational issues [3], and therefore, emotional modeling factors can serve as a foundation for modeling student motivation.

The design of ITSs should account for the cognitive characteristics of students to develop an effective learning model. These cognitive aspects encompass attention, knowledge acquisition, comprehension, memory, perception, concentration, collaboration, problem-solving, decision-making, analytical skills, and critical reasoning. Additionally, students' cognitive abilities and self-assessment of their knowledge are crucial for determining their focused areas of study [58]. In another case, the research by Tourtoglou and Virvou [92] is an example of how student modeling can be applied to target soft skills. Their approach utilizes simulated annealing to form optimal groups of learners in a UML learning environment. The focus is on balancing both technical and soft skills, such as collaboration and communication, to enhance group dynamics and learning outcomes. While not the ultimate approach, this method illustrates how computational techniques can be used to address the complex task of fostering soft skills in educational settings.

Adaptive and personalized educational systems should also consider students' metacognitive abilities, which involve monitoring, regulating, and coordinating information processes related to their cognitive activities [79]. Metacognition refers to students' capacity to recognize and manage their reasoning consciously, such as selecting learning objectives, utilizing prior knowledge, or deliberately choosing problem-solving approaches [5]. Specific metacognitive skills include reflection, self-awareness, self-control, self-regulation, self-explanation, self-evaluation, and self-management [74]. Furthermore, research has shown significant connections between students' motivational beliefs, online self-regulated learning, engagement, and performance [135].

A well-designed tutoring system has two main functions: Firstly, it serves as a diagnostic tool, evaluating the depth and nature of the student's knowledge. Secondly, it acts as a strategist, creating a personalized approach based on the student's comprehension [56]. The student model primarily fulfills this second function, serving as the basis for customization in ITSs [25]. Using data from the student model, the system dynamically adjusts its responses to each learner, offering personalized tutoring, assistance, and feedback.

The student model is designed to analyze students' errors, determine their educational requirements, and adjust teaching content and methods to align with the student's learning pace. It aims to precisely evaluate the student's skill level and cognitive state to deliver the most appropriate educational materials. An adaptive teaching system can also evaluate the student's profile to grasp their preferences and learning style, allowing it to determine the most effective approach to learning. Furthermore, the system can select suitable teaching methods to enhance tutoring sessions, improve learning outcomes, and recognize the student's emotional state to boost motivation. In addition, a student model can pinpoint the student's strengths and weaknesses to offer personalized guidance and constructive feedback. It can also present the student with more challenging tasks and suitable learning strategies to foster comprehensive learning and assist them in achieving educational goals by acknowledging their metacognitive skills.

The continuous progress in this field also considers privacy considerations, as emphasized in [59, 72], ensuring adherence to regulations such as GDPR. The assessment of student modeling approaches has also been a significant area of study to guarantee effectiveness, practicality, and educational outcomes in comparison to traditional educational methods or as stand-alone methods [23, 114]. These methodologies collectively contribute to customizing tutoring in AI-based interactive environments, thus optimizing learning results and user experiences [110].

3.2.1 Knowledge-Based Student Modeling

In many cases, knowledge-based and intelligent information systems and engineering [95] have been used. Knowledge-based systems employ expert knowledge and domain-specific information to enhance the accuracy and relevance of user models. These systems can create more comprehensive and insightful user representations by incorporating expert knowledge about user behaviors, preferences, and domain-specific factors.

A. The overlay and perturbation model

The overlay model, also known as the overlay student model, is one of the most recognized and commonly used student models in adaptive education systems [8]. Its widespread use stems from its core principle of separating the domain into concepts and representing the student's knowledge of each concept separately. In the overlay model, student knowledge is represented as a concept-value pair. Initially, this value was binary, indicating whether the student knows the domain concept. However, most modern educational systems now employ a weighted overlay model to capture different degrees of student knowledge. These systems use either qualitative measurements (e.g., characterizing knowledge levels as good, average, or poor) or quantitative measurements (e.g., using an integer value between 0 and 100) to reflect the level of student knowledge [18]. This enables the model to account for situations where a student may possess incomplete knowledge of a concept. Thus, the finer the domain breakdown into concepts, the more detailed the information about the student's knowledge level.

Brusilovsky and Vassileva devised two models for integrating traditional sequencing technology into ITSs [10]. The dynamic course creation (DCG) system facilitates domain authoring and automatically generates personalized online courses. DCG employs a numerical overlay model on concept structures to dynamically produce personalized courses, representing the learner's knowledge of each concept within a specific range. The concept-based course maintenance system (CoCoA) evaluates the consistency and quality of the course. It begins with an empty overlay model, scans the learning objects in the author-defined sequence, updates the student's model, and assesses the current state of the model and each subsequent object.

The ELM-ART (ELM Adaptive Remote Tutor) is a web-based ITS designed to assist with learning programming in Lisp [11]. It features support for example-based programming, intelligent problem-solving exercises, and advanced auditing and debugging capabilities. The system utilizes a weighted overlay model to create and maintain a personalized model for each user. This overlay model enables the system to provide adaptive features, such as informing students about their progress concerning every aspect of the course material.

In their work, Khan developed a curriculum overlay model to create an adaptive curriculum using digital library resources [45]. This model facilitates the representation of crucial curriculum components, such as courses, activities, learning goals, resources, and interdependencies. The research by Khan involves the development of a curriculum overlay model aimed at representing an adaptive curriculum using digital library resources. This model encompasses the critical components of the curriculum, such as courses, activities, learning objectives or target concepts, resources, and the relationships among these components.

The article authored by Triantafillou et al. outlines the design and development of the Adaptive Educational System based on Cognitive Styles (AEA-CS) [93]. This system maintains data on students' preferred cognitive styles to enhance their learning experience and outcomes. The architecture of AEA-CS includes a student modeling component with three types of profiles: personal (e.g., name and password), cognitive (e.g., cognitive style preferences), and knowledge profile based on an overlay model representing the student's knowledge of a subject. The overlay model uses qualitative measures to characterize the student's knowledge as "Unknown," "Know," "Learned," and "Well-Learned."

The overlay model assumes that a student's knowledge is a subset of an expert's knowledge. Consequently, errors resulting from a lack of understanding or knowledge are viewed as misconceptions. However, various factors can contribute to a student's incorrect answer. The perturbation model expands on the overlay model to address these errors and the student's knowledge, which may not be within the experts' domain knowledge [7, 84]. It utilizes a bug library that contains definitions of common student mistakes and their underlying reasons. This involves enumerating all possible bugs through experimental analyses (enumerative process) or generating bugs based on prevalent misunderstandings, often derived from an underlying cognitive theory (generative process). As a result, this model can help identify students' misconceptions and faulty rules underlying their incorrect answers. Subsequently, the system corrects students' misconceptions by providing relevant learning materials, guidance, and feedback.

Many ITSs utilize the perturbation approach to model student behavior. For instance, Lu et al. utilize a perturbation model encompassing 31 types of addition errors and 51 types of subtraction errors to help the system analyze and comprehend students' mistakes [51]. Additionally, LeCo-EAD applies the perturbation model to identify students' incorrect knowledge and provide personalized, real-time feedback and support to remote learners [26]. Moreover, Surjano and Maltby employed the perturbation model in combination with stereotypes and the overlay model to address student errors more effectively [74]. In 2010, Baschera and Gross utilized a perturbation student model to enhance the identification of student mistakes in spelling practice [6].

B. The constraint-based approach

The Constraint-Based approach to student modeling suggests representing knowledge in sets of constraints [64]. A student violating one of these constraints implies insufficient or incorrect knowledge. This approach also proposes that simple pattern matching is sufficient for the computations, significantly reducing the effort required to implement an ITS. There is no need to create a runnable expert model; thus, large-scale empirical studies of domain experts can be omitted. Additionally, there is no need to create a bug library, as it provides a much cheaper alternative, especially regarding the computational overheads.

Moreover, it fails to consider the student's problem-solving approach, as it should recognize innovative solutions as valid and supervise the student's independent exploration. Following each instance of a rule violation or a series of such instances, the system can provide appropriate guidance, particularly suitable for this subject area. An advantageous aspect is the system's capability to adapt to the level of instructional support. This enhanced flexibility allows the expert model to intervene and provide guidance promptly so the system can offer only suggestions or tips or monitor the student for incorrect actions before intervening.

The benefits mentioned previously have prompted researchers to develop various ITSs using the Cognitive-Based Model (CBM) approach. For instance, SQLT-Web is a web-based system designed to teach the SQL database language [57]. This system observes students' actions and adjusts to their learning capabilities and knowledge by analyzing their responses using the CBM component. Similarly, KERMIT is another ITS that enables university-level students to practice database design using the Entity-Correlations data model [90]. It represents domain knowledge as a set of constraints to assess students' responses to syntactic errors and compare them to the system's ideal solution. The knowledge base assists the system in determining correct alternatives that deviate from the system's solution.

Furthermore, Karaci discusses an ITS designed to teach Turkish punctuation. The ITS includes a Constraint-Based Model (CBM) that utilizes a pattern matcher to identify student errors and breached constraints [41]. It also analyses all the words in the student's responses to determine their relevance to the constraints. The CBM stores student information for a single session, including correct answers, errors, response times, and the usage of hints.

There are, however, potential drawbacks. For instance, the system may be unable to identify critical characteristics of problem instances that are highly indicative of the student's understanding of certain subjects. Moreover, the constraints may not offer the appropriate level of abstraction, leading to the acceptance of numerous incorrect solutions without breaching any constraints. As a result, the system may not effectively address the student's errors or connect their actions to relevant educational concepts.

C. User Stereotypes and Personas

According to Kay, stereotype-based syllogism entails forming a detailed model of a student based on initial impressions and default conjectures [44]. The initial interaction between the system and the user allows the system to make preliminary assessments of the student's knowledge level, which may involve asking questions or assigning a task. The data collected during this interaction is the foundation for determining various aspects of the student model. The system's capacity to infer student characteristics from the outset while avoiding a one-size-fits-all approach is an appealing aspect of stereotype-based modeling.

A stereotype is a collection of characteristics that co-exist in individuals, allowing the system to draw logical conclusions using only a small number of observations. A stereotype has a set of trigger conditions, t_{Mi}, where each t_{Mi} is a Boolean expression based upon components of the student model. Any tMi may be a single component c_j of the user model or a function of several components, $f(c_k)$. For instance, let us consider a stereotype designed to evaluate the understanding of Python concepts by an experienced C++ programmer. Initially, the system utilizes a component to gauge the student's proficiency as a C++ expert. Subsequently, a specific condition may refer to this component to trigger the stereotype asserting that the programmer possesses some knowledge of Python. The following equation defines the activation of the stereotype:

$$if \exists i, t_{Mi} = true \implies active(M) \tag{3.1}$$

When any trigger t_{Mi} becomes true, the stereotype M becomes active. A set of retraction conditions, r_{Mi}, deactivate the corresponding stereotype when they become true. Consider the following scenario: if the system identifies that the student lacks knowledge of Python's while loop and the if clause, it can be noted that these concepts are almost identical in Python and C++. Therefore, the absence of familiarity with these concepts in Python serves as a triggering condition for stereotype M. A stereotype is deactivated when any of the retraction conditions, r_{Mi}, becomes true:

$$\exists j, r_{Mj} = true \rightarrow notactive(M) \tag{3.2}$$

Grundy is a system designed to simulate the role of a librarian at a public library [80]. The system engages users by asking them questions to categorize them into specific stereotypes. For instance, if a user is new to the system, it assigns the ANY-PERSON stereotype. It then prompts users to describe themselves, allowing the system to assign appropriate stereotypes. For example, if a user is characterized as humorous or friendly, the system activates the optimism-related stereotype and recommends books with optimistic themes.

Suleman, Mizoguchi, and Ikeda introduce a system that employs a behavioral model containing information about the student's behavioral state [89]. The model generates a behavioral stereotype for each student interacting with the system and facilitates negotiation-driven learning. The ultimate goal is to enhance student engagement and promote cognitive and metacognitive skills by familiarizing students with the model and questioning its accuracy.

DEPTHS (Design Pattern Teaching Help System) is an ITS designed for teaching software design patterns [39]. It incorporates a student model that integrates stereotype and overlay modeling. When students first engage with the system, they choose a beginner, intermediate, or advanced stereotype to signify their level of knowledge in the subject area. The system then generates a student model with default attribute values based on the selected stereotype and tailors the learning materials until the initial assessment is completed. Subsequently, the system adjusts the values of other attributes in the student model based on the student's performance.

Another technique that has been associated with user modeling is the creation of user personas [46]. Personas embody typical users with well-defined attributes, facilitating designers' focus on the needs and objectives of potential users during system or interface development [15]. They are invented, especially when direct user trials or feedback are unavailable. These personas embody characteristics, needs, preferences, and behaviors representative of the anticipated user base, providing a tangible framework for design decisions and ensuring alignment with user expectations [53]. Pruitt and Adlin [78] point out that Personas offer a more captivating alternative to design primarily rooted in scenarios. Personas are advocated as a means to maintain user-centricity throughout the software design process, offering a tool for modeling software system users in a manner that is readily comprehensible to all design team members. User Personas, as a method, facilitate the identification and delineation of a system's archetype users. Armed with this understanding, the system can adapt accordingly, intuiting user expertise to deliver an optimal user experience. Madureira and colleagues introduce an architecture that integrates User Personas with a dynamic, evolving system, complemented by an evaluation conducted by its target users [53].

Probes may be used to create User Personas. Probes, in this context, refer to sets of prompts or tools distributed to participants to gather insights into their daily lives, experiences, and perspectives. These probes typically consist of various stimuli such as questionnaires, diaries, cameras, or other creative artifacts designed to elicit responses and reflections from participants. Probes aim to uncover rich qualitative data about individuals' behaviors, routines, preferences, and emotions in their natural environments without imposing researcher bias or constraints. For example, Campos and Paiva [13] aim to understand how people share memories and perform daily activities, using cultural probes to gather personal insights.

Recently, user stereotypes were introduced to automate the persona creation process, addressing deficiencies observed in both manual and semi-automated methodologies [46]. Through empirical comparisons, the investigation of Korsgaard and colleagues demonstrates that while manual techniques display variability among human designers, the proposed algorithm ensures consistency and rigor. Furthermore, the algorithm adeptly identifies optimal clusters within the dataset, diminishing the labor-intensive nature associated with manual cluster identification. Additionally, the algorithm's output mirrors the most substantial variances in the dataset, which aligns with patterns identified by multiple correspondence analysis. This methodological advancement shows potential for streamlining persona development and enhancing the efficiency and efficacy of user-centered design practices.

3.2.2 *Machine Learning for Student Modeling*

Many recent approaches in student modeling incorporate advancements in Machine Learning (ML) and Deep Learning (DL) [99]. ML, a subset of AI, focuses on developing algorithms that enable computers to learn from data and make predictions or decisions without being explicitly programmed. For instance, ML algorithms can recognize patterns in user behavior and preferences, contributing to creating more sophisticated user profiles. Other approaches involve the fusion of AI and ML methodologies [34]. This blending of techniques not only facilitates the creation of more comprehensive user models but also enhances the adaptability and responsiveness of personalized systems to individual users' needs and preferences.

ML entails acquiring and organizing knowledge, typically improving an agent's ability to classify items or solve issues within its particular area of expertise [85]. Consequently, a student modeling element within an educational framework can expand and consolidate its foundational knowledge, such as its error database, to enhance subsequent modeling efforts' precision, efficiency, and feasibility. Furthermore, insights into user behavior can generate learning instances that a ML system can harness to construct a model capable of forecasting future behavior [131]. Student models can adopt various methodologies, including delineating the cognitive skills that manifest as a result of student activities, discerning disparities between student and expert skills, recognizing behavioral patterns and preferences, and understanding student characteristics. Moreover, an educational system may employ ML techniques for predictive student modeling, enabling the prediction of a student's future performance on a problem or test through analysis of their past interactions within a learning environment [29].

iList is an ITS designed for teaching linked lists, a fundamental data structure commonly covered in introductory computer science courses. The system utilizes ML to create a customized model based on students' interactions [27]. This model is portrayed as a directed graph comprising two types of vertices: state and action vertices. In this model, a state vertex can link to multiple action vertices, while an action vertex can only connect to a single state vertex. Moreover, each action is linked to a probability representing a student's likelihood of taking that specific action.

Web-EasyMath is an online ITS designed for teaching algebra. It combines the stereotype technique and the k-nearest neighbor algorithm to initialize a new student's model. The k-nearest neighbor algorithm is a commonly used ML technique [102]. Initially, the system assigns students to a stereotype category based on their performance on a pretest related to their knowledge level. Then, it builds the student's model by applying the k-nearest neighbor algorithm to students within the same stereotype category as the new student. Web-EasyMath determines the neighbors by considering the student's school class, level of attention during assignments, and proficiency in basic arithmetic operations.

FITS is an ITS that utilizes flowcharts and the Bayesian Network ML algorithm to assess the user's mastery of individual topics within a specific problem domain, as well as to model the structure of the problem domain itself [36]. It offers students a tailored and personalized environment for developing flowcharts to improve their problem-solving abilities.

In other cases, ML advances [125] have been used to pinpoint similarities of individual students with other students [103]. Advancements in learning analytics enable measuring student engagement and predicting performance through ML paradigms [65]. Using Latent Dirichlet Allocation (LDA), students' sentiments towards personalized feedback in educational settings can be analyzed, offering valuable insights for improving feedback mechanisms [38]. An approach to student modeling in blended learning environments incorporates activities from both in-class and at-home learning, facilitating a meaningful representation of students' knowledge [33].

3.2.3 Cognitive Aspects and Theories for Student Modeling

Integrating cognitive theories into educational technologies significantly advances comprehension and enhances student learning processes. Cognitive theories, ranging from memory and attention to problem-solving, offer valuable insights into how students perceive, process, and retain information. By infusing these theories into educational systems, developers can create more educationally efficient and engaging teaching and learning tools.

For instance, the utilization of user modeling based on the cognitive theory of human plausible reasoning is exemplified in the learning process of novice users or students for Unix operating systems by Virvou [104, 105, 108], and in training scenarios for graphical user interfaces [113, 120]. This approach incorporates a cognitive theory related to chains of human reasoning for problem-solving. It enables the emulation of the reasoning processes that users might employ when confronted with problematic situations, thus accounting for plausible errors stemming from imperfect human reasoning. Furthermore, cognitive theories related to problem-solving can guide the development of instructional strategies that scaffold students' problem-solving skills. For instance, furnishing students with opportunities to engage in problem-solving across diverse contexts and offering timely feedback aligns with the tenets of cognitive load theory, fostering deeper understanding and the transfer of knowledge.

Additionally, fuzzy-based reasoning has found utility in modeling students' cognitive learning states, including progress pace and forgetting aspects, particularly in the domain of web-based learning environments for learning the programming language C [17] and the logic of programming [19]. This integration furnishes a structured framework for apprehending and addressing the subtleties of cognitive processes within educational contexts.

In educational applications, agents' expression of specific emotional states is made feasible by integrating the OCC theory [66], which constitutes a cognitive model of emotions. Systems such as the one developed by Alepis and Virvou [1] encapsulate an affective authoring module reliant on the OCC theory to model potential emotional states of users–students and to automatically create strategies of virtual medical instructor agents to enhance interactions between tutoring agents and medical students. By incorporating the OCC model, these systems can suggest that a tutoring agent expresses a particular emotional state to motivate the student during learning, thereby enhancing teaching effectiveness. Another instance is affective user modeling, as discussed by Katsionis and Virvou [42].

Moreover, recent research by Gao et al. [28] has highlighted the important role of practice as a cornerstone of cognitive activities, with ITSS leading the charge in employing cognitive process modeling to optimize learning strategies. These systems furnish a platform for inspecting students' cognitive processes, amplifying knowledge acquisition.

In conclusion, combining cognitive theories with educational technologies holds immense potential for refining student learning experiences. By incorporating insights from cognitive science, developers can create more adaptive, personalized, and captivating learning environments that encourage students' cognitive development and facilitate knowledge acquisition [125].

3.2.4 Fuzzy Student Modeling

Understanding a student's knowledge is a complex process influenced by subjective human judgment [16]. To address the uncertainty inherent in knowledge computation, fuzzy logic has been proposed as a method for handling uncertainty [136]. Algorithms based on fuzzy reasoning offer a way to determine the optimal model by considering multiple criteria and model specifications [83]. Therefore, fuzzy logic approaches are well-suited for analyzing students' knowledge levels, needs, and specificities to provide personalized guidance. Fuzzy logic extends classical Boolean logic to accommodate partial truth, where truth values range from "absolutely true" to "absolutely false." The concept of partial truth is represented by a continuous interval of truth values using the membership function (MF), which accepts values within the bounded interval of [0,1]. Fuzzy inference, also known as approximate reasoning, utilizes fuzzy logic to simulate human reasoning based on imprecise data and uncertainty for decision-making, involving fuzzy rules, facts, and a conclusion.

In their research, Goel, Lallé, & Luengo utilized a fuzzy logic student model with two rule-based components: one for assessing the student's knowledge level and another for predicting their performance [30]. Chrysafiadi and Virvou employed fuzzy logic theory to diagnose and characterize students' knowledge levels through an ITS focused on learning computer programming [16]. The model links the knowledge level to verbal terms such as "Unknown," "Unsatisfactory Known," "Known," and "Learned." Their system evaluation demonstrated that integrating fuzzy reasoning into the student model enhances students' satisfaction and efficiency, improves the system's adaptive capability, and makes more accurate and reliable decisions. Additionally, Karaci proposed an ITS that integrates the overlay model and CBM [41]. This system utilizes fuzzy logic, with inputs including the MYCIN certainty factor (which reduces uncertainty) and the student's number of attempts to answer a question. The fuzzy logic decision system outputs the student's learning level using sets such as "Definitely Does Not Know," "Might Know," and "Most Probably Knows." Moreover, Hwang et al. developed an ITS that employs a fuzzy inference system (FIS) to assess students' learning and affective status based on their responses to exercises and online learning behaviors [37]. The system also uses fuzzy theory to provide students with appropriate learning materials based on their current learning status.

Whether through the exclusive use or fusion of cognitive theories, ML, or fuzzy-based reasoning, as exemplified in [21, 22], the evolution towards highly adaptive educational systems is evident.

3.3 Player Modeling

In the context of user modeling, player modeling is an important element, particularly within the realm of game design and development. By understanding the subtleties of player behavior, preferences, and decision-making processes, developers can create more tailored and engaging gaming experiences. Advanced player models enable the creation of systems that can adapt dynamically in real-time, responding to individual players' specific needs and preferences as they engage with the game. This adaptive approach to player modeling closely aligns with the broader objectives of user modeling, which seeks to create personalized and intuitive user experiences across various interactive systems and platforms. Thus, player modeling plays a significant role in enhancing user satisfaction and immersion within gaming environments, ultimately contributing to the overall success of the gaming experience.

Player modeling is an essential aspect of game design and development. Traditionally, player modeling has focused on tracking and classifying behavior based on functional metrics such as deaths, specific choices, and completion times. However, recent research has introduced innovative approaches that combine expert domain knowledge and theoretical frameworks linking behavior and game design patterns. For instance, Cowley and Charles propose the concept of "Behavlets," [24] which derive play features from sequences of actions informed by psychological behavior theories. This approach aims to create player models that are directly interpretable concerning psychological theory, thereby improving the theoretical validity of player models. Similarly, Burelli and Yannakakis [12] introduce a methodology for adapting virtual camera behavior through player modeling. By modeling player preferences on virtual camera movements, this approach tailors viewpoint movements to the player type and gameplay style, resulting in an adaptive cinematographic experience.

In computer games, decisions are often intricately tied to gameplay rules and mechanics. This means that players' choices and actions are significantly influenced by the structure and constraints set by the game itself. Researchers can infer the motivations driving their decisions by analyzing how players interact with the game environment and respond to different stimuli, such as challenges, rewards, and obstacles. This approach allows for a deeper understanding of player preferences and behaviors within the context of the game's rules, enabling developers to tailor experiences more effectively to individual player needs. Pfau and colleagues [75] demonstrate the application of deep player behavior modeling to automate game balancing across multiple difficulty dimensions. By modeling individual player strategies and playing styles, this approach identifies and regulates imbalances between in-game units, characters, and classes, enhancing game performance.

The technique of the creation of personas has also been used in player modeling in games. Persona implementation by researchers and game developers in user modeling for games involves the exploration of user behavior and decision-making processes. Holmgard and colleagues [35] proposed a solution to this challenge by employing user behaviour data in personas to extract underlying motivations. Moreover, Melhart and colleagues have explored player modeling using personas by investigating whether predicting players' motivations solely by observing their gameplay data is possible. To address this question, they collected a large dataset of gameplay data from players of the popular game Tom Clancy's The Division. Additionally, they asked players to report their levels of competence, autonomy, relatedness, and presence using the Ubisoft Perceived Experience Questionnaire [55].

Overall, player modeling research encompasses a variety of methodologies, including integrating AI techniques, theoretical frameworks, and domain-specific knowledge. These approaches aim to create more sophisticated and adaptive player models, ultimately enhancing the player experience in video games.

3.4 Combination of Student and Player Modeling

3.4.1 *Modeling Approaches*

Student and user modeling play important roles in personalizing educational games, enabling tailored learning experiences that cater to individual needs and preferences while engaging and motivating students.

Student modeling in personalized educational games has notably enhanced the adaptability of educational experiences, as exemplified by personalized e-learning in VR environments, such as VR-ENGAGE [43, 122], which integrate VR educational games with intelligence to offer highly personalized learning experiences. In the domain of affective user modeling, as discussed by Katsionis and Virvou [42], cognitive theories of affect play a significant role in understanding emotional aspects within educational game-playing scenarios. These theories provide a framework for analyzing how users' emotions influence their interactions with educational games, particularly when their objectives involve both learning and winning. By integrating cognitive theories of affect, developers and researchers can gain insights into how users' emotional states impact their learning experiences and gameplay strategies. This understanding enables them to design educational games that effectively balance the dual objectives of fostering learning and engagement, enhancing the overall user experience.

3.4.2 *Types of Game Targeting Students' Learning*

When examining player modeling, one must consider the type of game being analyzed. Video game genres are distinct categories of games that share similar gameplay characteristics. Numerous studies have indicated that adventure and simulation games are particularly beneficial for educational purposes. Adventure games, in particular, can be seamlessly integrated into the educational process due to their narrative nature. Many researchers have developed educational adventure games and emphasized their advantages. However, most games lack adaptive elements or have limited capabilities, often providing students with hints, directions, or adjusting the difficulty level. While the game scenario is an essential component of an adventure game, there is a lack of games in the literature that automatically adjust the game's progression based on the student's cognitive level. Modifying the scenario expands the game with additional objectives, benefiting both the gaming and educational contexts. This leads to increased motivation and enjoyment for the player and additional exercises for the student to acquire the desired knowledge and master their skills.

To this end, Papadimitriou and Virvou [70] discuss adaptivity in scenarios within educational adventure games, highlighting the importance of dynamically adjusting gameplay based on user actions and progress. Building upon this concept, Papadimitriou, Chrysafiadi, and Virvou [71] propose FuzzEG, a fuzzy logic-based approach for adaptive scenarios in educational adventure games, emphasizing the use of fuzzy reasoning to enhance student modeling and adaptivity. Further, advancing cognitive-based adaptivity, Chrysafiadi, Papadimitriou, and Virvou [20] introduce cognitive-based adaptive scenarios in educational games, employing fuzzy reasoning techniques to tailor gameplay based on learners' cognitive states dynamically.

Addressing user modeling in young students, particularly those in elementary school, involves designing educational games and applications that cater to their specific learning needs and preferences. One approach to achieve this is to develop interactive and adaptive educational games tailored to teach fundamental subjects like math.

For example, researchers Papadimitriou and Virvou [69] created an online adventure game designed to teach young students mathematical concepts. This game incorporated engaging storytelling elements and interactive challenges to make learning math more enjoyable and accessible to children.

In another instance, Virvou and Papadimitriou developed a simple arithmetic lessons game using an adaptive snake game [117]. This approach involved designing a game where the difficulty level adapts based on the student's performance, ensuring that the learning experience remains challenging yet manageable as the student progresses.

The snake board game, often used as an educational tool for teaching arithmetic, is played similarly to the traditional game of Snake but with a mathematical twist. Here is how it typically works:

Setup: The game board consists of a grid with cells, similar to a traditional board game. Each cell may contain a number or a mathematical operation.

Objective: The objective of the game is to guide the snake (represented by a token or marker) around the board, collecting numbers or performing mathematical operations to reach a specific goal.

Movement: Players take turns rolling a die or spinning a spinner to determine how many spaces the snake can move. The snake moves horizontally or vertically, not diagonally, to adjacent cells.

Mathematical Operations: When the snake lands on a cell containing a mathematical operation (such as addition, subtraction, multiplication, or division), the player must perform the operation using the number on the cell and another number obtained from a previous cell. For example, if the snake lands on a cell with "+3" and the previous number collected was 5, the player adds 3–5 to get 8.

Goal: The game may have a specific goal or target number that players must reach by collecting numbers and performing operations. For example, the goal may be to reach a target of 20 by collecting and combining numbers through mathematical operations.

Winning: The game continues until a player reaches the target number or completes a specific objective. The first player to achieve the goal wins the game. The snake board game offers an interactive and engaging way for students to practice arithmetic skills while having fun playing a game. It encourages strategic thinking, mental math, and problem-solving abilities in a playful and educational context.

Similarly, in another project by Virvou and Papadimitriou, they created an educational game involving simple arithmetic additions using virtual coins [116]. This game provided a hands-on learning experience where students could visually interact with the mathematical concepts, making abstract ideas more tangible and easier to grasp.

In another domain area, that of language learning, student modeling in digital board games is exemplified by teaching English through a "Guess Who" game [118, 119]. In this game, students engage in language learning activities, particularly focused on English language skills. The game mechanics concern guessing the identity of characters based on specific characteristics, which include vocabulary related to appearance, personality traits, or other descriptors.

Student modeling in this scenario functions as follows:

Profile Creation: The game collects data on each student's performance, preferences, and learning style as they play. This data forms the basis of the student model.

Adaptive Gameplay: Using the student model, the game adapts its challenges, content, and feedback to suit each student's individual needs. For example, it may adjust the difficulty level of vocabulary or provide hints and prompts tailored to the student's proficiency level.

Personalized Learning: By incorporating student modeling, the game offers a personalized learning experience for each student. This personalization enhances engagement and motivation by catering to the individual interests and abilities of the learner.

Effective Learning Experiences: Incorporating student modeling in educational games like the "Guess Who" adaptation helps create more effective learning experiences. By customizing the gameplay to match each student's unique profile, the game maximizes learning outcomes and fosters skill development in a fun and interactive way.

3.4.3 Types of Game Targeting Students' Entertaining Preferences

The type of game is not only important for the particular students and player modeling to be created but plays a role on the actual entertaining aspect, as student may have diverse preferences o certain types of games. Different types of games appeal to different students based on their preferences, interests, and learning styles. For example, some students may prefer adventure games with narrative-driven storytelling, while others may enjoy puzzle games that challenge their problem-solving skills. The type of game chosen for educational purposes can significantly impact how engaged and motivated students are to participate in learning activities.

Different types of games influence player modeling in the following ways:

Creation of Player Models

Player modeling involves understanding individual players' characteristics, preferences, and behaviors within a game environment. The type of game influences the design and implementation of player models because different games require different approaches to adapt to player needs and preferences. For instance, an adventure game may track a player's progress through a narrative storyline, while a puzzle game may assess problem-solving strategies.

Entertainment Aspect

In addition to learning objectives, the entertainment value of educational games is crucial for engaging students and maintaining their interest over time. The type of game significantly contributes to its entertainment aspect, as certain game genres may inherently appeal more to students than others. By selecting game types that align with students' interests, educators can enhance the overall enjoyment and engagement of learning experiences.

For these reasons, Papadimitriou, Kamitsios, Chrysafiadi, and Virvou [73] explore how personalized reasoning and educational experiences in educational games can be achieved across various gaming formats, ranging from traditional point-and-click games to immersive VR mobile platforms. By examining different game formats, the research demonstrates the versatility of personalized learning approaches and their applicability across diverse gaming contexts.

In summary, the statement underscores the importance of considering the type of game in educational settings, both in terms of creating effective player models and ensuring the entertainment value of learning experiences. Additionally, it highlights ongoing research efforts to explore personalized learning approaches across different gaming formats to cater to diverse student preferences and needs.

3.5 Privacy and General Data Protection Regulation in Student and Player Modeling

Advances in AI and learning analytics, student and player modeling are significantly transforming education and decision support systems, but they also raise critical ethical issues. One major concern is privacy and data security, as AI systems require large datasets that often include sensitive information, necessitating robust security measures and transparent data governance [98, 125]. Bias and fairness are also pressing issues, as AI models can perpetuate existing biases in training data, leading to unfair outcomes, particularly for marginalized groups. It is crucial to train AI on diverse datasets and conduct regular audits to mitigate these biases [101, 127]. Furthermore, accountability and transparency are essential, as the complexity of AI algorithms can obscure decision-making processes, necessitating clear explanations and accountability frameworks to build trust [100, 121]. In the educational context, there are concerns about AI's impact on teaching, where over-reliance on technology might undermine human educators' roles, highlighting the need to balance AI use with fostering critical thinking and problem-solving skills [111, 124]. Additionally, AI's automation capabilities raise concerns about job displacement, requiring efforts to reskill workers and create new job opportunities to address potential economic inequalities [97, 126]. Addressing these ethical challenges is crucial to ensure that AI advancements align with societal values and contribute positively to human development.

Integrating AI in various applications has necessitated a focus on explainability, transparency, consent, trust, and compliance with privacy regulations such as the General Data Protection Regulation (GDPR). Explainability is very important in AI systems, as it enhances users' understanding of the decisions made by AI models, thus fostering trust and acceptance [67, 68]. Transparency in AI systems involves providing clear and accessible information about how AI models function, essential for building user trust and ensuring ethical compliance. Trust dynamics are particularly significant, as highlighted by Virvou, Tsihrintzis, and Tsichrintzi [129], who developed a model to enhance AI collaboration with human users by evaluating trust in systems like ChatGPT. Virvou and Tsihrintzis further explore the importance of consent and trust in AI [121], who propose a trust state-chart model to guide requirements engineering for AI-empowered software. Moreover, GDPR compliance is essential for protecting user privacy and data security, particularly in distributed systems where data is vulnerable to breaches [77]. The GDPR framework emphasizes user consent, transparency, and data protection, ensuring that personal information is handled responsibly [59]. Together, these elements highlight the ethical imperative to design AI systems that prioritize user privacy and trust while adhering to regulatory standards.

Student modeling techniques in educational games offer personalized experiences by customizing content to the unique needs of each learner. Furthermore, incorporating adaptive elements in educational or game contexts improves students' acceptance of ITSs and games as valuable learning tools, encourages their utilization, and increases engagement [72].

Hence, ITS and adaptive educational games are excellent data protection research subjects, particularly in the European Commission's General Data Protection Regulation (GDPR) context. The GDPR, introduced in response to a significant increase in online personal data sharing, seeks to empower individuals, including system users and students, using educational apps by strengthening their control over the collection and processing of their data. Implemented on May 25, 2018, across the EU, the GDPR applies to EU residents and visitors who access services or goods provided by entities within or outside the EU.

Replacing the outdated Data Protection Directive (DPD), the GDPR enhances and standardizes privacy protection across EU members and extends its jurisdiction to non-EU businesses and organizations under specific conditions. It also encompasses EU visitors, citizens, and residents, regardless of the geographical location of the data controller. Non-compliance with GDPR may result in financial penalties, necessitating the restructuring of data management practices among EU users.

Wachter and colleagues argue for a "right to explanation" to improve accountability and transparency in automated decision-making [130]. However, this right appears intentionally excluded regarding the right to avoid automated decision-making and access rights, creating complexities when comparing legal obligations with non-binding provisions.

Article 22 of the GDPR, which focuses on automated decision-making and profiling, poses potential challenges to industries such as credit, insurance, advertising, and social media [32]. This can disrupt underlying algorithmic structures and ML models, necessitating the development of more interpretable alternatives.

Sobolewski, Mazur, and Paliski analyze users' rights under GDPR, covering data erasure, processing objections, profiling objections, and consent for personal data collection and processing [86]. Whitley emphasizes consent as crucial for privacy [133]. At the same time, Politou, Alepis, Patsakis, and Virvou underscore the importance of consent with proof for effective profiling and personalization [76] in an extended examination of Privacy and Data Protection Challenges in the Distributed Era. Cenaite and Kosta explore children's data processing consent under both DPD and GDPR, noting the continued importance of consent under GDPR [52].

The book examines the intersections of ITSs and adaptive educational games with GDPR, using a specific educational game as a case study. It also offers recommendations on the necessary functionalities that designers of such systems should incorporate to obtain explicit and unambiguous consent from students. This allows the system to legally leverage its strengths, including data collection and processing and providing personalized experiences. The guidelines are designed for both adults and minors, addressing the points of conflict between GDPR and intelligent educational games (IEG). Identifying these conflicts arose from collaborating with Eirini Mougiakou, combining computer science and law as part of her research. In this research, Mougiakou, Papadimitriou, and Virvou [62] explore the implications of GDPR on synchronous and asynchronous learning methods, shedding light on the need for educational games to comply with strict privacy regulations. In their 2019 study, Papadimitriou, Mougiakou, and Virvou [72] shift the focus to the consent aspect within educational games concerning GDPR. This work underscores the importance of obtaining explicit consent from users, especially in the context of collecting and processing personal data within smart educational games. Mougiakou, Papadimitriou, and Virvou [60] examine transparency issues surrounding ITSs, particularly concerning children and adolescents. This highlights the ethical obligation to ensure transparency and accountability in educational technologies, especially when catering to vulnerable populations. Furthermore, Mougiakou and Papadimitriou [61] provide design guidelines for ITS concerning automated decision-making and personal data protection. This emphasizes the importance of incorporating privacy-enhancing measures and ethical considerations into the design and implementation of educational technologies. Finally, Mougiakou and Virvou [59] discuss GDPR privacy considerations in Unified Modeling Language (UML), using an e-learning program as a case study. This underscores the necessity of integrating privacy considerations into the development lifecycle of educational games, from conceptualization to deployment.

Overall, the synthesis of these works highlights the critical need for educational game developers to navigate and address complex ethical and privacy issues, especially concerning GDPR compliance. By incorporating transparency, consent mechanisms, and privacy-by-design principles, educational games can uphold ethical standards while providing engaging and effective learning experiences.

3.6 Initialization of User, Student and Player Models

The initialization of user, student, and player models in the context of ITSs and educational games directly addresses the cold start problem in ML. When a new student or player joins the system, there is typically limited to no historical data available regarding their preferences, behaviors, or skill levels. This scarcity of data presents a classic cold start scenario.

The cold start problem arises because the system needs to customize each user's learning or gaming experience, tailoring it to their individual needs and abilities. However, without adequate data about the new user, student, or player, the system may struggle to provide effective recommendations or guidance.

To tackle this challenge, initializing student models becomes paramount. The system makes informed estimates or assumptions about the new user's knowledge level, learning style, and error tendencies based on the available information by initializing the student model. This initialization process is a foundation for the model's comprehension of the new user's characteristics, akin to providing a starting point for personalized adaptation. Two approaches in the literature involve the following:

Generative Adversarial Imitation Learning (GAIL): The cold-start problem in ITSs is addressed through algorithms like generative adversarial imitation learning (GAIL), simulating real students' behavior trajectories [50]. As applied to ITSs, this algorithm simulates real students' behavior trajectories. By learning from the behavior patterns of real students, the model can better initialize itself and adapt to new students, even without extensive historical data specific to those students.

Generic Initialization Approach: Additionally, a generic approach has been proposed to initialize student models in Web-based educational applications, based on initial estimations of new students' knowledge level and error proneness [103]. This approach involves initializing student models based on initial estimations of new students' knowledge levels and error proneness. This initialization process likely leverages general educational principles or assessments to make educated guesses about the new student's characteristics and learning needs.

Both approaches aim to tackle the cold start problem by providing some level of initialization or adaptation mechanism for student models when they encounter new students. By doing so, these methods help improve the effectiveness of ITSs in catering to individual learners' diverse needs and characteristics, even when data specific to those learners is limited or unavailable.

3.7 Conclusions

3.7.1 Concluding Remarks of the Literature and Open Problems

The synthesis of the literature on user-player and student modeling highlights the paramount importance of integrating these approaches to tackle the complexities of personalization in educational games. Recognizing users' dual roles as students and players highlights the imperative for sophisticated models capable of harmonizing educational objectives with gaming mechanics, fostering engagement and effective learning.

Advancements in AI and Intelligent Applications play a central role in user modeling, enabling tailored experiences in educational software. These advancements empower systems to comprehensively understand user preferences, behaviors, and reactions and adapt content delivery in ways that optimize personalized learning experiences.

Despite the evident benefits of educational games in providing motivation and adaptive learning experiences, challenges persist in accurately modeling student knowledge, gaming strategies, and preferences. However, it is fundamental to recognize that both ITSs and educational games share a common objective: engaging learners through interactive and personalized experiences. Both combine technology to tailor content delivery based on individual learning needs and preferences, thus enhancing engagement and driving more effective learning outcomes.

For example, while ITSs simulate personalized guidance analogous to a teacher providing one-on-one instruction, educational games aim to immerse students in interactive experiences that adapt based on their progress and preferences. By emphasizing this shared objective, we can better appreciate how advancements in ITS can inform the design of educational games and vice versa.

Moreover, integrating elements of ITS into educational games and vice versa can lead to innovative approaches that enhance learning experiences. For instance, educational games may incorporate adaptive content delivery mechanisms inspired by ITS principles, providing tailored feedback and guidance to students as they progress through gameplay.

The review shows how advancements in AI and Intelligent Applications contribute to user modeling, facilitating personalized experiences in educational software. These advancements enable systems to comprehensively understand user preferences, behaviors, and reactions, thereby enhancing the efficacy of personalized learning experiences.

Despite educational games' benefits in providing motivation and adaptive learning experiences, challenges remain in effectively modeling student knowledge, gaming strategies, and preferences. The review identifies the need for flexible and sophisticated approaches to characterizing student knowledge levels accurately, particularly in scenarios where ambiguity exists.

The review highlights the importance of adaptive educational games in tailoring learning experiences to individual student's needs and preferences. By incorporating player modeling and adaptive content delivery elements, educational games can mitigate frustration, enhance enjoyment, and stimulate continued engagement in learning activities.

Developers create educational applications and games targeted at many students of different ages and levels of education. These applications are tailored to the diverse needs and learning styles of students. It is essential to recognize that each student possesses unique personality traits and learning preferences, which influence how they perceive and engage with educational content. Educational game developers strive to create immersive experiences that educate and cater to individual learning needs and preferences.

In the realm of educational gaming, the objectives of ITSs extend beyond mere content delivery. These systems aim to discern the specificities of each student, offering personalized guidance and support whenever needed. Similar to a teacher providing one-on-one instruction in a classroom setting, ITSs simulate this personalized guidance, aiding students in comprehending complex concepts and mastering the educational material at their own pace. This personalized approach significantly enhances the learning experience, fostering deeper understanding and retention of the subject matter.

Assessing students' knowledge levels can be achieved through student modeling techniques, which serve as the foundation for all ITSs to tailor their educational content accordingly. Although various student modeling approaches exist, most acknowledge the complexity of evaluating student knowledge. While in some cases, it may be clear whether a student thoroughly understands a lesson, there are instances where this assessment could be more precise, making it difficult for the teacher to form an opinion. The techniques mentioned provide flexibility in characterizing a student's knowledge level, such as overlay models that offer a nuanced understanding of a student's grasp of a subject. However, it is essential to recognize that this characterization may not always be straightforward and may contain elements of ambiguity.

Furthermore, we have demonstrated that educational games can significantly enhance student motivation by combining educational content with enjoyable, interactive experiences. These games create a stress-free environment where students can learn and play at their own pace, free from the fear of criticism. This reassurance about the positive impact on student learning can be a source of comfort for educators. Additionally, educational games can incorporate intelligent features to adapt to each student's unique characteristics, adjusting the content and game elements accordingly. Many studies on educational games have highlighted the incorporation of adaptive elements that modify the educational content or game components.

3.7.2 Advances of our Research to Address Open Problems

The work showcased underscores the advantages of adaptive educational games. These games help prevent student frustration and increase enjoyment and motivation to continue playing and learning.

Numerous studies underscore the distinct characteristics of students and utilize appropriate student models to deliver personalized learning experiences. However, these studies are often custom-built and not easily transferable to other contexts. Consequently, incorporating a robust student modeling technique and automated customization of educational materials tailored to individual student needs into a new application, whether an ITS or an adaptive educational game, demands significant time and effort. Additionally, each student possesses unique preferences for various digital game genres. For example, some may lean towards adventure games, while others may gravitate towards strategy games or favor action and exploration. Providing students the option to select their preferred game genre is crucial and advantageous as it can enhance the learning process by making it more engaging and impactful.

Multiple research studies focus on authoring tools for educators and individuals without programming expertise to easily import learning materials. These tools also enable users to track their students' progress and incorporate relevant exercises and theories. However, these systems' intelligent techniques are inaccessible for customisation, as they require programming skills or AI knowledge.

The GDPR has become integrated into people's everyday lives, mandating that data controllers obtain explicit consent from individuals to collect and process their data. Gathering and processing student data are fundamental to an ITS's operations, as they are essential for delivering personalized content. Nonetheless, research is needed to develop an ITS or adaptive educational game that aligns with GDPR.

This work addresses the gaps above by focusing on an educational adventure game with advanced adaptivity levels. The game dynamically adjusts the scenario based on the student's performance and knowledge level. As the player progresses, the game collects data related to their knowledge and introduces new puzzles and learning objectives accordingly. The student modeling component of the game applies fuzzy logic theory, which is well-suited for quantifying and assessing ambiguous and imprecise concepts like knowledge level. It also tailors questions and exercises to match the student's current level of knowledge. While fuzzy logic has been successfully used in educational applications, its use in educational games is unprecedented. The game described in this work features quizzes related to HTML, but it can accommodate content from any domain through an accompanying authoring tool. A notable aspect of this tool is that it allows teachers to configure fuzzy logic parameters more efficiently than other authoring tools. This can be achieved using simple verbal variables and appropriate forms, eliminating the need for programming skills. Additionally, the authoring tool provides an advanced option for educators with knowledge of fuzzy logic to define the mathematical functions and parameters governing the operation of the FLS in the current work.

The student modeling is performed online and independently from the presented adventure game. Each application can access the component, gather information about the student model, and even update it. As a result, a new application developer can avoid recreating the data collection and student model-building functionalities and instead retrieve the necessary information from this component and present it to the student. This saves the developer significant time and effort and allows them to focus solely on developing application-specific elements while optionally adding extra adaptive features. Furthermore, the presented game benefits from the autonomy of the student modeling component by using the student's current knowledge level to inform a rule-based system that determines the expansion of the game scenario. The following sections illustrate how a 3D exploration game utilizes the same student model and implements corresponding adaptive features.

We conclude by highlighting the potential conflicts between adventure games and GDPR and proposing recommendations for creating adaptive educational games that comply with GDPR regulations. As educational technologies continue to advance, it is crucial to consider ethical and legal frameworks like the General Data Protection Regulation (GDPR). The review underscores the significance of developing educational games and ITSs that align with GDPR guidelines while delivering personalized learning experiences.

Software user interfaces have to incorporate adaptivity and intelligence so that they may automatically adapt to the specific needs, interests, and knowledge of each individual user [107]. Moving forward, future research should focus on developing more sophisticated student modeling techniques that seamlessly integrate with educational games, enabling greater adaptivity and personalization. Additionally, efforts should be directed towards creating authoring tools that empower educators to customize educational content and adaptivity features without requiring extensive programming skills, facilitating wider adoption and use of personalized educational games.

In conclusion, this literature review provides valuable insights into the current state, challenges, and future directions of user-player and student modeling in personalized educational games. By addressing these complexities and considering ethical and regulatory considerations, researchers and developers can advance the effectiveness and engagement of personalized educational experiences for diverse learners.

In summary, the synergy between student and player modeling lies at the heart of effective personalization in educational games. By recognizing this synergy and exploring opportunities for integration, we can enhance engagement and effectiveness in educational software development.

References

1. Alepis, E., Virvou, M.: Automatic generation of emotions in tutoring agents for affective e-learning in medical education. Expert Syst. Appl. **38**(8), 9840–9847 (2011)
2. Alogogianni, E., Virvou, M.: Data mining for targeted inspections against undeclared work by applying the crisp-dm methodology. In: 2021 12th International Conference on Information, Intelligence, Systems & Applications (IISA), pp. 1–8. IEEE (2021)
3. Baker, R.S.: Modeling and understanding students' off-task behavior in intelligent tutoring systems. In: Proceedings of the SIGCHI Conference on Human Factors in Computing Systems, pp. 1059–1068 (2007)
4. Baker, R.S., Corbett, A.T., Koedinger, K.R., Wagner, A.Z.: Off-task behavior in the cognitive tutor classroom: When students "game the system". In: Proceedings of the SIGCHI Conference on Human Factors in Computing Systems, pp. 383–390 (2004)
5. Barak, M.: Motivating self-regulated learning in technology education. Int. J. Technol. Design Educ. **20**, 381–401 (2010)
6. Baschera, G.M., Gross, M.: Poisson-based inference for perturbation models in adaptive spelling training. Int. J. Artif. Intell. Educ. **20**(4), 333–360 (2010)
7. Brown, J.S., Burton, R.R.: Multiple representations of knowledge for tutorial reasoning. In: Representation and Understanding, pp. 311–349. Elsevier (1975)
8. Brusilovsky, P.: Developing adaptive educational hypermedia systems: from design models to authoring tools. In: Authoring Tools for Advanced Technology Learning Environments: Toward Cost-Effective Adaptive, Interactive and Intelligent Educational Software, pp. 377–409. Springer (2003)
9. Brusilovsky, P., Millán, E.: User models for adaptive hypermedia and adaptive educational systems. In: The Adaptive Web: Methods and Strategies of Web Personalization, pp. 3–53. Springer (2007)
10. Brusilovsky, P., Vassileva, J.: Course sequencing techniques for large-scale web-based education. Int. J. Contin. Eng. Educ. Life Long Learn. **13**(1–2), 75–94 (2003)
11. Brusilovsky, P., Schwarz, E., Weber, G.: Elm-art: an intelligent tutoring system on world wide web. In: Intelligent Tutoring Systems: Third International Conference, ITS'96 Montréal, Canada, 12–14 June 1996 Proceedings, vol. 3, pp. 261–269. Springer (1996)
12. Burelli, P., Yannakakis, G.N.: Adapting virtual camera behaviour through player modelling. User Model. User Adapt. Interact. **25**, 155–183 (2015)
13. Campos, J., Paiva, A.: A personal approach: the persona technique in a companion's design lifecycle. In: Human-Computer Interaction–INTERACT 2011: 13th IFIP TC 13 International Conference, Lisbon, Portugal, 5–9 Sept 2011, Proceedings, Part III, vol. 13, pp. 73–90. Springer (2011)

14. Cetintas, S., Si, L., Xin, Y.P.P., Hord, C.: Automatic detection of off-task behaviors in intelligent tutoring systems with machine learning techniques. IEEE Trans. Learn. Technol. **3**(3), 228–236 (2009)
15. Chang, Y.N., Lim, Y.K., Stolterman, E.: Personas: from theory to practices. In: Proceedings of the 5th Nordic Conference on Human-Computer Interaction: Building Bridges, pp. 439–442 (2008)
16. Chrysafiadi, K., Virvou, M.: Evaluating the integration of fuzzy logic into the student model of a web-based learning environment. Expert Syst. Appl. **39**(18), 13127–13134 (2012)
17. Chrysafiadi, K., Virvou, M.: Dynamically personalized e-training in computer programming and the language c. IEEE Trans. Educ. **56**(4), 385–392 (2013)
18. Chrysafiadi, K., Virvou, M.: Student modeling approaches: a literature review for the last decade. Expert Syst. Appl. **40**(11), 4715–4729 (2013)
19. Chrysafiadi, K., Virvou, M., et al.: Advances in Personalized Web-Based Education. Springer (2015)
20. Chrysafiadi, K., Papadimitriou, S., Virvou, M.: Cognitive-based adaptive scenarios in educational games using fuzzy reasoning. Knowl. Based Syst. **250**, 109111 (2022)
21. Chrysafiadi, K., Virvou, M., Tsihrintzis, G.A.: A fuzzy-based mechanism for automatic personalized assessment in an e-learning system for computer programming. Intell. Decis. Technol. (Preprint), 1–16 (2022)
22. Chrysafiadi, K., Virvou, M., Tsihrintzis, G., Hatzilygeroudis, I.: An adaptive learning environment for programming based on fuzzy logic and machine learning. Int. J. Artif. Intell, Tools (2023)
23. Chrysafiadi, K., Virvou, M., Tsihrintzis, G.A.: A fuzzy-based evaluation of e-learning acceptance and effectiveness by computer science students in greece in the period of covid-19. Electronics **12**(2), 428 (2023)
24. Cowley, B., Charles, D.: Behavlets: a method for practical player modelling using psychology-based player traits and domain specific features. User Model. User Adapt. Interact. **26**, 257–306 (2016)
25. Devedžic, V.: Semantic Web and Education, vol. 12. Springer Science & Business Media (2006)
26. Faraco, R.A., Rosatelli, M.C., Gauthier, F.A.: An approach of student modelling in a learning companion system. In: Advances in Artificial Intelligence–IBERAMIA 2004: 9th Ibero-American Conference on AI, Puebla, Mexico, 22–26 Nov 2004. Proceedings, vol. 9, pp. 891–900. Springer (2004)
27. Fossati, D., Di Eugenio, B., Ohlsson, S., Brown, C., Chen, L.: Data driven automatic feedback generation in the ilist intelligent tutoring system. Technol. Instr. Cogn. Learn. **10**(1), 5–26 (2015)
28. Gao, H., Zeng, Y., Ma, B., Pan, Y.: Improving knowledge learning through modelling students' practice-based cognitive processes. Cogn. Comput. 1–18 (2023)
29. Geden, M., Emerson, A., Rowe, J., Azevedo, R., Lester, J.: Predictive student modeling in educational games with multi-task learning. Proc. AAAI Conf. Artif. Intell. **34**, 654–661 (2020)
30. Goel, G., Lallé, S., Luengo, V.: Fuzzy logic representation for student modelling: case study on geometry. In: Intelligent Tutoring Systems: 11th International Conference, ITS 2012, Chania, Crete, Greece, 14–18 June 2012. Proceedings, vol. 11, pp. 428–433. Springer (2012)
31. González, C., Burguillo, J.C., Llamas, M.: A qualitative comparison of techniques for student modeling in intelligent tutoring systems. In: Proceedings. Frontiers in Education. 36th Annual Conference, pp. 13–18. IEEE (2006)
32. Goodman, B., Flaxman, S.: Eu regulations on algorithmic decision-making and a "right to explanation". In: ICML Workshop on Human Interpretability in Machine Learning (WHI 2016), New York, NY. (2016). http://arxiv.org/abs/1606.08813. (v1)
33. Hamzah, A., Sosnovsky, S.: Modelling student knowledge in blended learning. In: Adjunct Proceedings of the 31st ACM Conference on User Modeling, Adaptation and Personalization, pp. 76–80 (2023)

34. Hatzilygeroudis, I.K., Tsihrintzis, G.A., Jain, L.C.: Fusion of Machine Learning Paradigms: Theory and Applications, vol. 236. Springer Nature (2023)
35. Holmgard, C., Liapis, A., Togelius, J., Yannakakis, G.N.: Evolving personas for player decision modeling. In: 2014 IEEE Conference on Computational Intelligence and Games. IEEE (2014). https://doi.org/10.1109/cig.2014.6932911
36. Hooshyar, D., Ahmad, R.B., Yousefi, M., Fathi, M., Horng, S.J., Lim, H.: Applying an online game-based formative assessment in a flowchart-based intelligent tutoring system for improving problem-solving skills. Comput. Educ. **94**, 18–36 (2016)
37. Hwang, G.J., Sung, H.Y., Chang, S.C., Huang, X.C.: A fuzzy expert system-based adaptive learning approach to improving students' learning performances by considering affective and cognitive factors. Comput. Educ. Artif. Intell. **1**, 100003 (2020)
38. Ishmael, O., Kiely, E., Quigley, C., McGinty, D.: Topic modelling using latent dirichlet allocation (lda) and analysis of students sentiments. In: 2023 20th International Joint Conference on Computer Science and Software Engineering (JCSSE), pp. 1–6. IEEE (2023)
39. Jeremić, Z., Jovanović, J., Gašević, D.: Student modeling and assessment in intelligent tutoring of software patterns. Expert Syst. Appl. **39**(1), 210–222 (2012)
40. Jonassen, D.H., Grabowski, B.L.: Handbook of Individual Differences, Learning, and Instruction. Routledge (2012)
41. Karaci, A.: Intelligent tutoring system model based on fuzzy logic and constraint-based student model. Neural Comput. Appl. **31**(8), 3619–3628 (2019)
42. Katsionis, G., Virvou, M.: A cognitive theory for affective user modelling in a virtual reality educational game. In: 2004 IEEE International Conference on Systems, Man and Cybernetics (IEEE Cat. No. 04CH37583), vol. 2, pp. 1209–1213. IEEE (2004)
43. Katsionis, G., Virvou, M.: Personalised e-learning through an educational virtual reality game using web services. Multimed. Tools Appl. **39**, 47–71 (2008)
44. Kay, J.: Stereotypes, student models and scrutability. In: Intelligent Tutoring Systems: 5th International Conference, ITS 2000 Montréal, Canada, 19–23 June 2000 Proceedings, vol. 5, pp. 19–30. Springer (2000)
45. Khan, H., Maull, K., Sumner, T.: Curriculum overlay model for embedding digital resources. In: Proceedings of the 8th ACM/IEEE-CS Joint Conference on Digital Libraries, pp. 74–84 (2008)
46. Korsgaard, D., Bjørner, T., Sørensen, P.K., Burelli, P.: Creating user stereotypes for persona development from qualitative data through semi-automatic subspace clustering. User Model. User Adapt. Interact. **30**, 81–125 (2020)
47. Kyriacou, D.: A scrutable user modelling infrastructure for enabling life-long user modelling. In: International Conference on Adaptive Hypermedia and Adaptive Web-Based Systems, pp. 421–425. Springer (2008)
48. Lehman, B., Matthews, M., D'Mello, S., Person, N.: What are you feeling? Investigating student affective states during expert human tutoring sessions. In: International Conference on Intelligent Tutoring Systems, pp. 50–59. Springer (2008)
49. Li, N., Cohen, W.W., Koedinger, K.R., Matsuda, N.: (2011) A machine learning approach for automatic student model discovery. In: Edm, ERIC, pp 31–40
50. Li, Z., Shi, L., Wang, J., Cristea, A.I., Zhou, Y.: Sim-gail: A generative adversarial imitation learning approach of student modelling for intelligent tutoring systems. Neural Comput. Appl. **35**(34), 24369–24388 (2023). https://doi.org/10.1007/s00521-023-08989-w
51. Lu, C.H., Wu, C.W., Wu, S.H., Chiou, G.F., Hsu, W.L.: Ontological support in modeling learners' problem solving process. J. Educ. Technol. Soc. **8**(4), 64–74 (2005)
52. Macenaite, M., Kosta, E.: Consent for processing children's personal data in the eu: following in us footsteps? Inf. Commun. Technol. Law **26**(2), 146–197 (2017)
53. Madureira, A., Cunha, B., Pereira, J.P., Gomes, S., Pereira, I., Santos, J.M., Abraham, A.: Using personas for supporting user modeling on scheduling systems. In: 2014 14th International Conference on Hybrid Intelligent Systems, pp. 279–284. IEEE (2014)
54. Mahendar, M., Malik, A., Batra, I.: Facial micro-expression modelling-based student learning rate evaluation using vgg-cnn transfer learning model. SN Comput. Sci. **5**(2), 204 (2024)

55. Melhart, D., Azadvar, A., Canossa, A., Liapis, A., Yannakakis, G.N.: Your gameplay says it all: Modelling motivation in tom clancy's the division. In: 2019 IEEE Conference on Games (CoG), IEE (2019). https://doi.org/10.1109/cig.2019.8848123
56. Michaud, L.N., McCoy, K.F.: Empirical derivation of a sequence of user stereotypes for language learning. User Model. User Adapt. Interact. **14**, 317–350 (2004)
57. Mitrovic, A.: An intelligent sql tutor on the web. Int. J. Artif. Intell. Educ. **13**(2–4), 173–197 (2003)
58. Mitrovic, A., Martin, B.: Evaluating the effects of open student models on learning. In: Adaptive Hypermedia and Adaptive Web-Based Systems: Second International Conference, AH 2002 Málaga, Spain, 29–31 May 2002 Proceedings, vol. 2, pp. 296–305. Springer (2002)
59. Mougiakou, E., Virvou, M.: Based on gdpr privacy in uml: Case of e-learning program. In: 2017 8th International Conference on Information, Intelligence, Systems & Applications (IISA), pp. 1–8. IEEE (2017)
60. Mougiakou, E., Papadimitriou, S., Virvou, M.: Intelligent tutoring systems and transparency: the case of children and adolescents. In: 2018 9th International Conference on Information. Intelligence, Systems and Applications (IISA), pp. 1–8. IEEE (2018)
61. Mougiakou, E., Papadimitriou, S., Virvou, M.: Automated decision making and personal data protection in intelligent tutoring systems: design guidelines. In: Knowledge-Based Software Engineering: 2018: Proceedings of the 12th Joint Conference on Knowledge-Based Software Engineering (JCKBSE 2018) Corfu, Greece, vol. 12, pp. 231–241. Springer (2019)
62. Mougiakou, E., Papadimitriou, S., Virvou, M.: Synchronous and asynchronous learning methods under the light of general data protection regulation. In: 2020 11th International Conference on Information. Intelligence, Systems and Applications (IISA), pp. 1–7. IEEE
63. Moundridou, M., Virvou, M.: Authoring intelligent tutoring systems over the world wide web. In: Proceedings First International IEEE Symposium Intelligent Systems, vol. 1, pp. 160–165. IEEE (2002)
64. Ohlsson, S.: Constraint-based student modeling. In: Student modelling: the key to individualized knowledge-based instruction, pp. 167–189. Springer (1994)
65. Orji, F.A., Fatahi, S., Vassileva, J.: Data-driven approach for student engagement modelling based on learning behaviour. In: International Conference on Human-Computer Interaction, pp. 334–342. Springer (2023)
66. Ortony, A., Clore, G., Collins, A.: The Cognitive Structure of Emotions. Cambridge University Press (1990)
67. Panagoulias, D.P., Virvou, M., Tsihrintzis, G.A.: Tailored explainability in medical artificial intelligence-empowered applications: Personalisation via the technology acceptance model. In: 2023 IEEE 35th International Conference on Tools with Artificial Intelligence (ICTAI), pp. 486–490. IEEE (2023)
68. Panagoulias, D.P., Virvou, M., Tsihrintzis, G.A.: A novel framework for artificial intelligence explainability via the technology acceptance model and rapid estimate of adult literacy in medicine using machine learning. Expert Syst. Appl. **248**, 123375 (2024)
69. Papadimitriou, S., Virvou, M.: An online adventure game for teaching math. In: 2016 7th International Conference on Information, Intelligence, Systems & Applications (IISA), pp. 1–5. IEEE (2016)
70. Papadimitriou, S., Virvou, M.: Adaptivity in scenarios in an educational adventure game. In: 2017 8th International Conference on Information, Intelligence, Systems & Applications (IISA), pp. 1–6. IEEE (2017)
71. Papadimitriou, S., Chrysafiadi, K., Virvou, M.: Fuzzeg: Fuzzy logic for adaptive scenarios in an educational adventure game. Multimed. Tools Appl. **78**(22), 32023–3205 (2019). https://doi.org/10.1007/s11042-019-07955-w
72. Papadimitriou, S., Mougiakou, E., Virvou, M.: Smart educational games and consent under the scope of general data protection regulation. In: 2019 10th International Conference on Information, pp. 1–8. Intelligence, Systems and Applications (IISA), IEEE (2019)
73. Papadimitriou, S., Kamitsios, M., Chrysafiadi, K., Virvou, M.: Learn-and-play personalised reasoning from point-and-click to virtual reality mobile educational games. Intelligent Decision Technologies **15**(2), 321–332 (2021)

74. Peña, A., Kayashima, M., Mizoguchi, R., Dominguez, R.: Improving students' meta-cognitive skills within intelligent educational systems: a review. In: Foundations of Augmented Cognition. Directing the Future of Adaptive Systems: 6th International Conference, FAC 2011, Held as Part of HCI International 2011, Orlando, FL, USA, 9–14 July 2011. Proceedings, vol. 6, pp. 442–451. Springer (2011)
75. Pfau, J., Liapis, A., Yannakakis, G.N., Malaka, R.: Dungeons & replicants ii: automated game balancing across multiple difficulty dimensions via deep player behavior modeling. IEEE Trans, Games (2022)
76. Politou, E., Alepis, E., Patsakis, C.: Forgetting personal data and revoking consent under the gdpr: Challenges and proposed solutions. J. Cybersecur. **4**(1), tyy001 (2018)
77. Politou, E., Alepis, E., Virvou, M., Patsakis, C.: Privacy and Data Protection Challenges in the Distributed Era. Springer International Publishing (2022)
78. Pruitt, J., Adlin, T.: The Persona Lifecycle: A Field Guide for Interaction Designers (2006)
79. Resnick, L.B.: The Nature of Intelligence (1976)
80. Rich, E.: User modeling via stereotypes. Cogn. Sci. **3**(4), 329–354 (1979)
81. Rodrigo, M.M.T., Baker, R.S., Lagud, M.C., Lim, S.A.L., Macapanpan, A.F., Pascua, S., Santillano, J.Q., Sevilla, L.R., Sugay, J.O., Tep, S., et al.: Affect and usage choices in simulation problem solving environments. Front. Artif. Intell. Appl. **158**, 145 (2007)
82. Self, J.: Bypassing the intractable problem of student modelling. In: Proceedings of ITS'88, Montreal, pp. 18–24 (1988)
83. Shakouri, G.H., Tavassoli, N.Y.: Implementation of a hybrid fuzzy system as a decision support process: a fahp-fmcdm-fis composition. Expert Syst. Appl. **39**(3), 3682–3691 (2012)
84. Shyamala, R., Sunitha, R., Aghila, G.: Towards learner model sharing among heterogeneous e-learning environments. Int. J. Comput. Sci. Eng. **3**(5), 2034–2040 (2011)
85. Sison, R., Shimura, M.: Student modeling and machine learning. Int. J. Artif. Intell. Educ. (IJAIED) **9**, 128–158 (1998)
86. Sobolewski, M., Mazur, J., Paliński, M.: Gdpr: a step towards a user-centric internet? Intereconomics **52**, 207–213 (2017)
87. Stathopoulou, I., Alepis, E., Tsihrintzis, G., Virvou, M.: On assisting a visual-facial affect recognition system with keyboard-stroke pattern. In: Research and Development in Intelligent Systems XXVI: Incorporating Applications and Innovations in Intelligent Systems XVII, p. 451 (2009)
88. Sucar, L.E., Noguez, J.: Student Modeling, pp. 173–185. A Practical Guide to Applications, Bayesian Networks (2008)
89. Suleman, R.M., Mizoguchi, R., Ikeda, M.: Negotiation-driven learning. In: Artificial Intelligence in Education: 17th International Conference, AIED 2015, Madrid, Spain, 22–26 June 2015. Proceedings, vol. 17, pp. 470–479. Springer (2015)
90. Suraweera, P., Mitrovic, A.: An intelligent tutoring system for entity relationship modelling. Int. J. Artif. Intell. Educ. **14**(3–4), 375–417 (2004)
91. Thomson, D.: Towards a negotiable student model for constraint-based itss (2008)
92. Tourtoglou, K., Virvou, M.: Simulated annealing in finding optimum groups of learners of uml. In: Intelligent Interactive Multimedia Systems and Services, pp. 147–156. Springer (2010)
93. Triantafillou, E., Pomportsis, A., Demetriadis, S.: The design and the formative evaluation of an adaptive educational system based on cognitive styles. Comput. Educ. **41**(1), 87–103 (2003)
94. Triantafyllou, A.M., Tsihrintzis, G.A., Virvou, M., Alepis, E.: A bimodal system for emotion recognition via computer of known or unknown persons in normal or fatigue situations. Advances in Core Computer Science-Based Technologies: Papers in Honor of Professor Nikolaos Alexandris, pp. 9–35 (2021)
95. Tsihrintzis, G., Toro, C., Rios, S., Howlett, R., Jain, L.: 27th kes International Conference on Knowledge-Based and Intelligent Information & Engineering Systems kes2023 (2023)
96. Tsihrintzis, G.A., Virvou, M., Alepis, E., Stathopoulou, I.O.: Towards improving visual-facial emotion recognition through use of complementary keyboard-stroke pattern information. In: Fifth International Conference on Information Technology: New Generations (itng 2008), pp. 32–37. IEEE (2008)

97. Tsihrintzis, G.A., Virvou, M., Jain, L.C.: Multimedia Services in Intelligent Environments: Recommendation Services. Springer (2013)
98. Tsihrintzis, G.A., Virvou, M., Sakkopoulos, E., Jain, L.C.: Machine Learning Paradigms: Applications of Learning and Analytics in Intelligent Systems, vol. 1. Springer (2019)
99. Tsihrintzis, G.A., Virvou, M., Jain, L.C.: Advances in Machine Learning/Deep Learning-based Technologies. Springer (2022)
100. Tsihrintzis, G.A., Virvou, M., Hatzilygeroudis, I.: Special Issue on Selected Papers from the 33rd Annual IEEE International Conference on Tools with Artificial Intelligence (ictai-2021) (2023)
101. Tsihrintzis, G.A., Virvou, M., Doukas, H., Jain, L.C.: Introduction to advances in artificial intelligence-empowered decision support systems. In: Advances in Artificial Intelligence-Empowered Decision Support Systems: Papers in Honour of Professor John Psarras, pp. 1–11. Springer (2024)
102. Tsiriga, V., Virvou, M.: Initializing the student model using stereotypes and machine learning. In: Proceedings of the IEEE International Conference on Systems, Man and Cybernetics, Citeseer (2002)
103. Tsiriga, V., Virvou, M.: Initializing student models in web-based itss: a generic approach. In: Proceedings 3rd IEEE International Conference on Advanced Technologies, pp. 42–46. IEEE (2003)
104. Virvou, M.: A human plausible reasoning theory in the context of an active help system for unix1 users. Ph.D. thesis, University of Sussex, Falmer, East Sussex, UK (1993). https://ethos.bl.uk/OrderDetails.do?uin=uk.bl.ethos.787162
105. Virvou, M.: Automatic reasoning and help about human errors in using an operating system. Interact. Comput. **11**(5), 545–573 (1999)
106. Virvou, M.: A domain-independent reasoning mechanism for an its authoring tool. In: EdMedia+ Innovate Learning, Association for the Advancement of Computing in Education (AACE), pp. 1762–1764 (2000)
107. Virvou, M.: Intelligence and adaptivity in human-computer interaction concerning biotechnological software users. In: Scattering And Biomedical Engineering: Modeling and Applications, World Scientific, pp. 448–458 (2002)
108. Virvou, M.: Modelling the knowledge and reasoning of users in a knowledge-based authoring tool. Int. J. Bus. Perform. Manag. **5**(2–3), 399–412 (2003)
109. Virvou, M.: On the evaluation of the combined role of virtual reality games and animated agents in edutainment. Intell. Comput. Graph. **2011**, 79–95 (2012)
110. Virvou, M.: Artificial intelligence and user experience in reciprocity: contributions and state of the art. Intell. Dec. Technol. **17**(Preprint), 1–53 (2023)
111. Virvou, M.: The emerging role of artificial intelligence in teaching 21st century skills in software engineering. In: Keynote 1 in Proceedings of the 35th IEEE Conference on Software Engineering Education and Training, CSEE&T 2023 (2023)
112. Virvou, M., Alepis, E.: Mobile educational features in authoring tools for personalised tutoring. Comput. Educ. **44**(1), 53–68 (2005)
113. Virvou, M., Kabassi, K.: Reasoning about users' actions in a graphical user interface. Hum. Comput. Interact. **17**(4), 369–398 (2002)
114. Virvou, M., Kabassi, K.: Evaluating an intelligent graphical user interface by comparison with human experts. Knowl. Based Syst. **17**(1), 31–37 (2004)
115. Virvou, M., Moundridou, M.: An authoring tool for algebra-related domains. In: HCI (2), pp. 647–651. Citeseer (1999)
116. Virvou, M., Papadimitriou, S.: Learning addition by playing an adaptive game with coins. In: IISA 2013, pp. 1–4. IEEE (2013)
117. Virvou, M., Papadimitriou, S.: Simple arithmetic lessons through an adaptive snake game. In: 2013 International Conference on Computer, Information and Telecommunication Systems (CITS), pp. 1–4. IEEE (2013)
118. Virvou, M., Papadimitriou, S.: Use of digital 'guess who board game' in teaching english. In: IISA 2014, The 5th International Conference on Information. Intelligence, Systems and Applications, pp. 439–443. IEEE (2014)

119. Virvou, M., Papadimitriou, S.: Playing, learning, extending: educational "guess who" game that is renewable by a teacher. In: 2015 6th International Conference on Information. Intelligence, Systems and Applications (IISA), pp. 1–5. IEEE (2015)
120. Virvou, M., Stavrianou, A.: User modelling in a gui. In: Proceedings of HCI International (The 8th International Conference on Human-Computer Interaction) on Human-Computer Interaction: Ergonomics and User Interfaces, vol. I, pp. 261–265 (1999)
121. Virvou, M., Tsihrintzis, G.A.: A novel trust state-chart model for requirements engineering of trustful ai-empowered software. In: 2023 14th International Conference on Information, Intelligence, Systems & Applications (IISA), pp, 1–6. IEEE (2023)
122. Virvou, M., Manos, C., Katsionis, G., Tourtoglou, K.: Vr-engage: A virtual reality educational game that incorporates intelligence. In: Proceedings of IEEE International Conference on Advanced Learning Technologies, pp. 16–19 (2002)
123. Virvou, M., Manos, K., Katsionis, G., Tourtoglou, K.: Vr-integate: a knowledge-based authoring tool for virtual reality educational games. In: Proceedings of ED-MEDIA 2002 World Conferences on Educational Multimedia, Hypermedia and Telecommunications (2002)
124. Virvou, M., Alepis, E., Sidiropoulos, S.C.: A learning analytics tool for supporting teacher decision. In: 2015 6th International Conference on Information, Intelligence, Systems And Applications (IISA), pp. 1–3. IEEE (2015)
125. Virvou, M., Alepis, E., Tsihrintzis, G.A., Jain, L.C.: Machine Learning Paradigms: Advances in Learning Analytics. Springer (2020)
126. Virvou, M., Tsihrintzis, G.A., Bourbakis, N.G., Jain, L.C.: Handbook on Artificial Intelligence-Empowered Applied Software Engineering, vol. 2. Smart Software Applications in Cyber-Physical Systems, vol. 3. Springer (2022)
127. Virvou, M., Tsihrintzis, G.A., Bourbakis, N.G., Jain, L.C.: Introduction to handbook on artificial intelligence-empowered applied software engineering: novel methodologies to engineering smart software systems. In: Handbook on Artificial Intelligence-Empowered Applied Software Engineering: Novel Methodologies to Engineering Smart Software Systems, vol. 1, pp. 1–8. Springer (2022)
128. Virvou, M., Tsihrintzis, G.A., Tsoukalas, L.H., Jain, L.C.: Introduction to advances in artificial intelligence-based technologies. In: Advances in Artificial Intelligence-based Technologies: Selected Papers in Honour of Professor Nikolaos G Bourbakis, vol. 1, pp. 1–5 (2022)
129. Virvou, M., Tsihrintzis, G.A., Tsichrintzi, E.A.: Virtsi: A novel trust dynamics model enhancing artificial intelligence collaboration with human users-insights from a chatgpt evaluation study. Inf. Sci. **675**, 120759 (2024)
130. Wachter, S., Mittelstadt, B., Floridi, L.: Why a right to explanation of automated decision-making does not exist in the general data protection regulation. Int. Data Privacy Law **7**(2), 76–99 (2017)
131. Webb, G.I., Pazzani, M.J., Billsus, D.: Machine learning for user modeling. User Model. User Adapt. Interact. **11**, 19–29 (2001)
132. Weng, S., Shieh, C.S., Tsihrintzis, G.A.: Advances in Intelligent Information Hiding and Multimedia Signal Processing: Proceeding of the 18th IIH-MSP 2022 Kitakyushu. Springer, Japan (2023)
133. Whitley, E.A.: Informational privacy, consent and the "control" of personal data. Inf. Secur. Tech. Rep. **14**(3), 154–159 (2009)
134. Yang, G., Graf, S., et al.: A practical student model for a location-aware and context-sensitive personalized adaptive learning system. In: 2010 International Conference on Technology for Education, pp. 130–133. IEEE (2010)
135. Yossatorn, Y., Binali, T., Weng, C., Awuor, N.O.: Relating university students' online self-regulated english learning to motivational beliefs: a structural equation modelling analysis. Behav. Inf. Technol. **42**(7), 888–903 (2023)
136. Zadeh, L.A.: Fuzzy logic. Computer **21**(4), 83–93 (1988)

Chapter 4
Fuzzy Logic and Applications in Education and Games: Theory, Practical Implementations and a Literature Review

Abstract In computer science, fuzzy logic is a powerful tool that expands upon classical binary logic to represent ambiguity and imprecision mathematically. It emulates human thinking and effectively tackles issues involving uncertainty and vagueness. The concepts of fuzzy sets, membership functions, and fuzziness play a crucial role in developing fuzzy inference systems, which make decisions based on fuzzy logic. Furthermore, the application of interval type-2 fuzzy sets can better handle problems with significant uncertainty than traditional fuzzy sets. Fuzzy finite state machines (FFSMs) are another application of fuzzy logic that provides a more versatile approach to modeling complex systems. The chapter describes all of those above and highlights the use of fuzzy reasoning techniques in educational applications and games, demonstrating the potential of this technology to significantly enhance the way we learn and interact with technology. In addition to providing an overview of fuzzy logic principles and their applications, this chapter presents a review of educational applications and games where fuzzy logic has been utilized. The review examines various ways in which fuzzy logic has been integrated into educational contexts, showing its potential to enhance learning experiences. Additionally, this chapter identifies open problems and challenges in using fuzzy logic within educational settings, highlighting areas for future research and development. By synthesizing existing literature and identifying gaps in knowledge, this review contributes to a deeper understanding of the role of fuzzy logic in educational technology and points towards new directions for further exploration and innovation.

Keywords Educational software · Educational games · Artificial Intelligence in education · Fuzzy logic in education · User experience · AI-empowered software games · User modeling · Human-AI interaction · e-learning games · 21st-century skills

S. Papadimitriou and M. Virvou, *Artificial Intelligence—Based Games as Novel Holistic Educational Environments to Teach 21st Century Skills*, Intelligent Systems Reference Library 93, https://doi.org/10.1007/978-3-031-77464-5_4

4.1 An Overview

Fuzzy logic, a tool designed to handle issues with ambiguous or uncertain solutions, offers a sense of confidence in our decision-making. In situations where we struggle to determine the accuracy of a statement, it provides valuable adaptability by acknowledging the uncertainties present. This adaptability aids in resolving problems by thoroughly assessing all gathered data and arriving at a suitable conclusion. The fuzzy logic approach mirrors humans' decision-making process, effectively confidently managing unclear and imprecise information. Instead of relying on the traditional true/false or 1/0 values of boolean logic, it simplifies the complexities of the real world by employing degrees of truth. This method provides flexibility and enables a system to reason with a more extensive set of data and variables.

Lotfi Zadeh developed fuzzy logic after recognizing significant differences between human and computer decision-making processes [157]. He observed that conventional computer logic could not effectively process data representing vague concepts, a capability inherent in human thinking. He concluded that people do not always rely on precise numerical information but instead possess adept adaptive control. Consequently, he suggested that feedback controllers, which are mathematical systems analyzing input values, could be more effective and easier to develop if they were programmed to handle noisy or imprecise data. Fuzzy logic offers the following advantages:

- The methodology of fuzzy logic works similarly to human reasoning.
- Building fuzzy logic systems (FLSs) is simple and comprehensible.
- An FLS can handle imprecise, distorted, or noisy input.
- It finds extensive use in all sectors of life and yields effortlessly beneficial solutions to high-complexity challenges.
- The time to develop fuzzy logic is reduced compared to the time conventional methods require.
- Its flexibility allows users to insert and remove rules within the FLS conveniently.
- Fuzzy logic goes hand in hand with the mathematical concepts of set theory, making fuzzy syllogism uncomplicated.

However, fuzzy logic has some disadvantages and could be more helpful in some cases. When conventional logic arrives at a solution, there is no need to apply fuzzy logic as it overcomplicates everything. The disadvantages are the following:

- It is not recommended for problems that need high precision.
- An FLS needs much testing for verification and validation.
- Sometimes, the definition of exact rules and membership functions during the fuzzy logic setting is complicated.
- The results may not be widely accepted because they rely on assumptions.

Fuzzy logic finds applications in areas that range from process control to medical diagnosis. Following are some of the application areas where the fuzzy logic concept applies [6, 14–17, 107, 116]:

- In the aerospace sector, it controls the altitude of spacecraft and satellites.
- It controls traffic and speed in an automotive system.
- It works in large company businesses as a decision-making support system for evaluating personnel.
- It has been widely applied in the chemical industry for pH control, drying, and chemical distillation.
- It has applications in natural language processing [110] and several intensive Artificial Intelligence (AI) operations.
- It has applications in modern control systems such as expert systems.
- It supports neural networks because it imitates how a human makes decisions more quickly. It aggregates data and converts it into more meaningful data by generating partial truths as fuzzy sets.
- It works in medicine as a diagnostic system for interpreting medical findings.
- It has applications in energy modeling and renewable energy systems to map energy systems precisely.

4.2 Fuzzy Logic Systems

Fuzzy logic extends classical (conventional) logic and uses an efficient mathematical model to model information that approximates natural human logic and confronts the system's uncertainty and vagueness. A set of fuzzy rules in the form "if <antecedent> then <consequent>" that use linguistics variables with corresponding linguistic values describe the behavior of an FLS. Each linguistic variable represents a fuzzy set. For instance, the fuzzy set HEIGHT can take one or more of the following linguistic values: short, medium, and tall, while the variable FOOD_QUALITY in a restaurant can take the values tasteless, medium, and tasty. Thus, an FLS is an expert system that utilizes a set of fuzzy membership functions (MFs) and a base of fuzzy IF-THEN rules to reason about the data.

The most popular types of fuzzy systems are Mamdani and Takagi-Sugeno-Kang (TSK). For example, a Mamdani fuzzy system follows the following process:

1. It fuzzifies all input values into fuzzy membership functions (fuzzification).
2. It executes all the applied rules in the rule base to calculate fuzzy output functions (inference engine).
3. It de-fuzzifies the fuzzy output functions to get crisp output values (defuzzification).

Figure 4.1 depicts the system's architecture where the abovementioned steps occur.

The fuzzifier is the component where the fuzzification process (step 1) occurs. It transforms the raw inputs into fuzzy sets with membership varying in the range [0,1]. Should it be 0, the value does not participate in the given fuzzy set, whereas if it is 1, it belongs entirely to the fuzzy set. A value between 0 and 1 corresponds to the degree

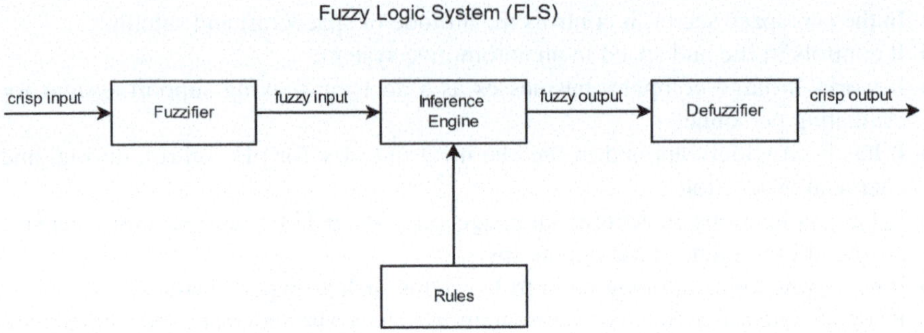

Fig. 4.1 The architecture of an FLS

of participation to which the value is a set member. Hence, the membership function is a graph that maps each point in the input space (called Universe U) to membership values between 0 and 1. The following equation defines the membership function:

$$\mu A : X \to [0, 1] \tag{4.1}$$

Here, X represents the Universe, and Y corresponds to any value between 0 and 1. The triangular membership function shapes are the most common among several other membership function shapes, such as trapezoidal, singleton, and Gaussian. For example, Fig. 4.2 illustrates trapezoidal membership functions for the fuzzy variable FOOD_QUALITY.

Once the fuzzification process is completed, the fuzzy sets are passed on to the inference engine for further processing with the rule-base component. This component consists of rules and IF-THEN conditions provided by experts to govern the decision-making system using linguistic information. The antecedent value is the input during the inference process, and a fuzzy set is generated as the output. Each rule is then executed, and the results are combined into a single fuzzy set to determine the final output, thus initiating an aggregation process. Consequently, the aggregation process takes the output function list generated from each rule during the inference process and produces a fuzzy set for each output variable.

Given the above, a rule-based component may contain the following rules:

Fig. 4.2 Membership functions of the linguistic variable FOOD_QUALITY

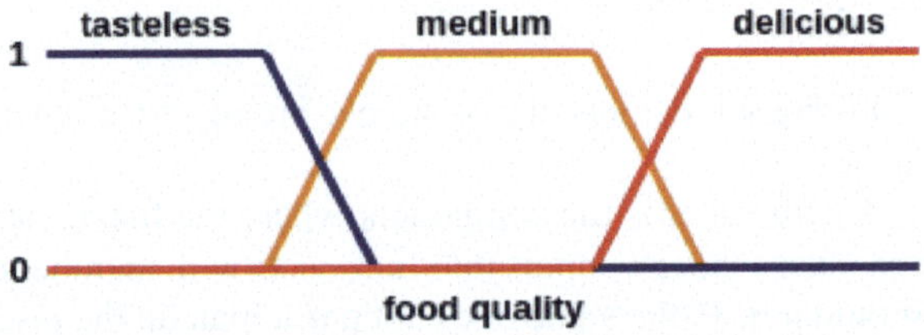

1. IF food_quality IS tasteless, THEN the tip is cheap.
2. IF food_quality IS delicious, THEN the tip is generous.

Given a predetermined numerical value representing food quality, the fuzzy set "tasteless" is assigned a specific degree of truth, which is then transferred to the fuzzy set "cheap" concerning the waiter's tip. If a fuzzy output set appears in multiple THEN parts, the values from the corresponding IF parts are combined using appropriate operators.

The final stage of an FLS involves the defuzzification process. This step takes the cumulative fuzzy output set as input and produces a single numeric value as output. As part of this process, the aggregated output, consisting of a range of values, must be de-fuzzified to obtain a crisp output value. Several defuzzification methods are available, such as the Max-Membership Principle, Centroid Method, Weighted Average Method, Mean-Max Membership, and others. FLS developers often seek expert guidance to determine the most suitable method.

Given the above, the algorithm for a Mamdani FLS contains the following steps:

1. Define all the fuzzy variables and fuzzy sets (input and output).
2. Create membership functions for the above fuzzy sets.
3. Construct the knowledge base of rules.
4. Transform crisp data into fuzzy sets using membership functions (fuzzification process).
5. Evaluate the rules in the rule knowledge base (inference engine).
6. Unite all the results acquired from the rule-based component (inference engine).
7. Convert the fuzzy output set into crisp output (defuzzification).

Regarding the TSK system, the defuzzification process occurs in the execution phase of the fuzzy rules. Rules differentiate in the part of consequents where polynomial functions (usually constant or linear) represent them. An example of a rule with a constant output value is the following:

IF food_quality IS tasteless, then the tip = 2

If the rule base contains two or more rules, the output will be the average of the consequent of each rule i (Y_i), weighted according to the membership value of its antecedent (h_i):

$$\frac{\Sigma_i(h_i \times Y_i)}{\Sigma_i h_i} \tag{4.2}$$

An example of a rule with a linear output would be:

IF food_quality IS delicious AND service IS excellent, then the tip = 2 * food_quality + 1 * service

The primary advantage of using TSK over Mamdani is that it is computationally efficient and works well within other algorithms, such as optimization algorithms. However, Mamdani has more expressive power and interpretability, making working with people more intuitive and easier.

4.3 Interval Type-2 Fuzzy Sets

Fuzzy logic effectively addresses decision-making problems that involve vagueness and uncertainty. In fuzzy sets, each element's participation is determined by a membership function that assigns values within the range of 0 to 1, denoting the level of uncertainty. However, there are scenarios where defining an accurate membership function for a fuzzy set is complicated [61]. In cases with heightened uncertainty, type-2 fuzzy sets present an alternative solution. They excel in handling linguistic ambiguities and modeling the inherent fuzziness and uncertainty of the underlying information, therefore reducing uncertainty [89, 90, 149]. Unlike standard fuzzy sets, type-2 fuzzy sets do not yield precise numbers but employ membership grades. Moreover, they utilize three-dimensional membership functions instead of the two-dimensional ones used in type-1 fuzzy sets, allowing additional degrees of freedom to represent uncertainties directly.

Zadeh introduced the concept of general type-2 fuzzy sets as an extension of the standard fuzzy set, the type-1 fuzzy set. Type-2 fuzzy sets have membership degrees that are themselves fuzzy. A type-2 membership, the primary one, has a grade in any subset in [0,1]. Additionally, a secondary membership function for each primary membership (which may also reside in [0,1]) represents the possibilities for the primary membership. However, general type-2 fuzzy logic sets are computationally intensive because type-reduction requires heavy computational resources. Therefore, the widespread application of generalized type-2 fuzzy systems has yet to occur. The solution is to use interval type-2 fuzzy sets due to their simplicity and decreased computational effort compared to the general type-2 fuzzy set [61]. The interval type-2 fuzzy sets utilize interval sets as secondary membership functions, which return either zero or one.

A type-2 FLS comprises a fuzzifier, a rule base element, a fuzzy inference engine, and an output processor (Fig. 4.3). The output processor generates a type-1 fuzzy set output from a type-reducer or a crisp number from a defuzzifier. The IF-THEN rules characterize a type-2 FLS, but its antecedent or consequent sets are now type-2. For example, a rule can have the following format:

If the temperature is moderate and the pressure is high, rotate the valve a bit to the right.

Uncertainty in the primary MF of a type-2 MF consists of a bounded region named the "footprint of uncertainty" (FOU) of a type-2 MF. An upper membership function (UMF) and a lower membership function (LMF) are two type-1 functions that include the bounds for the FOU of an interval type-2 MF (Fig. 4.4). The UMF is a subset with the maximum membership grade of the FOU, while the LMF is a subset with the minimum membership grade of the FOU [84, 92].

Type 2 sets and FLS have had applications, among other things, decision making, function approximation, time series prediction, survey processing, fuzzy relation equations solving, time-varying channel balancing, mobile robot control, image processing, video processing, computer vision, failure mode, and outcome analysis.

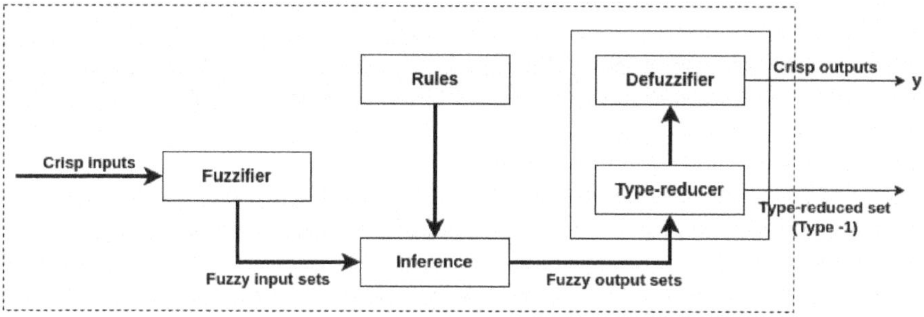

Fig. 4.3 The architecture of a Type-2 FLS

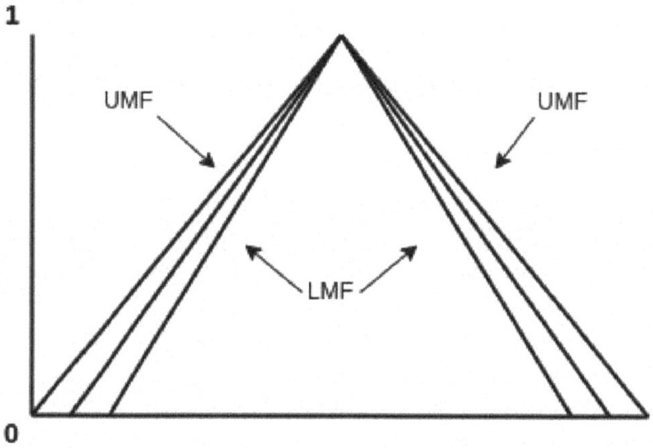

Fig. 4.4 Type-1 fuzzy sets contained in FOU

4.4 Fuzzy Finite State Machines

A finite state machine (FSM) is a mathematical model of computation that defines how a mathematical function computes an output based on an input. It is depicted as a graph with limited nodes representing possible system states and a finite number of arcs signifying transitions between states [52, 86]. At any given time, the machine exists in only one of the finite states, indicating the system's readiness to transition. FSMs are commonly used to model computer programs, sequential logic, and control execution flow. Two FSMs include Mealy machines, in which output values are based on the present state and input, and Moore machines, in which the output relies solely on the present state. They are applied to model problems in diverse domains such as games, linguistics, mathematics, and AI.

Incorporating fuzziness into the FSM allows for more precise modeling of processes that dynamically change over time [93]. The FFSM is an advanced extension of the standard FSM, using fuzzy values for inputs and outputs instead of crisp symbols. This facilitates seamless control and reasoning with degrees of truth, making the system more natural and adaptable. The FFSM's capability to handle and process input information with a degree of belonging provides a realistic model supported by a fuzzy reasoning mechanism. This mechanism utilizes degrees of truth associated with each state transition, enabling the system to have more than one active state based on each state's membership values.

The finite-state machine utilizes a series of fuzzy rules defined over the states and input/output values. However, when simulating the state machine, a fuzzy reasoning system must attentively consider these rules. Consequently, the representation does not offer much room for significant optimizations. In the end, the rule-base component contains fuzzy rules with the following format:

IF <state1> AND <transition> THEN <state2>

4.5 Fuzzy Reasoning in Education and Intelligent Tutoring Systems

Many intelligent techniques in education and Intelligent Tutoring Systems (ITSs) have been reviewed in [23]. These span all the areas of AI that have been at the center of research nowadays more than ever since AI is being applied to all social, scientific, and professional sectors [134] and in areas such as education [133] and even culture like Large Language Models (LLMs) [142] or Machine Learning (ML) [139] or artificial neural networks [87] in web-based education [96, 127, 137] or mobile [4]. Recent advancements in intelligent techniques have been extensively explored in recent publications. Tsihrintzis et al. [126] discuss the latest developments in the field, shedding light on the ongoing progress in AI tools. Furthermore, Weng, Shieh, and Tsihrintzis [148] highlight advancements in intelligent information hiding and multimedia signal processing, indicating the continuous evolution of these technologies. Hatzilygeroudis et al. [53] examine the fusion of ML paradigms, offering theoretical insights and practical applications that contribute to this area's expanding body of knowledge. Additionally, Tsihrintzis et al. [124] underscore ongoing research efforts in intelligent information and engineering systems, further emphasizing the dynamic nature of this field. Virvou and colleagues [140, 141] provide valuable insights into the application of AI in software engineering, particularly in cyber-physical systems, enriching the discourse on intelligent techniques. Collectively, these contributions highlight the relevance and impact of intelligent techniques across various domains, illustrating this field's dynamic and evolving landscape.

A special category of intelligent techniques that are being developed further and attracts a significant amount of research energy every year is fuzzy logic. In the context of intelligent applications, Chrysafiadi [15] explores many intelligent applications based on fuzzy logic. Recent advances include improving task scheduling through a fuzzy reasoner [13] or an intelligent fuzzy-based emergency alert generation for persons with episodic memory decline problems [18] or fuzzy logic in personalization of fire emergency alerts as in [5] and many other.

In the field of education and ITSs, fuzzy logic has emerged as a valuable tool for managing education and evaluating various factors influencing it [55, 75, 121, 146, 161]. It is also frequently employed as a method of reasoning integrated into the inference engine of ITS and educational systems [39, 94].

For instance, fuzzy logic finds application in evaluating the efficiency of educational programs [117], assessing the quality of higher education management using interval-valued neutrosophic sets [44], and establishing methodologies for evaluating the quality of innovation and entrepreneurship education [51].

There are many examples wherein fuzzy logic underpins the internal reasoning mechanisms of educational systems. Papadimitriou and colleagues proposed adaptive quizzes utilizing a Fuzzy Genetic Algorithm [105]. Similarly, Chrysafiadi and colleagues developed a fuzzy-based mechanism for automatic personalized assessment within an e-learning system focused on computer programming [32] and dynamic creation of tests [25]. It is worth noting that fuzzy logic has been used in many phases of the development process of educational systems, starting from usability requirements like in [47], where a Fuzzy Analytic Hierarchy Process (FAHP) is described for usability requirements of online education systems. In another instance, there has been an analysis of critical factors that affect early childhood education using touchscreen contrivances using a hybrid fuzzy AHP - fuzzy TOPSIS approach [68].

4.5.1 Decisions and Selection in Educational Contexts Using Fuzzy Logic

In educational contexts, many selection challenges arise when choosing among various options to meet specific objectives or needs. These challenges encompass selecting appropriate tools, platforms, models, or locations tailored to the unique requirements of educational programs, institutions, or initiatives. Educational stakeholders often face decisions regarding selecting distance education tools, platforms, or live platforms, where factors such as usability, functionality, and technological compatibility must be carefully considered. Additionally, selecting instructional design models, methodologies, or approaches for specialized domains like maritime education poses distinct challenges due to the need for alignment with specific learning objectives and industry requirements. Furthermore, broader decisions related to education center site selection necessitate considerations of geographic, infrastructural, and community-related factors to ensure optimal accessibility and effectiveness. Addressing these selection challenges requires robust decision-making frameworks capable of accommodating uncertainties, complexities, and diverse stakeholder perspectives inherent in educational contexts.

Selection processes in educational contexts often involve complex decision-making scenarios where various criteria need to be considered. Fuzzy logic, which can handle uncertainty and imprecision, has become valuable in aiding these selection tasks. For instance, Menekse and Camgöz-Akdag [91] utilize a novel spherical fuzzy Analytic Hierarchy Process (AHP) EDAS method for distance education tool selection, enhancing decision-making by considering multiple criteria. Adem et al. [1] focus on selecting suitable distance education platforms based on human-computer interaction criteria within a fuzzy environment, highlighting the importance of user experience in platform selection. Liu et al. [77] employ multi-attribute group decision-making techniques with linguistic intuitionistic cubic fuzzy aggregation operators for selecting online education live platforms, facilitating decision-making in a collaborative setting. Kandemir and Cicek [63] propose a fuzzy axiomatic design approach for selecting instructional design models in maritime education and training, emphasizing the need for tailored approaches in specialized educational domains. Additionally, Agrawal, Agarwal, and Bansal [2] explore the integration of fuzzy logic with metaheuristics for education center site selection, demonstrating the applicability of fuzzy methods in broader educational infrastructure decisions. These studies collectively showcase the versatility of fuzzy logic in addressing diverse selection challenges within educational settings, ultimately contributing to more informed and effective decision-making processes.

4.5.2 *Evaluation Strategies in Educational Contexts Using Fuzzy Logic*

Evaluation in educational contexts presents numerous challenges, particularly concerning the assessment of various facets such as user experience, adaptivity [96], learning outcomes, and overall quality. These may be extended to include evaluations concerning ethics and privacy [95]. Moreover, there are often methodologies that include comparisons, like, for example, comparisons of the automatic reasoning [26, 130–132, 135, 138] of intelligent systems with the reasoning of human experts [136].

Researchers address these challenges through many methodologies and frameworks using fuzzy logic. Chrysafiadi, Virvou, and Tsihrintzis, along with Hatzilygeroudis [36], evaluate the user experience, adaptivity, and learning outcomes of an ITS utilizing fuzzy logic. Similarly, Rao et al. [111] and Hu et al. [54] propose methodologies for evaluating physical education teaching quality, employing double-valued neutrosophic numbers and 2-tuple linguistic neutrosophic numbers, respectively. Other studies explore evaluation methods tailored to specific contexts, such as mental health education quality evaluation [145], regional culture integration into ideological education [150], and applied education in industrial universities [38].

Additionally, innovative approaches like cloud model-based assessment [154] and fuzzy decision-making for educational methods assessment [129] contribute to the diverse employment of evaluation methodologies. Technological advancements also influence evaluation practices, as evidenced by research evaluating ergonomics-related disorders in online education [128] and utilizing computer information technology for fuzzy evaluation in physical education [156]. Ayouni et al. [7] extend the fuzzy methodology to evaluate learning management systems in higher education, showing the adaptability of fuzzy logic across different educational domains. Furthermore, studies by Wang et al. [147], and Lv et al. [81] integrate AI with FLSs to develop evaluation frameworks for music teaching and learning management, respectively, highlighting the synergistic potential of fuzzy logic and AI technologies in educational assessment.

These endeavors collectively underscore the complexity and importance of evaluating various aspects of education, necessitating robust and adaptable evaluation methodologies to inform decision-making and drive continuous improvement.

4.5.3 Tutoring Strategies Using Fuzzy Logic

In educational contexts, tutoring strategies encounter many challenges that require careful navigation and innovative solutions. One significant hurdle lies in accommodating students' diverse learning styles, abilities, and preferences. To be effective, tutoring systems must be flexible and personalized, capable of adapting to individual needs while maintaining a cohesive learning environment. Additionally, the integration of emerging technologies presents both opportunities and complexities. While advancements such as multimedia-assistive technology offer exciting possibilities for enhancing learning experiences, their successful implementation requires meticulous planning and consideration of pedagogical principles to ensure meaningful integration into educational practices [71]. Moreover, the ongoing COVID-19 pandemic has highlighted the imperative for flexible and resilient tutoring strategies as institutions worldwide transition to online and hybrid learning models. This shift underscores the need for tutoring systems that seamlessly facilitate remote learning while maintaining engagement and effectiveness [114]. Furthermore, ensuring equitable access to tutoring resources and addressing socio-economic disparities pose significant challenges, necessitating proactive measures to promote inclusivity and accessibility in educational settings [39]. Despite these obstacles, advancements in fuzzy logic and AI offer promising avenues for overcoming challenges and optimizing tutoring strategies to meet the diverse needs of learners.

In the context of tutoring strategies, recent advancements have been driven by integrating personalized and adaptive approaches facilitated by fuzzy logic techniques. Konstantina Chrysafiadi and Maria Virvou have been pioneers in this domain, contributing significantly to the development of personalized web-based education systems [28]. Their work spans various aspects of adaptive instruction, including fuzzy logic-based methodologies for dynamic adaptation and modeling learner cognitive state transitions [21, 37]. Notably, their research explores the utilization of fuzzy cognitive maps for domain knowledge representation within adaptive e-learning environments, showcasing the versatility of fuzzy logic in capturing and managing complex knowledge structures [20]. Moreover, Papadimitriou and colleagues have contributed to the development of authoring tools that leverage fuzzy logic to generate personalized learning materials and track student progress across diverse educational platforms [101]. Their novel work exemplifies the transformative potential of fuzzy logic in enhancing tutoring strategies, offering tailored and adaptable learning experiences that cater to learners' individual needs and preferences.

4.5.4 Data Analytics and Management of Educational Contexts Using Fuzzy Logic

In the context of education, the effective utilization of data analytics and management presents a plethora of challenges. Educational information resources, particularly in peer-to-peer (P2P) sharing systems, require sophisticated solutions to ensure stability and efficiency [159]. Furthermore, the management of data in higher education demands advanced algorithms to handle complex datasets effectively. Guan [46] proposed an algorithm for data management in higher education based on fuzzy set theory and association rule mining, highlighting the need for innovative approaches in this domain. As the education sector transitions towards Education 4.0, stakeholders play pivotal roles, necessitating the development of modified picture fuzzy aggregation operators to address their diverse priorities [112]. Moreover, optimizing teaching management, especially in areas like college physical education, calls for innovative approaches such as fuzzy neural networks to enhance decision-making processes [11]. The advent of big data intelligence has marked significant progression in educational management systems, demanding the development of smart systems capable of processing vast amounts of data efficiently [41].

Thavi et al. [122] employed a fuzzy DEMATEL approach to identify critical factors for adopting cloud computing in developing countries' educational sectors, highlighting the complexity of integrating new technologies. Chen [12] explored the utilization of virtual crowd simulation and digital media to enhance the college physical education model, emphasizing the need for innovative approaches to engage students effectively.

Additionally, improving the effectiveness of physical education classroom teaching at the middle school level requires comprehensive strategies. Chen and Zhou [10] addressed this challenge by proposing methods to enhance teaching methodologies. Fei [40] introduced an LDA-based model for semantic annotation of Web English educational resources, demonstrating the potential for natural language processing techniques in educational data management.

Many studies have investigated various aspects of this domain, shedding light on the complexities involved. Huang et al. [151] investigated the effectiveness of ideological and political education reform in universities, showcasing the potential of data mining and AI technologies to enhance educational practices.

Juan [60] explored gesture recognition and information recommendation in distance education, highlighting the role of ML and Virtual Reality (VR) in improving remote learning experiences. Li [72] developed an AI agent technology-based web distance education system, demonstrating the integration of AI into educational platforms to facilitate distance learning.

Moreover, Liu and Wang [78] investigated ice and snow talent training using AI education informatization teaching models, illustrating innovative approaches to talent development. Lu [79] applied wireless networks and ML algorithms in entrepreneurship education, showcasing the potential of technology-enhanced learning environments in fostering entrepreneurial skills.

Additionally, Luo and Chen [80] proposed an adaptive entropy theory polymerization method for path optimization in edge computing educational systems, demonstrating advancements in optimizing educational processes through computational methods. Ma et al. [82] focused on improving traditional education effectiveness through computer AI and neural network systems, highlighting the transformative potential of technology in educational contexts.

Furthermore, Mao and Zhang [85] analyzed entrepreneurship education in colleges based on improved decision tree algorithms and fuzzy mathematics, providing insights into curriculum development strategies. Peng [106] conducted an effective analysis of the online education model based on AI, contributing to the understanding of AI-driven pedagogical approaches.

Sun and Gu [120] introduced a fuzzy knowledge graph system for AI-based smart education, illustrating the potential of fuzzy logic in educational data analytics. Wang and Li [56] proposed a systematic approach for English education models based on neural network algorithms, showcasing the integration of AI into language learning processes.

Furthermore, Li et al. [70] introduced a cross-dimension mining model of public opinion data in online education based on fuzzy association rules, demonstrating the applicability of fuzzy logic in analyzing educational data. Liu [76] designed and implemented an intelligent educational administration system using fuzzy clustering algorithms, showcasing the potential of fuzzy logic in administrative processes.

Lastly, Wang [143] explored music education's role in rescuing psychological stress in social crises based on fuzzy prediction algorithms, illustrating the potential of fuzzy logic in addressing socio-emotional needs through educational interventions.

These challenges underscore the need for robust data management strategies and advanced analytical tools to navigate the complexities of educational contexts effectively.

4.5.5 Big Data and Internet of Things in Educational Contexts Using Fuzzy Logic

In educational contexts, the integration of big data and Internet of Things (IoT) technologies presents both opportunities and challenges. Zhang et al. [160] explored big data analytics for intelligent online education, highlighting the potential for data-driven insights to enhance learning outcomes and educational experiences. Furthermore, the proliferation of educational big data necessitates intelligent platforms for real-time analytics, facilitating better insights into page view statistics and resource sharing [8].

Wang and Zhao [144] proposed an intelligent system for interactive online education based on cloud big data analytics, emphasizing the role of cloud computing in facilitating data analysis and personalized learning experiences. Fu et al. [41] investigated big data intelligence for smart educational management systems, showcasing the potential for data-driven decision-making in optimizing educational processes and resource allocation.

Moreover, Guo and Sun [58] focused on real-time monitoring of physical education classrooms in colleges and universities using open IoT and cloud computing technologies. Their study demonstrated the potential of IoT devices and cloud platforms in collecting real-time data on student activities and environmental conditions to enhance safety and efficiency in educational settings.

Furthermore, Xie and Yang [152] developed an IoT-based model for an intelligent innovation practice system in higher education institutions, illustrating the potential of IoT devices in fostering innovation and hands-on learning experiences among students.

Overall, while big data and IoT technologies offer promising opportunities to revolutionize educational practices, their implementation also poses challenges related to data privacy, security, interoperability, and ethical considerations. Addressing these challenges is important to fully integrate the potential of big data and IoT in educational contexts.

4.5.6 Integrating Artificial Intelligence into Educational Contexts Using Fuzzy Logic

Integrating AI and ML algorithms into educational systems poses its own set of challenges, requiring comprehensive analyses of critical factors and effective decision-making frameworks [146]. Moreover, the integration of AI into physical education pedagogy presents opportunities for enhancing teaching methods. Gaobin et al. [43] developed an AI fuzzy system to improve physical education teaching methods, emphasizing the role of technology in educational advancement. Guo and Li [48] discussed the application of AI in physical education, particularly in functional sports training, showcasing the potential for AI-driven approaches to optimize learning outcomes in sports education contexts. Furthermore, Li and Wang [73] assessed the effectiveness of physical education teaching in college through the lens of AI methods, indicating opportunities for AI-driven improvements in sports education. Lilingling [74] conducted empirical research on online mental health education teaching modes, emphasizing the role of AI in addressing mental health challenges among students.

Moreover, Saravanan [115] examined the impact of intelligence methodologies on education and training processes, emphasizing the transformative potential of innovative technologies in shaping learning experiences. Srivastava et al. [118] developed a smart learning assistance tool for inclusive education, highlighting the importance of leveraging technology to promote equitable access to education.

Additionally, Xu [153] developed an intelligent system for university legal education based on ML, highlighting the role of AI in legal pedagogy. Zhang [158] spearheaded reform and innovation of AI technology for information service in university physical education, contributing to advancements in sports education practices.

4.6 Fuzzy Reasoning in Educational Applications and Games

This section includes games and educational applications that utilize fuzzy reasoning mechanisms to offer a more personalized experience within the gaming or educational environment. For instance, Gokmen et al. employed fuzzy logic to evaluate students' performance in two specific exam tests [45]. The system utilizes suitable membership functions to describe students' final scores in each test using fuzzy sets such as "Very Low," "Low," "Average," "High," and "Very High." The assessment of the rules integrated into the fuzzy reasoning mechanism results in a fuzzy output that characterizes the final performance as "Very Unsuccessful," "Unsuccessful," "Average," "Successful," and "Very Successful" and subsequently computes the precise value of the performance. The authors noted that fuzzy logic offers more flexibility than conventional constant mathematical evaluation methods, as tutors can adjust membership functions and rules to provide the most accurate evaluation results.

Machado, Moreira, Gomes, Caldeira, and Santos developed an ITS to enhance the efficiency of the teaching process, catering to the specific needs of students, whether they are learning individually or in a group [83]. The methodology they devised effectively addresses linear programming problems and evaluations using variables, constraints, and objective functions. The system transforms the analyzed components (variables, constraints, and objective function) into fuzzy sets by employing triangular membership functions. Each variable represents the degree of success for the corresponding item and is characterized by fuzzy linguistic values such as "Very Low," "Low," "Medium," and "High." By applying predefined rules, the fuzzy inference engine generates a fuzzy output that reflects the problem's relevance to the student based on their level of knowledge. An assessment of the system demonstrated that the problems selected for each student using fuzzy logic techniques closely resembled those chosen by experts.

Chrysafiadi and Virvou designed and evaluated an educational module called Fuzzy Knowledge State Definer (FuzKSD) [24]. FuzKSD automatically identifies and adjusts the student's knowledge level for all domain knowledge concepts, employing fuzzy theory and fuzzy sets to map the student's knowledge level as a subset of the domain knowledge. This model also uses an inference engine to automatically update user stereotypes based on fuzzy sets. As a result, the system can more realistically replicate the student's knowledge level, offering better personalized adaptive guidance.

The Collaboration/Reflection Fuzzy Inference System (C/R-FIS) is an expert system based on fuzzy logic that automatically evaluates the cooperative activity of two users during their asynchronous, written, online collaboration [49]. This evaluation provides enhanced formative feedback for adaptive support during their asynchronous, written, and web-based collaboration, reducing the potential mismatch between individual cooperative skills by challenging self-adaptation processes. The system utilizes linguistic variables such as "initiative" and "creativity" and fuzzy values such as "satisfactory," "low," or "high." Consequently, the rule-based component of the system contains rules of the following format:

"IF the initiative is satisfactory AND (OR) the creativity is low, THEN collaboration is low."

The system's evaluation demonstrated its potential to enhance collaborative activity. Moreover, the integration of the system into a music education case study revealed that a fuzzy-logic-based system can be applied to other collaborative learning scenarios, including medicine, environmental engineering, law, and music education.

The Concept and Fuzzy Adaptive E-learning system (CaFAE) is an adaptive e-learning system that utilizes fuzzy logic techniques to assess the student's grasp of each domain concept and generate a prioritized list of concepts with learning resources to address any areas where the student's understanding may be deficient [3]. The system represents the student's level of knowledge using a weighted colored concept map. It consists of three main components: the FLS, the colored concept map, and the ranked concept list. The fuzzy logic component utilizes three primary linguistic variables: concept weight and concept error value as input variables and concept impact as an output variable. Concept weight determines how the student should study the related concepts within a topic. Meanwhile, concept error value evaluates the student's level of knowledge in each domain based on their responses to pre-tests or post-tests while engaging with the course materials. It employs four fuzzy values: "Learned," "Known," "Unsatisfied Known," and "Unknown," determined by appropriate membership functions.

The concept of impact value is based on the linguistic values "Small," "Medium," and "Large." A ranked list of concepts includes all impact values, starting with the highest ones. The system uses two methods to recommend suitable learning materials based on the student's performance. The first method prioritizes learning the concept with the highest impact value, followed by the highest concept weight value and the concept number. The second method prioritizes learning the concept with the highest impact value, followed by the highest concept weight value, the highest concept error value, and the concept number. The authors plan to conduct a pilot study in a classroom setting to test the system and highlight the benefits of integrating fuzzy logic theory into educational applications.

Karaci implemented fuzzy logic in an ITS designed to teach punctuation in Turkish [64]. The fuzzy logic module takes two input variables: the MYCIN CF (Certainty Factor) and the time taken for all attempts. MYCIN is an expert system developed in 1976 at Stanford University by a team of expert doctors to provide computer-based medical advice. It calculates a CF to reduce ambiguity and draw conclusions based on available information. The CF ranges between -1 and 1, with -1 indicating the student's unfamiliarity with the topic and 1 indicating familiarity. A CF closer to 0 indicates insufficient information to determine the student's knowledge. The MYCIN CF score related to a specific topic reflects the student's mastery and learning deficiency. Fuzzy sets related to the MYCIN CF include "DDNK" (definitely does not know), "MPDNK" (most probably does not know), "PDNK" (probably does not know), "MK" (might know), "PK" (probably knows), "MPK" (most probably knows), and "DK" (definitely knows). The fuzzy input variable for the number of attempts can take values such as "VL" (very low), "L" (low), "M" (moderate), "H" (high), and "VH" (very high). Through the fuzzy logic decision system, the ITS offers a dynamic learning environment that adapts to the student's needs, providing instant and intelligent feedback to facilitate learning from mistakes. Additionally, it caters to students who are struggling to learn, thereby supporting successful learning outcomes.

Proletool 3.0 is an ITS that teaches students predictive parsing techniques for compiler design [9]. The system focuses on the analytical aspect of the compilation process, with a specific emphasis on syntactic analysis to provide foundational knowledge about compiler theory. Additionally, it includes an exercise evaluation component that compares the student's solution with the correct answer, using fuzzy rules to assess the student's understanding of each intended learning outcome (ILO). These rules consider the student's mastery levels and the errors or correct answers from the student's last completed exercise related to the intended learning outcome. E.g., a fuzzy rule is the following:

IF recent_scores are Bad AND level_mastery is NOT Very Good, THEN achievement_level is not achieved.

The learning instructor component utilizes the fuzzy inference mechanism of the exercises corrector component to guide the student's progress toward reaching an ILO. A preliminary assessment of the system's efficacy and usefulness revealed that incorporating fuzzy logic produced an effective learning tool that enhances the student's learning process.

Megahed and Mohammed have pioneered a hybrid approach to modeling intelligent adaptive e-learning environments, focusing on students' emotional states [88]. This approach integrates convolutional neural networks (CNN) and fuzzy logic and consists of four key components: facial expression recognition, test and exam management, learning modeling, and a fuzzy inference system (FIS). The CNN identifies the student's emotional state by analyzing their facial expressions, while the FIS manages the ambiguity in learner responses during their interaction with the learning environment. Furthermore, the FIS utilizes the learning model's precise values of learning activity variables to generate recommendations for the next learning level. These learning activity variables include the current learning level, elapsed test time, answer validity ratio, and seven summarised facial expression classes representing aggregated emotional states. An example of a fuzzy rule in this approach is as follows:

If HappyExpr IS VeryHigh AND AngryExpr IS VeryLow, AND DisgustExpr IS VeryLow, AND FearExpr IS VeryLow, AND SadExpr IS VeryLow, AND SurpriseExpr IS VeryLow, AND NeutralExpr IS VeryLow, AND AnswerValidity IS DefinitelyValid, AND TestElapsedTime IS VeryLow, AND CurrentLevel IS Intermediate THEN NextLevel IS VeryAdvanced.

The authors showed that when an approach combines fuzzy logic with other methods, such as CNN, the final system offers the student a better learning experience and adaptation. It also considers their emotional states, whose identification is imprecise and challenging.

AI in gaming involves various techniques and methods to introduce intelligence into video games [108]. Game AI utilizes approaches from the broader field of AI, ranging from simple finite-state machines to evolutionary algorithms. Many video game AI developers opt for fuzzy logic to imbue games with realistic behaviors due to their simple formulation and descriptive nature. Numerous games integrate fuzzy logic techniques to enhance the realism of the game environment. For instance, the FPS game "Unreal" uses FFSMs to create more lifelike and natural enemy behaviors [59].

Similarly, the action game "SWAT 2" utilizes fuzzy logic to determine the tactical responses of enemy units based on the situation and the personality of their adversaries. There are even games like "Fuzzy Tactics," designed to teach players the concept of fuzzy logic by requiring them to command troops and destroy enemies using predefined linguistic fuzzy rules [109]. Furthermore, Renkas and Niewiadomski developed a hierarchical architecture of an FLS to control military vehicles in a 2D single-player game [113].

Chrysafiadi and Virvou have developed a web-based educational environment for teaching the programming language C, which uses fuzzy logic techniques to enhance the student model. Their system's student model rests on an overlay model representing the student's knowledge level. The fuzzy logic technique is incorporated into the student model to determine the student's knowledge level of each domain concept, update the student model, and make decisions about the instruction model that the system should follow for each student [19]. According to the evaluation results, the system's embedded fuzzy-based mechanism successfully adapts the sequence and content of lessons to meet the learners' requirements. As a result, students consider the system a practical, pleasant, and helpful tool for learning computer programming [27].

Chrysafiadi and Virvou have developed an innovative e-learning system to teach computer programming through fuzzy logic. This approach involves reasoning that is approximate rather than exact. The system employs fuzzy cognitive maps (FCMs) to accurately represent students' knowledge and the interdependencies between the various concepts in the learning material. This advanced system can deliver highly customized learning experiences based on the individual student's unique learning needs and pace. By adapting to the student's specific needs, the system can provide a more effective and engaging learning experience, improving the student's learning outcomes in computer programming [22].

Chrysafiadi et al. improved the ITS for teaching C programming. The system uses the distance-weighted k-nearest neighbor algorithm to assess the student's initial knowledge level and fuzzy logic to identify their current level of understanding and potential misconceptions in subsequent interactions [34]. Combining ML and fuzzy logic results in a more personalized learning experience that boosts student engagement and reduces dropout rates.

Fuzzy logic is not only used to assess the cognitive level of students but also to evaluate various education-related concepts. In recent times, fuzzy logic has been used to evaluate the effectiveness of e-learning courses during the COVID-19 pandemic. Researchers Chrysafiadi, Virvou, and Tsihrintzis applied fuzzy logic to analyze survey data to evaluate the acceptance and effectiveness of e-learning courses during the pandemic in Greece. This innovative technique helped produce more interpretable results using fuzzy rules to analyze survey data. The approach proved to be a valuable tool, providing new insights into the effectiveness of e-learning courses, which could be further used to improve online education [35].

The simplicity of development and realistic results produced by fuzzy logic make it an attractive option for many game developers and designers. Its capabilities are not limited to classical gaming; educational games benefit from fuzzy logic. "Select Box-Solve Question," a mobile educational game for learning computer programming, is an example of a game that uses fuzzy logic capabilities. It presents students with code snippets and challenges them to answer using multiple-choice questions [97, 98]. Additionally, the game employs a fuzzy logic approach to assess students' attitudes toward mobile learning. The FIS takes satisfaction, effect on learning, motivation, and usability as inputs of the attitude scale and produces corresponding output using the same fuzzy variables. This allows for a more nuanced assessment, where a high initial satisfaction may be offset by low values in other inputs, resulting in lower final satisfaction output. Through this approach, fuzzy logic effectively addresses uncertainty in the attitude determination process, outperforming traditional methods.

In their work, Ivan et al. explore the application of fuzzy logic in adaptive multiplayer math games to dynamically adjust the difficulty level of learning activities based on the student's performance and response times [57]. The student's performance is classified into "Low," "Medium," and "High" fuzzy sets, while response speed is categorized as "Slow" and "Fast." The system then makes inferences about the student's performance level and adjusts the difficulty of subsequent learning activities accordingly. The performance level is characterized as "Bad," "Standard," "Great," and "Perfect," while difficulty levels are designated as "Easy," "Normal," "Hard," and "Ultra-hard."

Following this, a rule-based system categorizes students into groups with similar difficulty levels to enhance competition and maintain motivation. Feedback from students indicates that the game effectively covers the school curriculum and facilitates understanding of complex Math and Science concepts, demonstrating the practical application of fuzzy logic in educational games.

Chrysafiadi, Papadimitriou, and Virvou described an educational adventure game that offers knowledge-based adaptivity in its educational material and scenario. The game dynamically customizes the difficulty level of the educational content and story to respond to each student's knowledge level and learning needs. This adaptation occurs through a fuzzy student model that detects the student's cognitive state and determines how the game story must evolve. The model determines whether the game scenario will expand to help students overcome weaknesses and improve their learning performance. The advantage of this approach is that by adapting the scenario, the game stimulates the student to participate actively in the learning process and provides educational content tailored to their knowledge level and weaknesses, leading to enhanced educational outcomes [31].

In their study, Papadimitriou, Chrysafiadi, and Virvou compared two versions of the same educational game [102]. The two versions took different approaches in deciding which quizzes to provide the students. The first version used the student's success grade on the previous quiz to determine the difficulty of the following quiz. The second version considered the student's overall progress using a fuzzy rule-based method to identify their knowledge level and learning needs when deciding the quiz's difficulty. The results of the evaluation indicated that using the fuzzy reasoning method had a positive impact on educational outcomes.

4.6.1 Fuzzy Logic in Player and Student Modeling of Educational Games

Fuzzy logic plays a significant role in student and player modeling within educational games, offering a flexible approach to handling uncertainty and imprecision. However, several challenges persist in employing fuzzy logic for effective dynamic difficulty adjustment, personalized reasoning, and adaptive scenarios in educational gaming environments.

One prominent challenge lies in dynamically adjusting the difficulty of educational games based on player performance. Chrysafiadi et al. [33] proposed a fuzzy-based dynamic difficulty adjustment system for an educational 3D game. However, implementing such systems requires precise modeling of player behavior and learning progress, which can be challenging due to human learning processes' inherent complexity and variability. Chrysafiadi, Kamitsios, and Virvou developed a 3D game that teaches HTML language using a fuzzy Dynamic Difficulty Adjustment (DDA) mechanism. The game adaptively adjusts the difficulty of battles and maze navigation according to individual player's skills. The game's uniqueness lies in using fuzzy DDA in the two dimensions of an educational game - educational content and play environment. It uses fuzzy logic to balance the game's difficulty with the user's playing skills. Fuzzy logic also permits the application of DDA without the need for extensive learning data, unlike other AI techniques for DDA, such as ML techniques.

Moreover, personalizing reasoning mechanisms in educational games presents another set of challenges. Papadimitriou et al. [104] discussed the transition from point-and-click to VR mobile educational games and emphasized the importance of personalized learning experiences. However, designing effective personalized reasoning systems using fuzzy logic requires robust student modeling techniques to accurately capture individual preferences, cognitive abilities, and learning styles [62].

In the context of student modeling, integrating fuzzy reasoning techniques presents opportunities and challenges. Papadimitriou [99] explored adaptive scenarios and tutoring using fuzzy reasoning in an intelligent educational adventure game, highlighting the potential for personalized learning experiences. However, developing cognitive-based adaptive scenarios using fuzzy reasoning requires extensive domain knowledge and careful consideration of various factors influencing student learning [31]. Papadimitriou, Chrysafiadi, and Virvou have recognized the effectiveness of adaptive educational games in catering to individual learning needs and increasing students' interest in learning. To this end, they have developed an adaptive educational adventure game for teaching HTML. The game utilizes fuzzy logic to assess the student's knowledge level and generate corresponding fuzzy states. The fuzzy states where a student enters during gameplay determine the progression of the game's storyline and provide the player with additional motivation to continue playing and ultimately acquire the required knowledge [30]. Papadimitriou, Chrysafiadi, Kamitsios, and Virvou applied web services technology to make a fuzzy mechanism accessible to any external educational application or game. The fuzzy mechanism exploits fuzzy logic to identify the student's knowledge level and create their model. Two games (a 3d exploration game and a 2d adventure game) and a web application also used the mechanism, all three accessing the same student model. By maintaining a profile for each student and sharing it across multiple applications, the system offers a unique advantage that leads to better learning outcomes. With the ability to choose their preferred application, students can personalize their learning experience and achieve better educational outcomes [100].

Furthermore, incorporating gameplay data into learner modeling introduces additional complexities. Hamdaoui et al. [50] proposed a learner modeling approach in educational games based on fuzzy logic and gameplay data, aiming to enhance the adaptivity and effectiveness of educational experiences. Nonetheless, extracting meaningful insights from gameplay data and integrating them into fuzzy logic-based models remains challenging due to the inherent ambiguity and noise in real-world data [67].

In conclusion, while fuzzy logic offers a flexible framework for student and player modeling in educational games, addressing the challenges associated with dynamic difficulty adjustment, personalized reasoning, adaptive scenarios, and integration of gameplay data is essential for realizing its full potential. Collaborative efforts from researchers, educators, and game developers are crucial in overcoming these challenges and creating engaging, effective, and personalized educational gaming experiences.

4.6.2 *Fuzzy Logic in Tutoring Strategies of Educational Games*

In educational games, designing effective tutoring strategies using fuzzy logic presents several challenges. Fuzzy logic offers a flexible framework for adapting tutoring strategies based on learners' progress and needs, but its implementation requires addressing various complexities.

One challenge is dynamically adapting the plot of an educational game in response to the learner's progress. Chrysafiadi, Papadimitriou, and Virvou [30] proposed using fuzzy states for dynamic plot adaptation, allowing the game to adjust its storyline based on the learner's performance. However, implementing such a system requires precise modeling of learner progress and behavior to ensure the adaptation aligns with educational objectives.

Moreover, incorporating fuzzy logic for adaptive scenarios in educational adventure games presents its own set of challenges. Papadimitriou, Chrysafiadi, and Virvou [103] introduced FuzzEG, an adaptive educational game utilizing fuzzy logic for adaptive scenarios. While this approach offers personalized learning experiences, designing effective fuzzy logic rules and algorithms to adjust game scenarios dynamically requires careful consideration of various factors influencing learner engagement and comprehension.

Additionally, integrating fuzzy cognitive maps into game-based learning models poses challenges. Ying and Yang [155] proposed a game-based learning system using the ARCS model and fuzzy logic to enhance learner motivation and engagement. However, effectively incorporating fuzzy cognitive maps into educational games requires robust mapping of knowledge domains and learner interactions, as well as fine-tuning the fuzzy logic parameters to ensure optimal learning outcomes.

Furthermore, modeling collaborative learning scenarios, such as the Beer Game, using fuzzy constraint-directed agent negotiation presents challenges. Tseng, Lan, and Lai [123] explored modeling the Beer Game as role-play collaborative learning via fuzzy constraint-directed agent negotiation. Implementing such models necessitates sophisticated algorithms for agent negotiation and decision-making and clear guidelines for incorporating fuzzy constraints to enhance collaborative learning experiences effectively.

In conclusion, while fuzzy logic offers promising opportunities for designing adaptive tutoring strategies in educational games, addressing challenges such as dynamic plot adaptation, adaptive scenario design, integration of fuzzy cognitive maps, and modeling collaborative learning scenarios is essential. Collaborative efforts from researchers, educators, and game developers are crucial in overcoming these challenges and creating engaging, effective, and personalized educational gaming experiences.

4.6.3 Fuzzy Logic in Affective Reasoning in Educational Games

Affective computing, the study and development of systems that can recognize, interpret, and simulate human emotions, holds significant promise for enhancing educational games. However, it also presents several challenges that must be addressed for effective implementation.

One primary challenge lies in the induction of emotional states within educational video games. While affective computing aims to imbue games with emotional elements to enhance engagement and learning, achieving this in a precise and controlled manner is complex. Researchers like Lara-Alvarez et al. [69] have explored methods such as fuzzy control systems to induce emotional states in educational video games. However, implementing these systems effectively requires a deep understanding of both the educational content and the nuances of human emotions.

Another challenge is accurately recognizing facial expressions, a crucial aspect of affective computing in educational games. Gaman et al. [42] proposed an adaptive learning method utilizing fuzzy-ART for facial expression recognition. Despite technological advancements, recognizing facial expressions accurately in real-time within a dynamic educational game environment remains a significant hurdle.

Moreover, integrating affective user modeling into educational games poses its own set of challenges. Katsionis and Virvou [65] proposed a cognitive theory for affective user modeling in VR reality educational games, highlighting the importance of personalization. However, implementing personalized e-learning experiences through educational games using web services introduces complexities related to data privacy, security, and interoperability [66].

Furthermore, enhancing visual-facial emotion recognition systems with complementary keyboard-stroke pattern information is an area of ongoing research. Tsihrintzis et al. [125] and Stathopoulou et al. [119] explored methods to improve emotion recognition by incorporating keyboard-stroke patterns. While such approaches show promise, they also raise concerns about user privacy and the need for seamless integration within the educational game environment.

In conclusion, while affective computing holds great potential for enhancing educational games, addressing the challenges surrounding the induction of emotional states, accurate facial expression recognition, personalized user modeling, and privacy considerations is essential for realizing its full benefits. Collaborative efforts from researchers, educators, and game developers are crucial in overcoming these challenges and creating immersive, emotionally engaging educational experiences.

4.6.4 Evaluation of Fuzzy Logic in Educational Games

Evaluating the effectiveness of fuzzy logic in educational games poses several challenges, including methodological considerations and the comparison of fuzzy logic-based approaches with other instructional methods.

One challenge lies in designing appropriate evaluation methodologies to assess the impact of fuzzy logic on learning outcomes within educational games. Papadimitriou, Chrysafiadi, and Virvou [102] addressed this challenge by evaluating the use of fuzzy logic in an educational game for adaptation. However, determining suitable metrics and criteria for evaluating the efficacy of fuzzy logic-based adaptations requires careful consideration of factors such as learner engagement, knowledge retention, and transferability of skills.

Furthermore, comparing the effectiveness of fuzzy logic-based educational games with other instructional methods presents challenges. Chrysafiadi, Papadimitriou, and Virvou [29] conducted a study comparing web-based educational applications with educational games. However, designing rigorous experimental designs and controlling for confounding variables while ensuring ecological validity is essential for drawing meaningful conclusions about the relative effectiveness of different instructional approaches.

Moreover, assessing the generalizability and scalability of fuzzy logic-based adaptations across diverse learner populations and educational contexts is challenging. Educational games incorporating fuzzy logic may yield varying results based on factors such as learner characteristics, subject matter, and instructional design. Therefore, conducting multi-site studies and replicating findings across different settings is crucial for establishing the robustness of fuzzy logic-based approaches in educational games.

Additionally, addressing the limitations and biases inherent in evaluation methods used in educational game research is essential. Common challenges include reliance on self-report measures, limited sample sizes, and lack of long-term follow-up assessments. Overcoming these challenges requires employing mixed-methods approaches, triangulating data from multiple sources, and adopting longitudinal study designs to provide a comprehensive understanding of the impact of fuzzy logic in educational games over time.

In conclusion, evaluating the effectiveness of fuzzy logic in educational games involves addressing methodological challenges, comparing fuzzy logic-based approaches with other instructional methods, assessing generalizability and scalability, and overcoming limitations inherent in evaluation methods. Collaborative efforts from researchers, educators, and game developers are essential in overcoming these challenges and advancing the field of educational game evaluation.

References

1. Adem, A., Çakıt, E., Dağdeviren, M.: Selection of suitable distance education platforms based on human-computer interaction criteria under fuzzy environment. Neural Comput. Appl. **34**(10), 7919–7931 (2022)
2. Agrawal, A., Agarwal, A., Bansal, P.: Integration of fuzzy logic with metaheuristics for education center site selection. Educ. Inf. Technol. **26**, 103–124 (2021)
3. Al Duhayyim, M., Newbury, P.: Concept-based and fuzzy adaptive e-learning. In: Proceedings of the 3rd International Conference on Information and Education Innovations, pp. 49–56 (2018)
4. Alepis, E., Virvou, M.: Object-Oriented User Interfaces for Personalized Mobile Learning. Springer (2014)
5. Alonistioti, N., Tsichrintzi, E.A., Chrysafiadi, K., Alepis, E.: Requirements for fuzzy logic in personalisation of fire emergency alerts. In: 2023 14th International Conference on Information, Intelligence, Systems & Applications (IISA), pp. 1–8. IEEE (2023)
6. Anand, A.S., et al.: Applications of fuzzy logic in renewable energy systems-a review. Renew. Sustain, Energy Rev (2015)
7. Ayouni, S., Menzli, L.J., Hajjej, F., Maddeh, M., Al-Otaibi, S.: Fuzzy vikor application for learning management systems evaluation in higher education. Int. J. Inf. Commun. Technol. Educ. (IJICTE) **17**(2), 17–35 (2021)
8. Bai, X., Li, J.: Intelligent platform for real-time page view statistics using educational big data digital resource sharing. J. Intell. Fuzzy Syst. **40**(2), 2851–2860 (2021). https://doi.org/10.3233/jifs-189325
9. Castro-Schez, J.J., Glez-Morcillo, C., Albusac, J., Vallejo, D.: An intelligent tutoring system for supporting active learning: a case study on predictive parsing learning. Inf. Sci. **544**, 446–468 (2021)
10. Chen, C., Zhou, Q.: Improving the effectiveness of middle school physical education classroom teaching. J. Intell. Fuzzy Syst. **40**(2), 3433–3444 (2021). https://doi.org/10.3233/jifs-189381
11. Chen, R., Wang, T., Kim, S.: Optimizing teaching management in college physical education: a fuzzy neural network approach. Soft Comput. **27**(24), 19299–19315 (2023). https://doi.org/10.1007/s00500-023-09318-y
12. Chen, Y.: Research on college physical education model based on virtual crowd simulation and digital media. J. Intell. Fuzzy Syst. **40**(4), 7195–7207 (2021). https://doi.org/10.3233/jifs-189547
13. Chrysafiadi, K.: Improving task scheduling by using a fuzzy reasoner. Intell. Decis. Technol. **14**(2), 165–170 (2020)
14. Chrysafiadi, K.: Special collection of papers on 'current fuzzy logic-based software applications and systems'. Intell. Decis. Technol. **14**(2), 161–163 (2020)
15. Chrysafiadi, K.: Fuzzy logic-based software systems. In: Fuzzy Logic-Based Software Systems, pp. 31–129. Springer (2023)
16. Chrysafiadi, K.: The role of fuzzy logic in artificial intelligence and smart applications. In: Fuzzy Logic-Based Software Systems, pp. 25–29. Springer (2023)
17. Chrysafiadi, K., Tsichrintzi, E.A.: An intelligent fuzzy-based emergency alert generation for persons with episodic memory decline problems. Intell. Decis. Technol. **15**(4), 681–690 (2021)
18. Chrysafiadi, K., Tsichrintzi, E.A.: Smart home assistance for humans with episodic memory decline problems using a fuzzy rule-based mechanism: the case of stove usage. In: 2021 12th International Conference on Information, Intelligence, Systems & Applications (IISA), pp. 1–5. IEEE (2021)
19. Chrysafiadi, K., Virvou, M.: Evaluating the integration of fuzzy logic into the student model of a web-based learning environment. Expert Syst. Appl. **39**(18), 13127–13134 (2012)
20. Chrysafiadi, K., Virvou, M.: Using fuzzy cognitive maps for the domain knowledge representation of an adaptive e-learning system. In: JCKBSE, pp. 257–265 (2012)

21. Chrysafiadi, K., Virvou, M.: Fuzzy logic for dynamic adaptation: Represent and manage changeable user states. In: IISA 2013, pp. 1–6. IEEE (2013)
22. Chrysafiadi, K., Virvou, M.: A knowledge representation approach using fuzzy cognitive maps for better navigation support in an adaptive learning system. SpringerPlus **2**(1), 1–13 (2013)
23. Chrysafiadi, K., Virvou, M.: Student modeling approaches: a literature review for the last decade. Expert Syst. Appl. **40**(11), 4715–4729 (2013)
24. Chrysafiadi, K., Virvou, M.: Fuzzy logic for adaptive instruction in an e-learning environment for computer programming. IEEE Trans. Fuzzy Syst. **23**(1), 164–177 (2014)
25. Chrysafiadi, K., Virvou, M.: Create dynamically adaptive test on the fly using fuzzy logic. In: 2018 9th International Conference on Information, Intelligence, Systems and Applications (IISA), pp. 1–8. IEEE (2018)
26. Chrysafiadi, K., Virvou, M.: Enhancing automatic reasoning of human errors in an operating system using fuzzy logic. In: 2019 10th International Conference on Information, Intelligence, Systems and Applications (IISA), pp. 1–4. IEEE (2019)
27. Chrysafiadi, K., Virvou, M.: Evaluating the user experience of a fuzzy-based intelligent tutoring system. In: 2021 12th International Conference On Information, Intelligence, Systems & Applications (IISA), pp. 1–7. IEEE (2021)
28. Chrysafiadi, K., Virvou, M., et al.: Advances in Personalized Web-Based Education. Springer (2015)
29. Chrysafiadi, K., Papadimitriou, S., Virvou, M.: Which is better for learning: a web-based educational application or an educational game? In: 2019 International Symposium on Performance Evaluation of Computer and Telecommunication Systems (SPECTS), pp. 1–6. IEEE (2019)
30. Chrysafiadi, K., Papadimitriou, S., Virvou, M.: Fuzzy states for dynamic adaptation of the plot of an educational game in relation to the learner's progress. In: 2020 11th International Conference on Information, Intelligence, Systems and Applications (IISA), pp. 1–7. IEEE (2020)
31. Chrysafiadi, K., Papadimitriou, S., Virvou, M.: Cognitive-based adaptive scenarios in educational games using fuzzy reasoning. Knowl. Based Syst. **250**, 109111 (2022)
32. Chrysafiadi, K., Virvou, M., Tsihrintzis, G.A.: A fuzzy-based mechanism for automatic personalized assessment in an e-learning system for computer programming. Intell. Decis. Technol. (Preprint), 1–16 (2022)
33. Chrysafiadi, K., Kamitsios, M., Virvou, M.: Fuzzy-based dynamic difficulty adjustment of an educational 3d-game. Multimed. Tools Appl. 1–25 (2023)
34. Chrysafiadi, K., Virvou, M., Tsihrintzis, G., Hatzilygeroudis, I.: An adaptive learning environment for programming based on fuzzy logic and machine learning. Int. J. Artif. Intell, Tools (2023)
35. Chrysafiadi, K., Virvou, M., Tsihrintzis, G.A.: A fuzzy-based evaluation of e-learning acceptance and effectiveness by computer science students in greece in the period of covid-19. Electronics **12**(2), 428 (2023)
36. Chrysafiadi, K., Virvou, M., Tsihrintzis, G.A., Hatzilygeroudis, I.: Evaluating the user's experience, adaptivity and learning outcomes of a fuzzy-based intelligent tutoring system for computer programming for academic students in greece. Educ. Inf. Technol. **28**(6), 6453–6483 (2023)
37. Chrysafiadi, K.C.: Modeling the learner's cognitive state transitions using fuzzy logic techniques for adaptive learning. Ph.D. thesis, University of Piraeus (Greece) (2014)
38. Çiftçi, S., Pınarbaşı, M., Yazıcı, E., Alakaş, H.M.: Evaluation of the companies for applied education under fuzzy environment: a case in an industrial university. Neural Comput. Appl. **35**(29), 21675–21695 (2023)
39. Dutt, S., Ahuja, N.J., Kumar, M.: An intelligent tutoring system architecture based on fuzzy neural network (fnn) for special education of learning disabled learners. Educ. Inf. Technol. **27**(2), 2613–2633 (2022)

40. Fei, X.: An lda based model for semantic annotation of web english educational resources. J. Intell. Fuzzy Syst. **40**(2), 3445–3454 (2021). https://doi.org/10.3233/jifs-189382
41. Fu, C., Jiang, H., Chen, X.: Big data intelligence for smart educational management systems. J. Intell. Fuzzy Syst. **40**(2), 2881–2890 (2021)
42. Gaman, Y., Ishii, M., Kageyama, Y., Takahashi, T., Nishida, M., Shirasu, M.: Adaptive learning method in facial expression recognition model using fuzzy-art. In: 2019 IEEE 1st Global Conference on Life Sciences and Technologies (LifeTech), pp. 85–86. IEEE (2019)
43. Gaobin, Huan Nan, C., Zhen Zhong, L.: An artificial intelligence fuzzy system for improvement of physical education teaching method. J. Intell. Fuzzy Syst. **40**(2), 3595–3604 (2021). https://doi.org/10.3233/jifs-189395
44. Geng, X., Geng, X.: Combining grey relation analysis with mcgdm to evaluate the higher education management quality with interval-valued neutrosophic sets. J. Intell. Fuzzy Syst. **45**(3), 3791–3805 (2023)
45. Gokmen, G., Akinci, T.Ç., Tektaş, M., Onat, N., Kocyigit, G., Tektaş, N.: Evaluation of student performance in laboratory applications using fuzzy logic. Proc. Soc. Behav. Sci. **2**(2), 902–909 (2010)
46. Guan, Y.: An algorithm for data management of higher education based on fuzzy set theory-association rule mining algorithm. Informatica **47**(9) (2023). https://doi.org/10.31449/inf.v47i9.5222
47. Gulzar, K., Tariq, O., Mustafa, S., Mohsin, S.M., Kazmi, S.N., Akber, S.M.A., Abazeed, M., Ali, M.: A fuzzy analytic hierarchy process for usability requirements of online education systems. IEEE Access (2023)
48. Guo, Q., Li, B.: Role of ai physical education based on application of functional sports training. J. Intell. Fuzzy Syst. **40**(2), 3337–3345 (2021). https://doi.org/10.3233/jifs-189373
49. Hadjileontiadou, S.J., Nikolaidou, G.N., Hadjileontiadis, L.J., Balafoutas, G.N.: On enhancing on-line collaboration using fuzzy logic modeling. J. Educ. Technol. Soc. **7**(2), 68–81 (2004)
50. Hamdaoui, N., Idrissi, M.K., Bennani, S.: Learner modeling in educational games based on fuzzy logic and gameplay data. Int. J. Game Based Learn. (IJGBL) **11**(2), 38–60 (2021)
51. Han, Y., Yan, R., Gou, C.: An integrated multiple attribute decision making methodology for quality evaluation of innovation and entrepreneurship education with interval-valued intuitionistic fuzzy information. J. Intell. Fuzzy Syst. (Preprint), 1–19 (2023)
52. Hardy, Y., Steeb, W.H., Hardy, Y., Steeb, W.H.: Finite state machines. Classical and Quantum Computing: With C++ and Java Simulations, pp. 229–250 (2001)
53. Hatzilygeroudis, I.K., Tsihrintzis, G.A., Jain, L.C.: Fusion of Machine Learning Paradigms: Theory and Applications, vol. 236. Springer Nature (2023)
54. Hu, W., Shao, Y., Liu, Y.: A novel madm-based efficient methodology with 2-tuple linguistic neutrosophic numbers and applications to physical education teaching quality evaluation. J. Intell. Fuzzy Syst. (Preprint), 1–15 (2023)
55. Huang, J., Zhang, C.: An magdm approach with 2-tuple linguistic neutrosophic number for mental health education evaluation of college students. J. Intell. Fuzzy Syst. (Preprint), 1–12 (2023)
56. Hui, W., Aiyuan, L.: A systematic approach for english education model based on the neural network algorithm. J. Intell. Fuzzy Syst. **40**(2), 3455–3466 (2021). https://doi.org/10.3233/jifs-189383
57. Ivan, C., Rumondor, P.C., Ricky, M.Y., Yossy, E.H., Budiharto, W.: Help the math town: adaptive multiplayer math-science games using fuzzy logic. Proc. Comput. Sci. **116**, 309–317 (2017)
58. Jianbang, G., Changxin, S.: Real-time monitoring of physical education classroom in colleges and universities based on open iot and cloud computing. J. Intell. Fuzzy Syst. **40**(4), 7397–7409 (2021). https://doi.org/10.3233/jifs-189563
59. Johnson, D, Wiles, J.: Computer games with intelligence. In: 10th IEEE International Conference on Fuzzy Systems. (Cat. No. 01CH37297), vol. 3, pp. 1355–1358. IEEE (2001)
60. Juan, W.: Gesture recognition and information recommendation based on machine learning and virtual reality in distance education. J. Intell. Fuzzy Syst. **40**(4), 7509–7519 (2021). https://doi.org/10.3233/jifs-189572

61. Kahraman, C., Öztayşi, B., Sarı, I.U., Turanoğlu, E.: Fuzzy analytic hierarchy process with interval type-2 fuzzy sets. Knowl. Based Syst. **59**, 48–57 (2014)
62. Kamitsios, M., Papadimitriou, S., Chrysafiadi, K., Virvou, M.: Incorporating quizzes into a 3d educational game calling web services. In: EdMedia+ Innovate Learning, Association for the Advancement of Computing in Education (AACE), pp. 1170–1179 (2018)
63. Kandemir, I., Cicek, K.: Development an instructional design model selection approach for maritime education and training using fuzzy axiomatic design. Educ. Inf. Technol. 1–22 (2023)
64. Karaci, A.: Intelligent tutoring system model based on fuzzy logic and constraint-based student model. Neural Comput. Appl. **31**(8), 3619–3628 (2019)
65. Katsionis, G., Virvou, M.: A cognitive theory for affective user modelling in a virtual reality educational game. In: 2004 IEEE International Conference on Systems, Man and Cybernetics (IEEE Cat. No. 04CH37583), vol. 2, pp. 1209–1213. IEEE (2004)
66. Katsionis, G., Virvou, M.: Personalised e-learning through an educational virtual reality game using web services. Multimed. Tools Appl. **39**, 47–71 (2008)
67. Khenissi, M.A., Essalmi, F.: Automatic generation of fuzzy logic components for enhancing the mechanism of learner's modeling while using educational games. In: 2015 5th International Conference on Information & Communication Technology and Accessibility (ICTA), pp. 1–6. IEEE (2015)
68. Kukreja, V., Jain, A.K., Singh, A., Kaushal, R.K., Aggarwal, A.: Analysing moderators and critical factors that affect early childhood education with the usage of touchscreen contrivances: a hybrid fuzzy ahp-fuzzy topsis approach. Educ. Inf. Technol. **28**(5), 5621–5650 (2023)
69. Lara-Alvarez, C., Mitre-Hernandez, H., Flores, J.J., Pérez-Espinosa, H.: Induction of emotional states in educational video games through a fuzzy control system. IEEE Trans. Affect. Comput. **12**(1), 66–77 (2018)
70. Lx, Li., Huo, Y., Lin, J.C.W.: Cross-dimension mining model of public opinion data in online education based on fuzzy association rules. Mobile Netw. Appl. **26**(5), 2127–2140 (2021). https://doi.org/10.1007/s11036-021-01769-7
71. Li, N., Chen, X., Subramani, S., Kadry, S.N.: Improved fuzzy-assisted multimedia-assistive technology for engineering education. Comput. Appl. Eng. Educ. **29**(2), 453–464 (2021)
72. Li, R.: An artificial intelligence agent technology based web distance education system. J. Intell. Fuzzy Syst. **40**(2), 3289–3299 (2021). https://doi.org/10.3233/jifs-189369
73. Li, Z., Wang, H.: The effectiveness of physical education teaching in college based on artificial intelligence methods. J. Intell. Fuzzy Syst. **40**(2), 3301–3311 (2021). https://doi.org/10.3233/jifs-189370
74. Lilingling.: Online mental health education teaching mode and empirical research based on artificial intelligence. J. Intell. Fuzzy Syst. **40**(2), 3467–3476 (2021). https://doi.org/10.3233/jifs-189384
75. Liu, C.: (2023) Gra method for probabilistic simplified neutrosophic madm and application to talent training quality evaluation of segmented education. J. Intell. Fuzzy Syst. (Preprint), 1–11
76. Liu, F.: Design and implementation of intelligent educational administration system using fuzzy clustering algorithm. Sci. Program. **2021**, 1–14 (2021). https://doi.org/10.1155/2021/9485654
77. Liu, H., Liu, Y., Xu, L., Abdullah, S.: Multi-attribute group decision-making for online education live platform selection based on linguistic intuitionistic cubic fuzzy aggregation operators. Comput. Appl. Math. **40**, 1–34 (2021)
78. Liu, S., Wang, J.: Ice and snow talent training based on construction and analysis of artificial intelligence education informatization teaching model. J. Intell. Fuzzy Syst. **40**(2), 3421–3431 (2021). https://doi.org/10.3233/jifs-189380
79. Lu, H.: Application of wireless network and machine learning algorithm in entrepreneurship education of remote intelligent classroom. J. Intell. Fuzzy Syst. **40**(2), 2133–2144 (2021). https://doi.org/10.3233/jifs-189213

80. Luo, G., Chen, Z.: Adaptive entropy theory polymerization method for path optimization in edge computing educational systems. J. Intell. Fuzzy Syst. **40**(2), 2941–2951 (2021). https://doi.org/10.3233/jifs-189333
81. Lv, Z., Shen, H.: Artificial intelligence with fuzzy logic system for learning management evaluation in higher educational systems. J. Intell. Fuzzy Syst. **40**(2), 3501–3511 (2021)
82. Ma, W., Zhao, X., Guo, Y.: Improving the effectiveness of traditional education based on computer artificial intelligence and neural network system. J. Intell. Fuzzy Syst. **40**(2), 2565–2575 (2021)
83. Machado, M.A.S., Moreira, T.D.R.G., Gomes, L.F.A.M., Caldeira, A.M., Santos, D.J.: A fuzzy logic application in virtual education. Proc. Comput. Sci. **91**, 19–26 (2016)
84. Madasu, V.K., Hanmandlu, M., Vasikarla, S.: A novel approach for fuzzy edge detection using type ii fuzzy sets. Mathematics of Data/Image Pattern Recognition, Compression, and Encryption with Applications XI, SPIE **7075**, 157–164 (2008)
85. Mao, L., Zhang, W.: Analysis of entrepreneurship education in colleges and based on improved decision tree algorithm and fuzzy mathematics. J. Intell. Fuzzy Syst. **40**(2), 2095–2107 (2021). https://doi.org/10.3233/jifs-189210
86. Marinescu, R., Seceleanu, C., Le Guen, H., Pettersson, P.: A research overview of tool-supported model-based testing of requirements-based designs. Adv. Comput. **98**, 89–140 (2015)
87. Matzavela V, Chrysafiadi, K., Alepis, E.: Questionnaires and artificial neural networks: a literature review on modern techniques in education. In: 2017 IEEE Global Engineering Education Conference (EDUCON), pp. 1700–1704. IEEE (2017)
88. Megahed, M., Mohammed, A.: Modeling adaptive e-learning environment using facial expressions and fuzzy logic. Expert Syst. Appl. **157**, 113460 (2020)
89. Mendel, J.M.: Fuzzy sets for words: a new beginning. In: The 12th IEEE International Conference on Fuzzy Systems, 2003. FUZZ'03, vol. 1, pp. 37–42. IEEE (2003)
90. Mendel, J.M.: Computing with words: Zadeh, turing, popper and occam. IEEE Comput. Intell. Mag. **2**(4), 10–17 (2007)
91. Menekşe, A., Camgöz Akdağ, H.: Distance education tool selection using novel spherical fuzzy ahp edas. Soft Comput. **26**(4), 1617–1635 (2022)
92. Mo, H., Wang, F.Y., Zhou, M., Li, R., Xiao, Z.: Footprint of uncertainty for type-2 fuzzy sets. Inf. Sci. **272**, 96–110 (2014)
93. Mohmed, G., Lotfi, A., Pourabdollah, A.: Enhanced fuzzy finite state machine for human activity modelling and recognition. J. Ambient Intell. Hum. Comput. **11**(12), 6077–6091 (2020)
94. Montes, R., Zuheros, C., Morales, J., Zermeño, N., Duran, J., Herrera, F.: Design and consensus content validity of the questionnaire for b-learning education: a 2-tuple fuzzy linguistic delphi based decision support tool. Appl. Soft Comput. **147**, 110755 (2023)
95. Mougiakou, E., Virvou, M.: Based on gdpr privacy in uml: case of e-learning program. In: 2017 8th International Conference on Information, Intelligence, Systems & Applications (IISA), pp. 1–8. IEEE (2017)
96. Moundridou, M., Virvou, M.: Authoring and delivering adaptive web-based textbooks using wear. In: Proceedings IEEE International Conference on Advanced Learning Technologies, pp. 185–188. IEEE (2001)
97. Ozdemir, A., Balbal, K.F.: Fuzzy logic based performance analysis of educational mobile game for engineering students. Comput. Appl. Eng. Educ. **28**(6), 1536–1548 (2020)
98. Ozdemir, A., Balbal, K.F., Senel, B.C.: Mobile devices use in analyzing the engineering students attitude towards programming by using a fuzzy logic technique. Comput. Appl. Eng. Educ. **29**(3), 538–549 (2021)
99. Papadimitriou, S.: Adaptive scenarios and tutoring using fuzzy reasoning in an intelligent educational adventure game (2022)
100. Papadimitriou, S., Chrysafiadi, K., Kamitsios, M., Virvou, M.: Fsp creator: a novel web service api creator of fuzzy students progress profile. In: 2018 9th International Conference on Information, Intelligence, Systems and Applications (IISA), pp. 1–7. IEEE (2018)

101. Papadimitriou, S., Chrysafiadi, K., Virvou, M.: Authoring technological and platform independent learning material and student's progress profile using web services. In: Knowledge-Based Software Engineering: 2018: Proceedings of the 12th Joint Conference on Knowledge-Based Software Engineering (JCKBSE 2018) Corfu, Greece, vol. 12, pp. 242–251. Springer (2019)
102. Papadimitriou, S., Chrysafiadi, K., Virvou, M.: Evaluating the use of fuzzy logic in an educational game for offering adaptation. In: 2019 International Conference on Computer, Information and Telecommunication Systems (CITS), pp. 1–5. IEEE (2019)
103. Papadimitriou, S., Chrysafiadi, K., Virvou, M.: Fuzzeg: fuzzy logic for adaptive scenarios in an educational adventure game. Multimed. Tools Appl. **78**(22), 32023–32053 (2019). https://doi.org/10.1007/s11042-019-07955-w
104. Papadimitriou, S., Kamitsios, M., Chrysafiadi, K., Virvou, M.: Learn-and-play personalised reasoning from point-and-click to virtual reality mobile educational games. Intell. Decis. Technol. **15**(2), 321–332 (2021)
105. Papadimitriou, S., Chrysafiadi, K., Virvou, M.: Adaptive quizzes using fuzzy genetic algorithm. In: 2023 14th International Conference on Information, Intelligence, Systems & Applications (IISA), pp. 1–8. IEEE (2023)
106. Peng, L.: An effective analysis of online education model based on artificial intelligence. J. Intell. Fuzzy Syst. **40**(2), 3535–3546 (2021). https://doi.org/10.3233/jifs-189390
107. Phuong, N.H., Kreinovich, V.: Fuzzy logic and its applications in medicine. Int. J. Med. Inf. **62**(2–3), 165–173 (2001)
108. Pirovano, M.: The Use of Fuzzy Logic for Artificial Intelligence in Games. University of Milano, Milano (2012)
109. Pirovano, M., Lanzi, P.L.: Fuzzy tactics: a scripting game that leverages fuzzy logic as an engaging game mechanic. Expert Syst. Appl. **41**(13), 6029–6038 (2014)
110. Psarologou, A., Bourbakis, N., Virvou, M.: A mapping mechanism of nl sentences onto an spn state machine for understanding purposes. In: IISA 2014, The 5th International Conference on Information, Intelligence, Systems and Applications, pp. 321–324. IEEE (2014)
111. Rao, J., Peng, L., Rao, J., Cao, X.: Modified taxonomy method for double-valued neutrosophic number madm and applications to physical education teaching quality evaluation in colleges and universities. J. Intell. Fuzzy Syst. (Preprint), 1–10 (2023)
112. Razzaq, A., Riaz, M.: Some modified picture fuzzy average aggregation operators with priority roles of stakeholders in implementation of education 4.0. J. Intell. Fuzzy Syst. **44**(6), 10159–10181 (2023). https://doi.org/10.3233/jifs-224600
113. Renkas, K., Niewiadomski, A.: Hierarchical fuzzy logic systems and controlling vehicles in computer games. J. Appl. Comput. Sci. **22**(1), 201–212 (2014)
114. Saraji, M.K., Mardani, A., Köppen, M., Mishra, A.R., Rani, P.: An extended hesitant fuzzy set using swara-multimoora approach to adapt online education for the control of the pandemic spread of covid-19 in higher education institutions. Artif. Intell. Rev. **55**(1), 181–206 (2022)
115. Saravanan, V.: Impact of intelligence methodologies on education and training process. J. Intell. Fuzzy Syst. **40**(2), 3237–3238 (2021)
116. Singh, B., Mishra, A.K.: Fuzzy logic control system and its applications. Int. Res. J. Eng. Technol. **2**(8), 742–746 (2015)
117. Sirbiladze, G., Kacprzyk, J., Midodashvili, B., Khachidze, M., Midodashvili, L., Parshutkin, I.: Possibilistic simulation based interactive fuzzy magdm under discrimination q-rung picture linguistic information. Application in educational programs efficiency evaluation. Eng. Appl. Artif. Intell. **123**, 106278 (2023)
118. Srivastava, S., Varshney, A., Katyal, S., Kaur, R., Gaur, V.: A smart learning assistance tool for inclusive education. J. Intell. Fuzzy Syst. **40**(6), 11981–11994 (2021). https://doi.org/10.3233/jifs-210075
119. Stathopoulou, I., Alepis, E., Tsihrintzis, G., Virvou, M.: On assisting a visual-facial affect recognition system with keyboard-stroke pattern. Research and Development in Intelligent Systems XXVI: Incorporating Applications and Innovations in Intelligent Systems XVII, p. 451 (2009)

120. Sun, P., Gu, L.: Fuzzy knowledge graph system for artificial intelligence-based smart education. J. Intell. Fuzzy Syst. **40**(2), 2929–2940 (2021). https://doi.org/10.3233/jifs-189332
121. Tang, M., Si, X., Wang, R., Gao, W.: A new fuzzy comprehensive evaluation model for influencing factors of physical education. Int. J. Emerg. Technol. Learn. (iJET) **16**(17), 164–176 (2021)
122. Thavi, R.R., Narwane, V.S., Jhaveri, R.H., Raut, R.D.: To determine the critical factors for the adoption of cloud computing in the educational sector in developing countries-a fuzzy dematel approach. Kybernetes **51**(11), 3340–3365 (2021). https://doi.org/10.1108/k-12-2020-0864
123. Tseng, C.C., Lan, C.H., Lai, K.R.: Modeling beer game as role-play collaborative learning via fuzzy constraint-directed agent negotiation. In: 2008 Eighth IEEE International Conference on Advanced Learning Technologies, pp. 634–638. IEEE (2008)
124. Tsihrintzis, G., Toro, C., Rios, S., Howlett, R., Jain, L.: 27th kes International Conference on Knowledge-Based and Intelligent Information & Engineering Systems kes2023 (2023)
125. Tsihrintzis, G.A., Virvou, M., Alepis, E., Stathopoulou, I.O.: Towards improving visual-facial emotion recognition through use of complementary keyboard-stroke pattern information. In: Fifth International Conference on Information Technology: New Generations (itng 2008), pp. 32–37. IEEE (2008)
126. Tsihrintzis, G.A., Virvou, M., Hatzilygeroudis, I.: Special Issue on Selected Papers from the 33rd Annual IEEE International Conference on Tools with Artificial Intelligence (ictai-2021) (2023)
127. Tsiriga, V., Virvou, M.: Modelling the student to individualise tutoring in a web-based icall. Int. J. Contin. Eng. Educ. Life Long Learn. **13**(3–4), 350–365 (2003)
128. Upadhyay, H., Juneja, S., Juneja, A., Dhiman, G., Kautish, S.: Evaluation of ergonomics-related disorders in online education using fuzzy ahp. Comput. Intell. Neurosci. **2021**, 1–11 (2021)
129. Vazquez Reyes, B.O., Teixeira, T., Colmenero, J.C., Picinin, C.T.: Assessing educational methods for tomorrow's supply chain leaders with the integration of skill development priorities: a fuzzy decision-making approach. J. Enterp. Inf. Manag. **36**(2), 349–380 (2023)
130. Virvou, M.: A human plausible reasoning theory in the context of an active help system for unix1 users. Ph.D. thesis, University of Sussex, Falmer, East Sussex, UK (1993). https://ethos.bl.uk/OrderDetails.do?uin=uk.bl.ethos.787162
131. Virvou, M.: Automatic reasoning and help about human errors in using an operating system. Interact. Comput. **11**(5), 545–573 (1999)
132. Virvou, M.: Modelling the knowledge and reasoning of users in a knowledge-based authoring tool. Int. J. Bus. Perform. Manag. **5**(2–3), 399–412 (2003)
133. Virvou, M.: The emerging era of human-ai interaction: Keynote address. In: 2022 13th International Conference on Information, Intelligence, Systems & Applications (IISA), pp. 1–10. IEEE (2022)
134. Virvou, M.: Artificial intelligence and user experience in reciprocity: contributions and state of the art. Intell. Decis. Technol. **17**(Preprint), 1–53 (2023)
135. Virvou, M., Kabassi, K.: Reasoning about users' actions in a graphical user interface. Hum. Comput. Interact. **17**(4), 369–398 (2002)
136. Virvou, M., Kabassi, K.: Evaluating an intelligent graphical user interface by comparison with human experts. Knowl. Based Syst. **17**(1), 31–37 (2004)
137. Virvou, M., Moundridou, M.: Student and instructor models: two kinds of user model and their interaction in an its authoring tool. In: User Modeling 2001: 8th International Conference, UM 2001 Sonthofen, Germany, July 13–17, 2001 Proceedings, vol. 8, pp. 158–167. Springer (2001)
138. Virvou, M., Stavrianou, A.: User modelling in a gui. In: Proceedings of HCI International (The 8th International Conference on Human-Computer Interaction) on Human-Computer Interaction: Ergonomics and User Interfaces, vol. I, pp. 261–265 (1999)
139. Virvou, M., Alepis, E., Tsihrintzis, G.A., Jain, L.C.: Machine Learning Paradigms: Advances in Learning Analytics. Springer (2020)

140. Virvou, M., Tsihrintzis, G.A., Bourbakis, N.G., Jain, L.C.: Handbook on Artificial Intelligence-Empowered Applied Software Engineering: Smart Software Applications in Cyber-Physical Systems, vols. 2 and 3. Springer (2022)
141. Virvou, M., Tsihrintzis, G.A., Bourbakis, N.G., Jain, L.C.: Introduction to handbook on artificial intelligence-empowered applied software engineering—Novel methodologies to engineering smart software systems. In: Handbook on Artificial Intelligence-Empowered Applied Software Engineering: Novel Methodologies to Engineering Smart Software Systems, vol. 1, pp. 1–8. Springer (2022)
142. Virvou, M., Tsihrintzis, G.A., Sotiropouloss, D.N., Chrysafiadi, K., Sakkopoulos, E., Tsichrintzi, E.A.: Chatgpt in artificial intelligence-empowered e-learning for cultural heritage: the case of lyrics and poems. In: 2023 14th International Conference on Information, Intelligence, Systems & Applications (IISA), pp. 1–9. IEEE (2023)
143. Wang, J.: Music education to rescue psychological stress in social crisis based on fuzzy prediction algorithm. Sci. Program. **2021**, 1–6 (2021). https://doi.org/10.1155/2021/2039235
144. Wang, J., Zhao, B.: Intelligent system for interactive online education based on cloud big data analytics. J. Intell. Fuzzy Syst. **40**(2), 2839–2849 (2021). https://doi.org/10.3233/jifs-189324
145. Wang, R., Rong, X.: Extended group decision making method for quality evaluation of mental health education of college students with hesitant triangular fuzzy information. J. Intell. Fuzzy Syst. **45**(2), 2835–2845 (2023)
146. Wang, Y., Sun, C., Guo, Y.: A multi-attribute fuzzy evaluation model for the teaching quality of physical education in colleges and its implementation strategies. Int. J. Emerg. Technol. Learn. (iJET) **16**(2), 159–172 (2021)
147. Wang, Z., Lyu, D.: Design and realization of a fuzzy comprehensive evaluation system for music teaching in higher education. Int. J. Emerg. Technol. Learn. (iJET) **16**(22), 59–72 (2021)
148. Weng, S., Shieh, C.S., Tsihrintzis, G.A.: Advances in Intelligent Information Hiding and Multimedia Signal Processing: Proceeding of the 18th IIH-MSP 2022 Kitakyushu. Springer, Japan (2023)
149. Wu, D., Mendel, J.M.: Uncertainty measures for interval type-2 fuzzy sets. Inf. Sci. **177**(23), 5378–5393 (2007)
150. Xia, F.: Method for education value evaluation with regional culture integration into international students' ideological education under fuzzy number intuitionistic fuzzy environment. J. Intell. Fuzzy Syst. **44**(2), 1865–1880 (2023)
151. Xiaoyang, H., Junzhi, Z., Jingyuan, F., Xiuxia, Z.: Effectiveness of ideological and political education reform in universities based on data mining artificial intelligence technology. J. Intell. Fuzzy Syst. **40**(2), 3743–3754 (2021). https://doi.org/10.3233/jifs-189408
152. Xie, J., Yang, Y.: Iot-based model for intelligent innovation practice system in higher education institutions. J. Intell. Fuzzy Syst. **40**(2), 2861–2870 (2021). https://doi.org/10.3233/jifs-189326
153. Xu, H.: Intelligent system for university legal education based on machine learning. J. Intell. Fuzzy Syst. **40**(2), 2685–2696 (2021). https://doi.org/10.3233/jifs-189311
154. Xu, X., Xie, J., Wang, H., Lin, M.: Online education satisfaction assessment based on cloud model and fuzzy topsis. Appl. Intell. **52**(12), 13659–13674 (2022)
155. Ying, M.H., Yang, K.T.: A game-based learning system using the arcs model and fuzzy logic. J. Softw. **8**(9), 2155–2162 (2013)
156. Yu, S.: Application of computer information technology in college physical education using fuzzy evaluation theory. Comput. Intell. **37**(3), 1181–1198 (2021)
157. Zadeh, L.A.: Fuzzy logic=computing with words. IEEE Trans. Fuzzy Syst. **4**(2), 103–111 (1996)
158. Zhang, J.: Reform and innovation of artificial intelligence technology for information service in university physical education. J. Intell. Fuzzy Syst. **40**(2), 3325–3335 (2021)
159. Zhang, L.: Optimal symmetric interval solution of fuzzy relation inequality considering the stability in p2p educational information resources sharing system. Fuzzy Sets Syst. **478**, 108835 (2024). https://doi.org/10.1016/j.fss.2023.108835

160. Zhang, R., Zhao, W., Wang, Y.: Big data analytics for intelligent online education. J. Intell. Fuzzy Syst. **40**(2), 2815–2825 (2021). https://doi.org/10.3233/jifs-189322
161. Zhao, X., Zheng, C.: Fuzzy evaluation of physical education teaching quality in colleges based on analytic hierarchy process. Int. J. Emerg. Technol. Learn. (iJET) **16**(6), 217–230 (2021)

Chapter 5
EPATHLO Educational Game Suite and Fuzzy Reasoning

Abstract Recently, there has been a growing interest in utilizing computer games as a means of education. Educational computer games aim to boost students' motivation and engagement while catering to their unique needs and characteristics for better academic outcomes. To this end, a new AI-based educational game, Fuzzy EPATHLO, has been developed to teach the HTML programming language. Fuzzy EPATHLO operates under the umbrella of EPATHLO Educational Game Suite, which applies fuzzy sets to represent the student's knowledge level more realistically and provide a customized and engaging learning experience. The game utilizes this information to determine the appropriate difficulty level of quizzes and if the scenario should dynamically expand to provide a more challenging and immersive experience. By doing so, the game adapts to the student's learning pace and provides an optimal learning experience that is both challenging and engaging. This approach provides a highly personalized learning and gaming experience for the student. The game has undergone evaluation, and the outcomes are encouraging.

Keywords Educational software · Personalized educational games · Artificial Intelligence in educational games · Fuzzy logic in education · User experience · AI-empowered software games · User modeling · Human-AI interaction · e-learning games · 21st-century skills · Human-centred Artificial Intelligence

5.1 Overview of the EPATHLO Educational Game Suite

This section describes the innovative EPATHLO Educational Game Suite, an innovative suite of educational games. The word "EPATHLO" is Greek and means anything given as a reward or honorary distinction to someone who wins a sporting event or other competition. The suite's title indicates that learning through the use of the suite is a valuable reward for the student player, as it offers enhanced learning benefits through highly entertaining environments.

Figure 5.1 illustrates the different components that compose the educational suite. Each component has its essential role, and the whole system works like a well-wound clock that provides the student-player with the best possible educational and enter-

S. Papadimitriou and M. Virvou, *Artificial Intelligence—Based Games as Novel Holistic Educational Environments to Teach 21st Century Skills*, Intelligent Systems Reference Library 93, https://doi.org/10.1007/978-3-031-77464-5_5

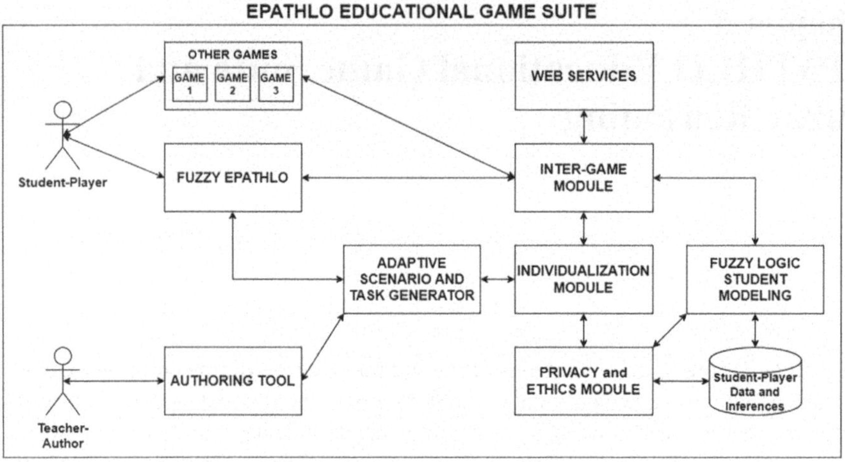

Fig. 5.1 The architecture of the EPATHLO educational game suite

taining experience. The Individualization and Fuzzy Modeling modules are essentially an Intelligent Tutoring System (ITS) that hosts the educational material (theory and exercises), builds student models, and generates exercises tailored to the student's knowledge level. The Fuzzy Student modeling component utilizes fuzzy reasoning to assess the student's knowledge level. The Individualization Module uses this information and produces personalized exercises. The Inter-Game Module communicates with the Individualization Module and Fuzzy Student Modelling Module. It uses web services technology to send personalized exercises and the student's knowledge level to different game environments.

The suite includes an adaptive educational game called Fuzzy EPATHLO, which we present in this chapter. It is a 2D adventure game whose story progresses through exploration and puzzle-solving. Fuzzy EPATHLO utilizes information from Fuzzy Student Modeling about the student's knowledge level. It uses a component of its own named Adaptive Scenario and Task Generator. The component extends the game's scenario by generating new tasks and enhances the student-player's soft and hard skills.

One of the suite's powerful elements is allowing any external game to communicate with it and exploit its educational components and advanced features. Chapter 6 presents Surviving Businessman, a game created by Chrysafiadi and Kamitsios [21], which uses the suite's components. It is an entirely different game from Fuzzy EPATHLO since its environment is 3D and supports Virtual Reality (VR). Fuzzy EPATHLO and Surviving Businesmman have access to the same educational material and student model built with the help of the Individualization Module.

According to the above, the suite's modular structure is beneficial, as any game can easily be integrated into the suite and turned into an educational game. In this way, the suite can transform a captivating digital game into an educational game and feed it with educational content suited to the student's educational needs. Therefore, with its unique gaming features, the educational game can keep the user's interest on the player's side at a high level and, at the same time, it can enhance the user's learning benefits on the student's side due to the intelligent techniques the suite applies for educational personalization. Also, when the player chooses to play a different game, it is unnecessary to rebuild the student model, as the suite retains the corresponding information. Thus, personalization continues from where it had stopped when playing the previous game.

The suite also offers a user-friendly tool for teachers and tutors. The EPATHLO Author is a web-based authoring tool that empowers educators to create and edit lessons, theories, and exercises. It also provides a comprehensive view of students' progress and performance during exercises. The EPATHLO Author allows quick intervention in the EPATHLO Fuzzy Student Modeling Module, even without prior knowledge of fuzzy reasoning. This flexibility in adjusting the suite's intelligent techniques can lead to a more accurate assessment of the student's knowledge level. The detailed features of the EPATHLO Authoring Tool are outlined in Chap. 7.

Explainability, transparency, consent, trust, GDPR, ethics, and privacy are essential for developing user-centered AI systems. Explainability, particularly in educational and medical contexts, aids user understanding and acceptance by clarifying AI decision-making processes, as Panagoulias et al. [45, 46] suggest. GDPR emphasizes the need for explicit consent and data protection, particularly in e-learning environments, with system designs incorporating these principles to protect user data [41, 42]. Trust in AI extends beyond technical performance to include ethical considerations, with Virvou et al. [65, 69] proposing trust models that focus on AI-human interaction dynamics. Privacy challenges in the distributed data era demand robust frameworks and strategies to secure personal data [57]. Collectively, these elements enhance AI systems by making them more user-centered, secure, and compliant with data protection regulations.

Given the above, the suite includes the Privacy and Ethics Module, a component under development. Its role is to ensure that an educational game that uses the suite to add personalized content complies with the General Data Protection Regulation (GDPR) rules. No game will collect or process user data without the user's consent. The game shall describe in full transparency all data processing operations when they take place and the benefits they provide to the learner. The personalized learning and gaming experience will only be possible with the learner's consent, which will be revocable at any time. The student can access all the data collected about them and ask for them to be deleted. All the module functionalities are related to the ethics an adaptive educational game should follow and are described in detail in Chap. 9.

5.2 The Game's Architecture

Figure 5.2 illustrates the system's architecture, a sophisticated gaming ITS with advanced capabilities. The right part is the Individualization Module and Fuzzy

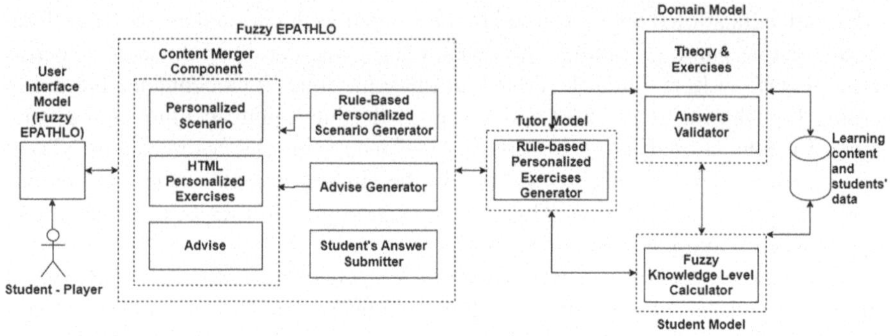

Fig. 5.2 The architecture of the system

Student Modeling Module, which we described in the previous section. It includes everything related to the educational part of the system, namely the domain model, the student model, and the tutor model. Thus, it adheres to the architecture of a typical ITS [3]. The domain model accommodates the educational material accompanying the learning object, such as theory, exercises, and questions of various types, e.g., single-choice or multiple-choice, answers, and difficulty levels. In addition, it includes a tool that verifies whether the students' answers to an exercise are correct. The student model includes all the information concerning the student's cognitive status and learning progress. The Fuzzy Knowledge Level Calculator integrated into the student model employs fuzzy logic techniques [70] to assess the student's knowledge level. All the data in the system, such as instructional materials, student data, and performance, are stored in a database. The tutor model is connected to the domain and student models and uses a rule-based system to create personalized exercises tailored to the student's knowledge level.

Fuzzy EPATHLO is the main game integrated into the EPATHLO Educational Game Suite. It provides the user with the user interface to interact with the whole system [50]. It is the subsystem of the suite that exchanges information with the tutor model. The tutor model sends the theory, the personalized exercises, and the student's current knowledge level to Fuzzy EPATHLO. Subsequently, the game uses a built-in rule system that adapts its scenario based on the student's educational needs. In addition, it maintains information regarding the student's failed quiz-solving attempts. It uses the advice generator to generate advice regarding the material students can study to overcome difficulties. If the student follows this advice, Fuzzy EPATHLO receives the corresponding material from the tutor model and displays it on the user's screen. The content merger component merges the personalized scenario, exercises, and hints and creates the content presented to the user through the game. In addition, when the user participates in a quiz through the game, the latter sends the answers to the tutor model to check their correctness and update the student's knowledge level. The functionality of the components of the system is as follows:

The right part of the system adjusts the educational content, while the left part adjusts the game context and elements. It is important to note that the right part, which includes the tutor, domain model, and student models, operates independently of the game [47]. This means that tasks such as updating a student model, creating customized quizzes, or validating quiz responses occur within this system. An external application can interact with the EPATHLO Inter-Game Module, trigger its tasks, and receive the relevant results [17, 35, 51]. In Chap. 6, the technical aspects of the communication between the EPATHLO Inter-Game Module and an external application and how the latter can leverage the capabilities of the former and integrate them into various graphical environments without the need for reimplementing are discussed. On the other hand, the tools for generating advice and personalizing the storyline are part of Fuzzy EPATHLO and cannot operate independently.

Finally, the privacy issues of students' data and inferences are analyzed further.

5.3 The Environment of the Fuzzy EPATHLO

In Fuzzy EPATHLO, players take on the protagonist role and find themselves imprisoned in a house with numerous rooms and a sprawling garden (Fig. 5.3). The primary educational goal of the game is to teach the HTML language. At the game's onset, a character appears on a wall screen in the first room, explaining the protagonist's predicament (Fig. 5.4). To escape, the protagonist must solve puzzles and complete side quests to uncover the path to freedom. Along the way, the protagonist gathers various items, combines them, and interacts with non-playable characters (NPCs)

Fig. 5.3 The title screen of the game

Fig. 5.4 The game environment

who entrust multiple quests to aid the protagonist's escape. Completing all quests (Table 5.1) is essential to progress the game's narrative and facilitate the player's exit from the building.

The user interface follows the standard design of point-and-click adventure games [10, 26, 27] and is easily navigable using touch gestures on personal computers or mobile devices. The main gameplay area covers three-quarters of the screen, where the player interacts with NPCs and objects. In contrast, the bottom one-quarter of the screen is dedicated to the inventory system for storing gathered items. Players can combine items from their inventory or surroundings and offer items to characters by selecting them from the inventory. Clicking on a character without selecting an inventory item initiates a dialogue, leading to the NPC's assignment of a new task, rewarding the player with a critical item necessary for advancing the game. These critical items are essential for progressing through the story and attempting to escape from the game. For example, players can create a refreshing beverage by combining a lemon slice with a soda in a glass, which they can then offer to a thirsty NPC (Fig. 5.5).

In Fig. 5.6, we can see the game map, which consists of a tile grid representing the game world. The blue tiles indicate the start and end rooms. White tiles are standard rooms where players navigate and solve puzzles. The green tiles are linked to learning tasks, where players complete quizzes customized to their expertise level. High-scoring players, yet with some weaknesses in the green rooms, may encounter red rooms with supplementary quizzes. Rooms with rounded corners are part of the scenario expansion and only appear to players struggling with quizzes. Solid lines

Table 5.1 The list of the main quests the NPCs assign to the hero

Assigner (NPC)	Quest	Result according to the student's knowledge level	Knowledge level
Witch	Find a dry crystal ball, money, and an air conditioner for the witch	The witch NPC assigns the hero a quiz.	Not applicable
	Solve quiz	Get a scythe	Learned
		Unlock a new quiz assigned by the zombie NPC	Known
Zombie	Solve quiz	Get a scythe	Learned
Thirsty man	Offer the thirsty NPC something refreshing and tasty	Get a bucket	Not applicable
Kid	Help the kid to play with his board game	Get a music album	Not applicable
Older man	Find a music album for the older NPC	Get a stool	Not applicable
NPC with thick hair	Solve quiz	Pull a heavy object from the ceiling and lift it	Learned
		Pull a heavy object from the ceiling but cannot lift it. Unlock a new quiz assigned by the muscle man NPC	Known
Muscle man	Solve quiz	Accompany the hero to help him lift the heavy object	Learned
Businessman	Solve quiz	Get a full oil can	Learned
		Get an empty oil can. Unlock a new quiz assigned by the chef NPC	Known
Chef	Solve quiz	Unlock the hidden room to fill the empty oil can	Learned
Final examiner	Solve quiz	Final escape	Learned

indicate pathways between rooms, while dotted lines require puzzle completion to reveal the corresponding paths.

We can visualize the hero's escape plan in the diagram labeled as Fig. 5.7. The ellipses in the diagram represent individual steps that the player can execute without any additional action required, such as "Take glow stick" or "Take battery." On the

Fig. 5.5 The hero offers an NPC a soda with lemon

other hand, the green boxes indicate dependent actions. For clarity, the figure does not illustrate the complete action graph and some of the dependencies of the displayed actions, like “Attach the hose to the magic hydra” or “Place the letter E.” The final step, “ESCAPE,” is depicted as a brown rectangle. According to the diagram, the player must first acquire the glowing stick to illuminate a dimly lit room. Upon doing so, the player discovers a switch and activates it to turn on the lights. Subsequently, while surveying the illuminated room, the player notices footprints on the wall that lead to a secret door. By carefully following the footprints in a specific sequence, the protagonist discovers a tile with the letter F. Placing the F tile, along with the I, R, and E tiles, in separate sockets opens a hidden passageway, bringing the hero one step closer to escaping.

The illustrated treeview does not include the quizzes but rather the steps associated with the game context. At the highest level of the hierarchy, the tree represents the game’s ultimate objective. As we move down the tree, we can see the sequence of actions the player needs to take to accomplish the game’s goal. The hierarchical structure of the steps helps to examine and confirm that there are no activity loops. Such loops may lead to the player getting stuck in a situation they cannot escape, making it impossible to complete the game. In turn, the inability to finish the game could lead to player frustration, resulting in a lack of interest in replaying the game.

Furthermore, the complete hierarchical tree illustrates that executing all actions and solving each puzzle is essential for completing the game. During the game development, there were instances in puzzle construction where the player could complete the game by taking only three actions. The action diagram aided in addressing certain

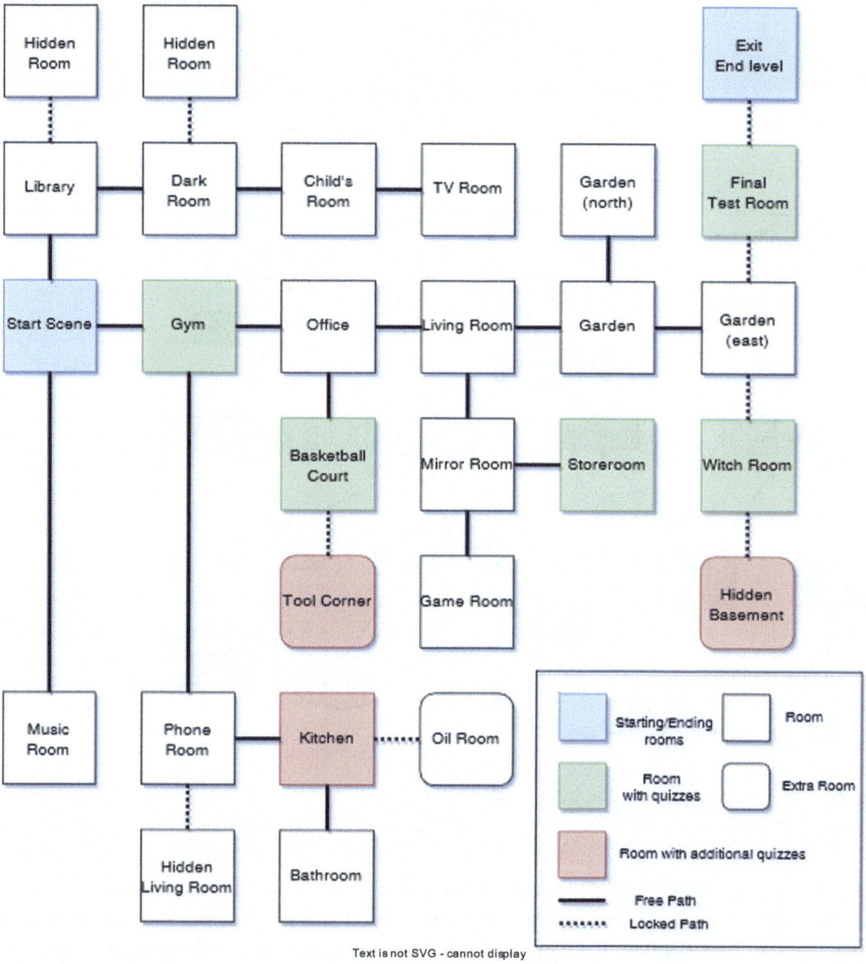

Fig. 5.6 The game map

puzzles and ensuring that all actions were necessary to complete the game. Allowing the game to be finished with only a few steps could diminish the player's motivation.

The game's narrative is non-linear, allowing players to move between rooms and tackle puzzles in any order to some extent. Certain puzzles are interconnected, as depicted in the action diagram. While some puzzles must be completed before others, the order in which they are solved varies for each player. Despite the different paths players can take, they must fulfill the same predefined tasks within the game's storyline. The game also incorporates additional tasks for players who require more time to grasp the learning material. These intermediate steps are seamlessly integrated to encourage continued play and provide tailored questions to help players overcome challenges, enhancing their understanding of the game's mechanics and storyline.

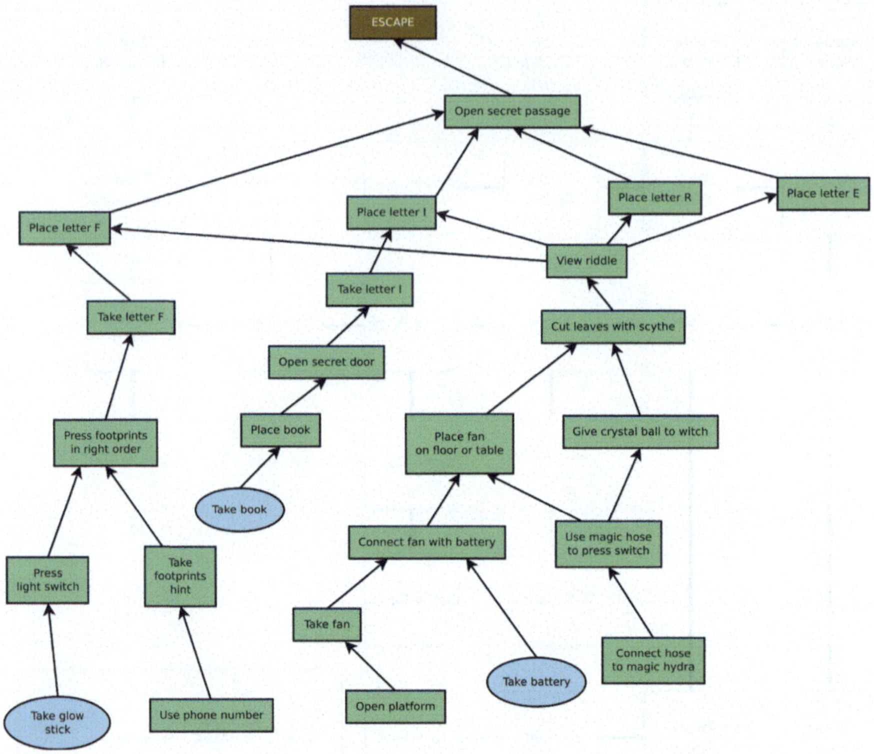

Fig. 5.7 A sample of actions the hero needs to perform

Although every player undergoes a similar game experience, they encounter different paths based on the order in which puzzles are solved and the number of questions they must master. The sequence of puzzle-solving is contingent on the player's actions, and the quantity of quests and tasks varies based on the player's learning progress throughout the game.

In the quests, players embark on missions to find specific objects (Fig. 5.8) or answer questions related to the teaching material, which may be in the form of single or multiple-choice exercises (Fig. 5.9). NPCs assign tasks in the first category by describing their problems, and players must locate items to help them. For example, a boy in his room might have a freeway board game but no car. Upon finding and delivering the car to the child, the boy may mention that the car needs oil to move on the track. The player then continues to explore the rooms to find the oil. The player is rewarded with essential items to aid in other quests upon task completion.

The game segments that require players to solve challenges serve as the areas where educational content is integrated into the game. During these segments, there is a two-way communication between Fuzzy EPATHLO and the EPATHLO Inter-Game Module. This communication retrieves questions tailored to the student's

Fig. 5.8 The NPC assigns a quest concerning discovering an object

Fig. 5.9 The NPC assigns a quest concerning solving a quiz

knowledge level and sends back the answers for performance calculation and profile updates. The questions in these segments come in four types: single-choice, multiple-choice, fill-in-the-blanks, and sorting, although the game currently only utilizes single-choice questions. Additionally, the questions are categorized based on difficulty levels, which include easy, moderate, and difficult. The player's success rate in these challenges determines the reward they receive and may impact the expansion of the game's storyline. Exceptional performance results in a higher reward for the player without affecting the predetermined game flow. Conversely, an acceptable performance point out the player's areas of improvement, provides a lower reward, and opens up additional game scenarios for the player.

The most recent scenario marks a turning point in the game, signifying the beginning of its expansion. Despite the students' flaws, the game recognizes and appropriately rewards their satisfactory performance. The evolving scenario involves introducing more tasks and puzzles, heightening the player's enthusiasm and eagerness to advance in the game. Additionally, expanding the scenario leads to supplementary quizzes tailored to the student's expertise and learning needs. This support assists and inspires them to overcome the challenges identified by the game, not by necessitating a game restart–a potential cause of boredom–but by offering additional incentives and objectives. Enriching the scenario with new game elements spurs the player to actively engage in the educational experience. Increased quiz participation effectively addresses weaknesses and enhances students' grasp of the learning material.

For example, an NPC in a storeroom presents the hero with a challenge. If the hero completes the challenge with a high success rate, the NPC will leave, allowing the hero to remove a heavy object suspended from the ceiling. By doing so, the hero can progress to the next quest. However, if the success rate is adequate but not high enough, the hero will not be able to lift the heavy object. In this scenario, a strong NPC appears in another room and presents the hero with a new challenge designed to address the weaknesses identified in the previous one. This tailored challenge ensures that the hero can overcome their weaknesses and progress. In this successive challenge, the hero must achieve a high success score so that the muscular NPC will offer assistance by carrying the heavy object (Fig. 5.10).

5.4 Student Modeling and Adaptivity Based on Fuzzy Logic

5.4.1 Student Modeling

The users of educational software, including those who use educational games, consist of a diverse group of learners with unique attributes, backgrounds, personalities, knowledge levels, skills, and learning needs. The system must recognize and cater to each student's knowledge level, learning requirements, and competencies by offering appropriate learning resources and methods. One effective way to calculate a student's proficiency level is through student modeling. Consequently, many adaptive educational applications, including games, utilize student modeling techniques to evaluate students' knowledge levels.

Fig. 5.10 The muscular man assigns a repetitive quiz if the player encounters difficulties

There are several methods of student modeling, including stereotypes, overlay models, Machine Learning (ML) techniques, fuzzy logic, neural networks, and others. Most ITSs employ the overlay model as their foundational principle. This model stores data that evaluates a student's knowledge of every concept in the domain model. In its simplest form, the student's knowledge is represented as a binary value (e.g., "Known" or "Unknown") overlaid on domain knowledge. While many successful adaptive educational systems use this classical format [24] of an overlay model, most systems employ a weighted overlay model, which can differentiate between different levels of student knowledge using qualitative values [6, 53] (e.g., "Good," "Average," and "Poor") [6, 53], quantitative integer values (e.g., 0 to 100) [7, 23], or the probability that the student knows the concept [30, 58]. Therefore, a weighted overlay model represents the student's knowledge as a collection of "concept-value" pairs, one for each domain concept.

Assessing students' knowledge is difficult as it often depends on characteristics that are not immediately observable or measurable. For instance, claiming that a student has fully grasped a concept in a particular subject area can be misleading. They might be unfamiliar with one concept aspect while being well-versed in another. Therefore, a student's knowledge cannot be treated as a clear-cut variable, as its evaluation involves uncertainty and human subjectivity. A potential solution to this challenge is the application of fuzzy reasoning to handle uncertainties in everyday problems stemming from incomplete or inaccurate data and human subjectivity.

Dr. Konstantina Chrysafiadi, an expert in designing fuzzy inference systems (FISs), collaborated on implementing the student model for the EPATHLO Inter-game Individualization Module. Utilizing the weighted overlay model and fuzzy logic technique, the student model assesses the student's current knowledge level for each concept. The model uses qualitative values such as "Unknown," "Unsatisfactorily Known," "Known," and "Learned" to characterize the student's knowledge level. This model was selected for its simplicity, broad applicability, and effectiveness in creating student models. Furthermore, it enables detailed user modeling and establishes a clear link between the User and Domain Model.

Furthermore, the student model is not only robust but also highly adaptable. It can autonomously assess students' understanding across a wide range of subjects. When students first engage in activities, their comprehension of each concept is initially categorized as 'Unknown.' This adaptability is a key feature of the student model, allowing it to intersect with the expert model and evolve as students refine their understanding.

The above quantitative values are challenging to calculate because they involve subjectivity and ambiguity. Fuzzy logic is used to quantify vague concepts and successfully calculate these values. It is important to note that the quartet $(\mu_{Un}, \mu_{UK}, \mu_K, \mu_L)$ describes the student's knowledge level related to a specific domain concept. The members of the quartet define the participation of the student's current knowledge level in each one of the following four fuzzy sets:

- Unknown (U_n): The student has little or no knowledge of the educational material.
- Unsatisfactorily Known (U_K): The student has moderate knowledge of the delivered educational material.
- Known (K): The student knows the delivered educational material well.
- Learned (L): The student's knowledge of the learning material delivered is excellent, and they have absorbed it.

The membership functions (MF) specified below, referred to as membership functions, accept the student's success percentage rate post-quiz as the input X and determine their knowledge level for each fuzzy set. All membership functions are trapezoidal (Fig. 5.11).

$$\mu_{Un}(x) = \begin{Bmatrix} 1, & x < 45 \\ 1 - \frac{x-45}{5}, & 45 < x < 50 \\ 0, & x \geq 50 \end{Bmatrix} \tag{5.1}$$

$$\mu_{Uk}(x) = \begin{Bmatrix} \frac{x-45}{5}, & 45 < x < 50 \\ 1, & 50 \leq x \leq 65 \\ 1 - \frac{x-65}{5}, & 65 < x < 70 \\ 0, & x \leq 45, x \geq 70 \end{Bmatrix} \tag{5.2}$$

$$\mu_K(x) = \begin{Bmatrix} \frac{x-65}{5}, & 65 < x < 70 \\ 1, & 70 \leq x \leq 85 \\ 1 - \frac{x-85}{5}, & 85 < x < 90 \\ 0, & x \leq 65, x \geq 90 \end{Bmatrix} \tag{5.3}$$

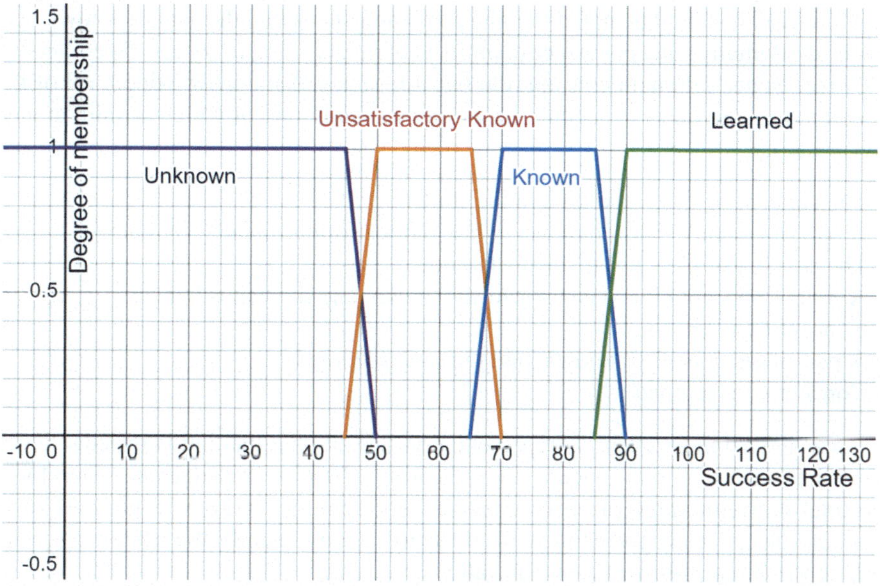

Fig. 5.11 The membership functions represent the student's knowledge level

$$\mu_L(x) = \begin{Bmatrix} \frac{x-85}{5}, & 85 < x < 90 \\ 1, & 90 \le x \le 100 \\ 0, & x \le 85 \end{Bmatrix} \tag{5.4}$$

The above membership functions can range from values of 0 to 1, i.e.:

$$\mu_{Un}(x), \mu_{UK}(x), \mu_K(x), \mu_L(x) \epsilon [0, 1]$$

As a result, based on the specified membership functions and their associated thresholds, the fuzzy sets (or the qualitative values within the weighted overlay model) represent the following interpretations:

- Unknown (U_n): The student does not know the domain concept. The success rate for the quiz on the domain concept is between 0 and 50%.
- Unsatisfactory known (U_K): The student has minor or moderate knowledge of the concept of the domain. The success rate on the domain concept quiz ranges from 45
- Known (K): The student knows the domain concept well. The success rate for the domain concept quiz varies from 65 to 90%.
- Learned (L): The student has learned the domain concept very well. The success rate on the domain concept quiz is between 85 and 100%.

Moreover, it should be noted that the constraint states that the various knowledge levels of a domain concept must be comprised of neighboring fuzzy sets. Specifically, if C_i represents a domain concept, the feasible states of the knowledge level of C_i as defined by the fuzzy sets could be as follows (only one of which is valid at any given time):

- The concept of C_i is unknown to the student.
- A part of concept C_i is unknown, while the rest is unsatisfactorily known.
- The concept of C_i is unsatisfactorily known.
- A part of concept C_i is unsatisfactorily known, and the rest is known.
- Concept C_i is known.
- A part of concept C_i is known, and the rest is learned.
- Concept C_i is learned.

To achieve a more accurate representation, we assume that the knowledge level associated with a domain concept is 1, indicating a 100% knowledge rate. Therefore, the total partition value of each fuzzy set representing the knowledge level of a domain concept at any given time will add up to 1. Consequently, the relation $\mu_{U(x)} + mu_{UK(x)} + mu_{K(x)} + mu_{L(x)} = 1$ is valid.

5.5 Adaptivity in Educational Gaming Environment and Content

Educational games should provide personalized learning content, continuous motivation, and challenges to enhance the learning and gaming experience. Despite this, many games in the literature lack adaptive capabilities and fail to offer additional motivation to players. In the past, personalized educational games have been developed to adapt dynamically to students' needs, preferences, and cognitive states [37, 61, 62, 64, 66, 67]. These earlier approaches focused on various student characteristics such as memory, reasoning, and affect, which were related to students' knowledge levels and primarily addressed the educational aspect and students' behavior in the games. In contrast, Fuzzy EPATHLO employs a mechanism that combines the overlay model with fuzzy logic to dynamically adjust scenarios based on the student's knowledge level. This approach enhances adaptivity by addressing both the gaming experience and educational content through scenario adjustments. In educational games, it is very important to achieve a good balance of the gaming part with the educational part because if the entertainment aspect is overlooked in favor of education, then the advantages of games, such as immersion and motivation, may be lost.

While there has been similar work in the context of ITSs that have effectively utilized a similar mechanism using Fuzzy Logic [11, 14], no educational game offers comparable adaptive support to student-players. EPATHLO adapts its content according to the student model generated using fuzzy logic. After each exercise, the

EPATHLO Inter-Game Module transmits the knowledge level to Fuzzy EPATHLO, which utilizes the Adaptive Scenario and Task Generator components to determine whether to expand the scenario. Moreover, when the game requests a new exercise, the tutor model collaborates with the domain and student model to create a new personalized exercise, which is then sent to Fuzzy EPATHLO. By linking the game scenario to the student's knowledge level through the rule-based personalized scenario generator, the game meets the requirement of keeping students motivated to persist until they overcome their learning challenges.

5.5.1 Adaptivity in the Educational Part of Fuzzy EPATHLO

The adaptivity of the educational component entails customizing the difficulty level of an exercise to align with the student's historical knowledge level, which is indicated by fuzzy sets. Questions are categorized as easy, moderate, or difficult. During the initial quiz attempt, the tutor model chooses fifteen easy questions. Subsequently, the student's performance in the quiz determines their placement in the relevant fuzzy set. In subsequent attempts, the tutor model retrieves the student's knowledge level from the previous quiz and selects appropriate questions for the current assessment. For each following exercise, the tutor model analyses the student's knowledge level classifications from the two most recent exercises and tailors question selection accordingly. This approach allows the system to utilize the student's most recent activities based on historical data.

The rules presented in Table 5.2 depict how the system adjusts exercise content to accommodate student's individual needs, thereby augmenting the learning experience by helping students address any deficiencies and advancing their knowledge and cognitive abilities.

5.5.2 Adaptivity in the Gaming Part of Fuzzy EPATHLO

The rule-based personalized scenario generator in Fuzzy EPATHLO plays a crucial role in adjusting the game's storyline to sustain the player's motivation and involvement at high levels. Following each quiz, the system assesses the student's latest performance and then makes decisions based on this classification. The game receives information about the student's current knowledge level to determine whether to offer rewards, what type to give, and whether to introduce additional rooms, puzzles, NPCs, and exercises. The accompanying diagrams depict the state transitions of key game elements (quests, rooms, NPCs) based on the student's knowledge level, which is evaluated after completing quizzes in specific game sections. This specifically pertains to elements that could potentially lead to the expansion of the game's scenario under certain conditions.

Table 5.2 The rules that adjust the difficulty level of the questions

Quiz number	Knowledge level before the current quiz	Knowledge level before the previous quiz	Number of questions			Rule number
			Easy	Moderate	Difficult	
1st	Not applicable	Not applicable	15	0	0	1
2nd	Unknown	Not applicable	15	0	0	2
	Unsatisfactorily known		7	8	0	3
	Known		0	7	8	4
	Learned		0	0	15	5
3rd, 4th, 5th, …, nth	Unknown or unsatisfacto-rily known	Unknown or unsatisfacto-rily known	15	0	0	6
		Known	7	8	0	7
		Learned	0	7	8	8
	Known	Unknown or unsatisfacto-rily known	7	8	0	9
		Known	0	7	8	10
		Learned	0	5	10	11
	Learned	Unknown	7	8	0	12
		Unsatisfactorily known	3	8	4	13
		Known	0	5	10	14
		Learned	0	0	15	15

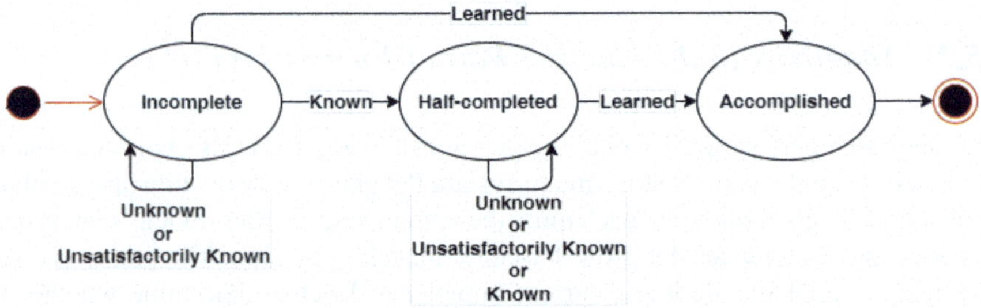

Fig. 5.12 The state chart diagram of the quests related to objects

Figure 5.12 shows the possible states of a quest to find an object. For example, the hero must find an oil can at some point in the game. Initially, he does not have the object in his possession. Therefore, the quest is in an "Incomplete" state. Then, the oil can is supplied by a specific NPC, who assigns the hero to solve a quiz. If the hero answers with an excellent success rate and his knowledge level becomes "Learned," the NPC hands him the oil can. Thus, the quest transitions to the "Accomplished" state.

On the other hand, if the hero answers satisfactorily, the NPC offers him the oil can, but the can is empty, i.e., the quest enters the "Half-completed" state. Then, the player must navigate to another part of the game, participate in a repetitive quiz, and improve his knowledge. This way, he will find a barrel of oil, fill his can, and transfer the quest to the "Accomplished" state. In any case, if the player has a knowledge level of "Unknown" or "Unsatisfactorily Known," the quest remains in the same state.

Figure 5.13 similarly depicts the possible states of the rooms involved in the search for an object and are part of the scenario extension. For example, the witch's room (Fig. 5.22) is initially "Existing." The witch assigns the hero a quiz. If the hero answers with a perfect success rate and his knowledge level becomes "Learned," the room transitions to the "Accomplished" state, and the player continues with other quests. If the success rate is slightly lower and the knowledge level becomes "Known," the room enters the "Expanded" state. Then, the witch points the player to a trapdoor that leads to a hidden basement, an expansion of the previous one. There, a new NPC will challenge him with a repetitive quiz, and he must raise his knowledge level to "Learned." At that point, the room will transition to the "Accomplished" state, and there will be no reason for the player to return to it.

Figure 5.14 indicates the state chart diagram of the NPCs that are part of the scenario expansion. For example, a muscular athlete is initially in the "Non-existent" state and is not visible anywhere in the game. When the hero goes to the storeroom, the NPC informs him that to allow him to lower a heavy object hanging from the ceiling, he must successfully answer the quiz he will be assigned. If, after completing the quiz, the player's knowledge level is "Learned," the NPC leaves, and the player lowers the object and takes possession of it. At that point, the muscular athlete's status becomes "Left," meaning he silently exits the game, and there is no chance of him appearing. If the student's proficiency level becomes "Known," he can lower the object but not lift it because it is too heavy. Then, the muscular athlete's status becomes "Appeared," meaning that he appears at some point in the game. When the

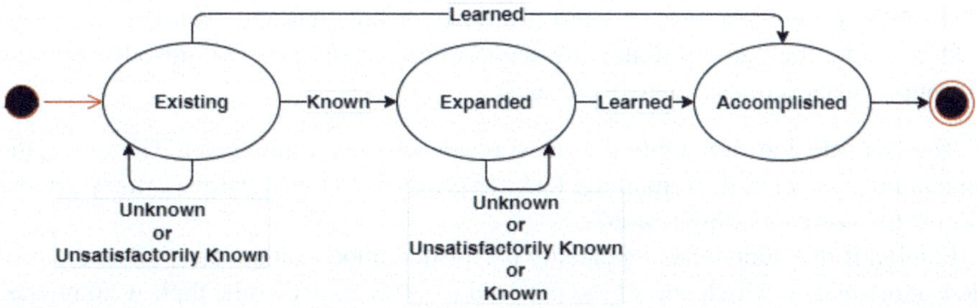

Fig. 5.13 The state chart diagram of the rooms related to objects

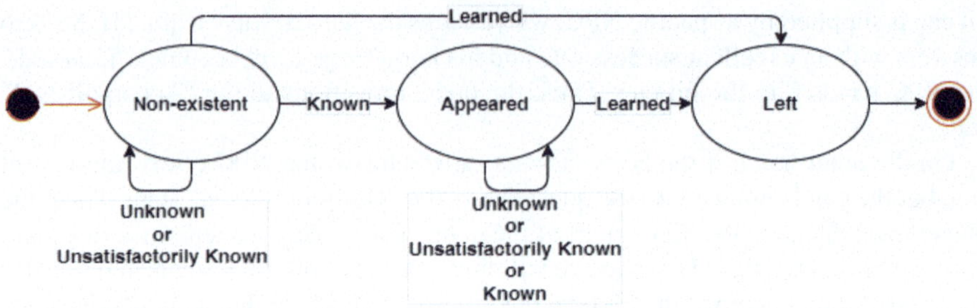

Fig. 5.14 The state chart diagram of the NPCs related to objects

player spots him, he will participate in a repeat quiz. If his knowledge level becomes "Learned," the NPC helps him pick up the item and leaves the game. He enters the "Left" state and will not appear again. Similarly, in this case, if the student's knowledge level is "Unknown" or "Unsatisfactorily Known," the NPC remains in the same state.

The intermediate states are part of the game extension in all three cases of the game elements described. They are only activated under certain circumstances and when students need to improve their knowledge and acquire new challenges to continue playing. Moreover, in the repetitive quizzes, if the knowledge level remains "Known," the state of the game objects remains the same.

Extending the game scenario elevates player motivation and fosters deeper engagement with the curriculum. The expansion seamlessly integrates with the experience, going unnoticed by the students. According to the following rules, the students' quiz performance and resultant knowledge level directly influence the game's progression:

- If the player belongs to one of the "Unknown" or "Unsatisfactorily Known" fuzzy sets, the game does not offer any rewards. Instead, the player needs to interact with the NPC assigned the quiz and participate in a revised exercise with adjusted questions according to Table 5.2 rules.
- The NPC provides the maximum reward if the player's knowledge level is part of the "Learned" fuzzy set. The game scenario remains unchanged.
- The NPC provides a lower reward if the player falls into the "Known" category. Meanwhile, the game initiates the scenario expansion process, introducing new rooms, puzzles, quests, and challenges.

The last rule does not apply if a quest reaches its maximum level. Therefore, the student must excel in the remaining tasks to progress to the "Learned" fuzzy set and receive the corresponding rewards.

Finally, if the student has repeatedly performed poorly on quizzes, the game provides guidance on which part of the material to study to overcome their weaknesses. The advice is provided by the appearance of either a new NPC (Fig. 5.15) or a paper lying on the floor.

Fig. 5.15 An agent is responsible for giving clues about the most recent questions that a student has answered incorrectly

5.5.3 Examples of Operation

In the context of the game, examples will be used to demonstrate how fuzzy logic impacts the educational content. Each example will involve quizzes that the player needs to solve. There are three key components to consider when engaging with a quiz: the description, the educational aspect, and the gaming aspect. The description will detail the player's movements, including the rooms they navigate and the interactive items or NPCs they encounter. The educational aspect will explain how fuzzy logic influences the selection of quiz questions. The final part will assess the player's success rate, their rankings within the fuzzy sets, and how these rankings directly influence the game's storyline, determining the player's next challenges and rewards.

5.5.3.1 Example 1

Interaction 1

Description

The player encounters a businessman in a room (Fig. 5.16). The character offers to help the hero if the hero successfully answers various HTML questions.

Educational part

The player takes the quiz for the first time, so their knowledge level is "Unknown." They need to answer fifteen randomized easy questions according to the rule for the first quiz (Fig. 5.17).

Fig. 5.16 The NPC offers to help the hero

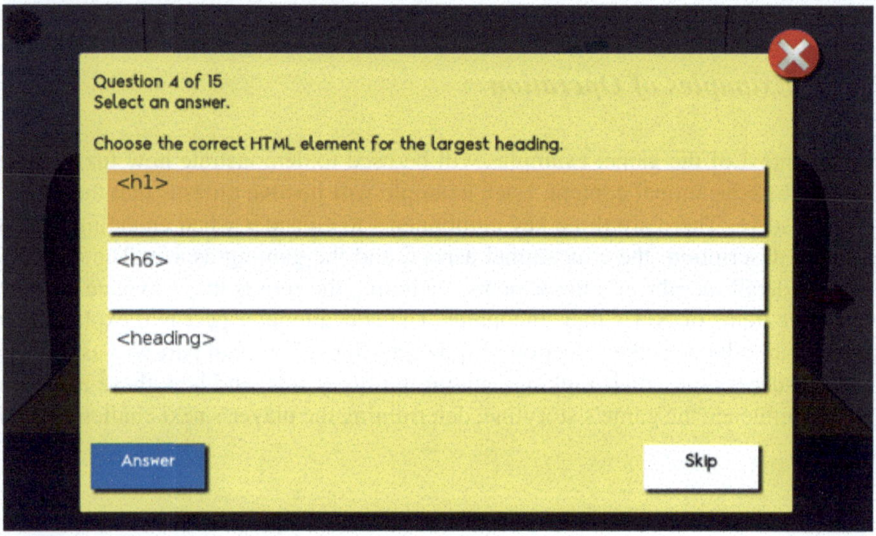

Fig. 5.17 The first quiz contains fifteen easy questions

Game part

The player correctly answers about 93% (14/15 questions), placing them firmly in the "Learned" category. Subsequently, based on the rule-based component of Fuzzy EPATHLO, the character rewards the player by presenting the hero with a can of oil (Fig. 5.18). This oil can is vital for the game's progression, as the hero must deliver it to a boy to pour oil on the highway board game.

Fig. 5.18 The NPC offers the hero a can full of oil as a reward

Fig. 5.19 A room where a weighted object is fixed to the room ceiling

Interaction 2

Description

Upon moving to the next room, the player encounters a heavy object secured to the ceiling (Fig. 5.19). An NPC obstructs the hero from cutting the rope to release the object but offers to leave the room if the hero completes an assigned quiz with a perfect success rate.

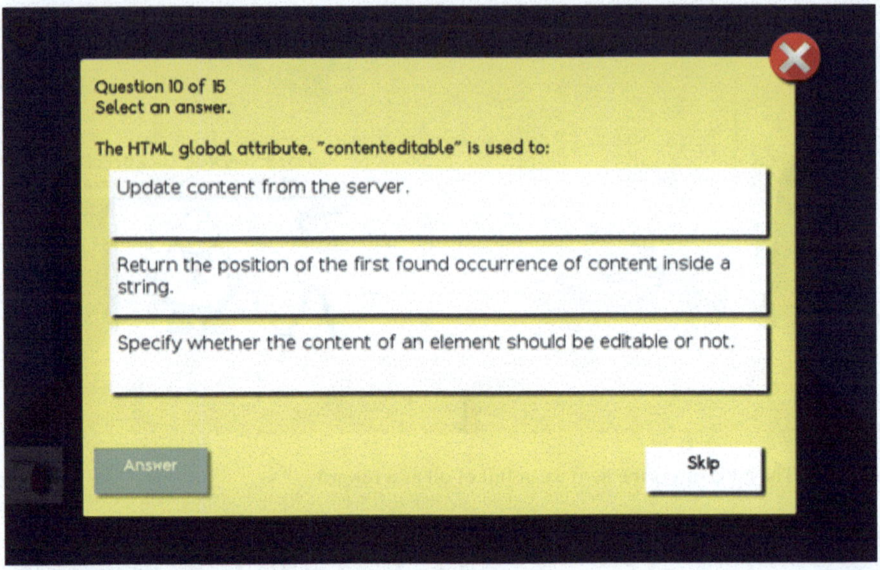

Fig. 5.20 The second quiz consists of fifteen difficult questions

Educational part

It is the second quiz that the player needs to complete. The player's current rank is "Learned," so they must answer fifteen difficult questions (Fig. 5.20).

Game part

The player responds correctly to all 15 questions, achieving a 100% success rate. Consequently, based on the membership functions, their classification remains within the fuzzy set "Learner." Subsequently, the NPC exits the room, allowing the hero to sever the rope and raise the heavy object (Fig. 5.21).

Interaction 3

Description

The player arrives in a room containing a witch. Upon presenting her with specific items (a crystal ball, money, and a fan), the player must answer a quiz (Fig. 5.22).

Educational part

The previous two categories were identified as "Learned," so the test will consist of fifteen randomly selected difficult questions.

Fig. 5.21 The NPC walks away, and the hero can retrieve the item

Fig. 5.22 The witch asks questions related to HTML concepts

Game part

The player achieves a score of 14/15, with a success rate of 93%, maintaining their position in the "Learned" category. After completing three consecutive quizzes with the "Learned" ranking, the player proceeds to the next quiz, which is the final challenge (Fig. 5.23) in the game. The game requires consistent high performance on several quizzes to demonstrate the player's understanding of the learning domain.

Fig. 5.23 The final test appears after enough classifications in the fuzzy "Learned" set

5.5.3.2 Example 2

Interaction 1

Description

The player enters the room with a heavy object, but the NPC prevents them from taking it. The first quiz requires the student to answer fifteen easy, randomly chosen questions.

Game part

The learner answers 8 out of 15 questions correctly, a success rate of almost 53%. As per the membership functions, this places them within the "Unsatisfactorily known" category with 100% certainty (Fig. 5.24).

Interaction 2

Description

If the player moves to a different room instead of staying in the same one, they will encounter an NPC wearing a tie. This NPC will present a new quiz featuring questions related to HTML.

Educational part

At the second encounter, with the player's knowledge level falling into the "Unsatisfactorily known" category, rule number 2 comes into play. This rule dictates that the quiz will comprise seven easy questions and eight average difficulty questions (Fig. 5.25).

Fig. 5.24 The hero does not get any help due to the "Unsatisfactorily Known" set

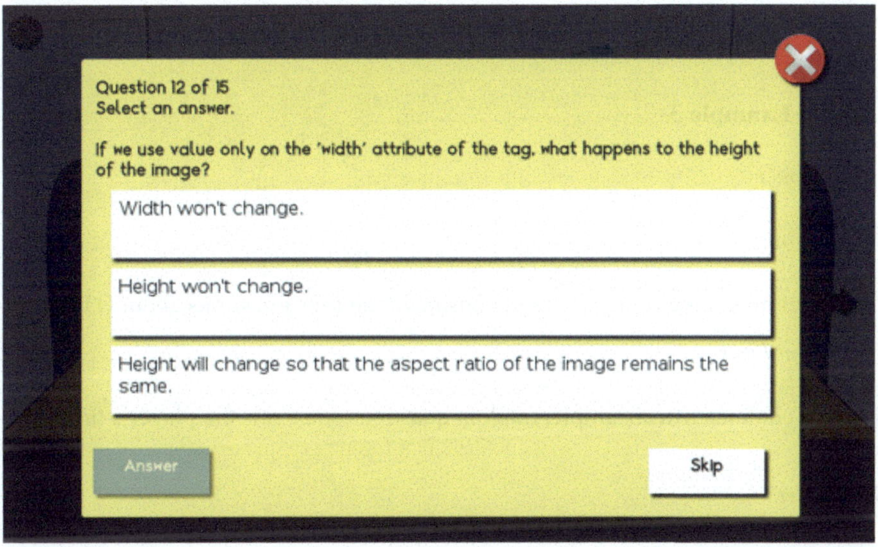

Fig. 5.25 The quiz consists of seven easy and eight moderate-difficulty questions

Game part

The player answers 14 out of 15 questions correctly, resulting in a success rate of 93%. As a result, the student falls into the "Learned" category with 100% participation. Consequently, the businessman presents the can filled with oil to the hero.

Interaction 3

Description

The player returns to the room with the weighty item to try the quiz again.

Educational part

This is the third instance of interaction. The previous two levels attained were "Not satisfactorily known" and "Acquired knowledge." Therefore, per Table 5.2, the quiz will consist of seven simple and eight moderately challenging questions.

Game part

The player answers with a 100% success rate (15/15), demonstrating a high level of knowledge and belonging to the "Learned" category. As a result, the character in the room departs, allowing the player to obtain the object. The player proceeds to engage with the next quiz on their journey.

5.5.3.3 Example 3

Interaction 1

Description

The player enters the room where the businessman asks questions about HTML.

Educational part

The quiz contains fifteen simple, random questions since it is the player's first interaction.

Game part

The player correctly answers 12 out of 15 questions, resulting in a success rate of 80%. They rank at 100% in the "Known" fuzzy set, triggering the application of rule 3 (basic quiz, set is "Known" but not "Learned"). In this scenario, the player's performance is deemed acceptable but with minor shortcomings. Consequently, the NPC character rewards the player with an item of lesser value compared to achieving a ranking in the "Learned" set. The player receives an empty can from the character and is tasked with finding the oil elsewhere (Fig. 5.26). Figure 5.27 illustrates the different game states and their connections. Simultaneously, the game sets in motion a background process that expands the scenario, becoming evident during subsequent interactions.

Fig. 5.26 Because of the "Known" set, the NPC rewards the player with an empty oilcan

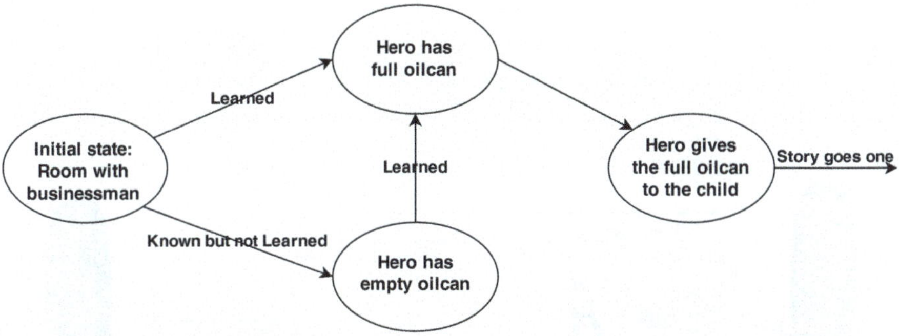

Fig. 5.27 The diagram of the possible game states in the room with the businessman

Interaction 2

Description

Upon entering the kitchen, the player finds no NPC character at first. However, based on the previous interaction and the resulting knowledge classification, the game introduces a chef into the kitchen and assigns the player a new quiz (Fig. 5.28). Essentially, a new character emerges with the specific goal of administering a recurring test to the player in order to assist them in overcoming their weaknesses.

Educational part

The quiz contains seven moderately difficult and eight challenging questions for this second interaction, per rule 4.

Fig. 5.28 When the player's performance is strong but with some minor challenges, the chef will appear

Fig. 5.29 A hidden door that leads to a new room appears

Game part

The player correctly answers 14 out of 15 questions, resulting in a 93% success rate, placing their knowledge level in the "Learned" category at 100%. Following this, the chef reveals a secret door (Fig. 5.29), leading the player to a previously undiscovered hidden room. An oil container exists inside the room, allowing the hero to refill their empty oilcan and continue their journey (Fig. 5.30). The quiz is optional and not mandatory for the hero to complete. The game generated it to help students overcome minor difficulties in previous interactions. Completing the quiz requires the student to partially classify their knowledge level in the "Learned" category.

Fig. 5.30 A container full of oil exists in the new room

Interaction 3

Description

The player enters the room with a heavy object.

Educational part

This marks the third interaction; therefore, we will consider the two most recent classifications, 100% "Learned" and 100% "Known." According to Table 5.2, the quiz comprises seven questions of average difficulty and eight challenging questions.

Game part

The player answers 10 out of 15 questions correctly, achieving a success rate of about 67%. Based on the membership functions, the knowledge level falls 60% into the "Unsatisfactorily known" set and 40% into the "Known" set. As the quiz is categorized as compulsory, rule 3 comes into play. According to this rule, the hero will be awarded a small prize and can continue the adventure. Subsequently, the NPC character exits the room, leaving the hero to discover that the item is too heavy to lift (Fig. 5.31). This silent expansion of the game scenario leads the hero to determine how to acquire the object.

Interaction 4

Description

When the character enters the businessman's room, they find another NPC present, a muscular athlete. The character initiates a conversation with the athlete, giving the character a new quiz to complete. (Fig. 5.32).

Fig. 5.31 The player has minor shortcomings; therefore, the object is quite heavy

Fig. 5.32 The muscular man offers to assist the player in carrying the weighty object

Educational part

The new quiz's difficulty level will be determined by the two most recent knowledge ranks. Based on the table, when the last rank is "Known," and the second-to-last rank is "Learned," the quiz will consist of fifteen difficult questions.

Game part

The player responds correctly to all 15 questions, achieving a flawless 100% success rate. As a result, the player is moved back to the "Learned" category with a perfect score of 100%. Subsequently, the muscular character joins forces with the hero to help carry the heavy object. Together, they place the heavy object on a platform, triggering the reveal and unlocking of a hidden trapdoor. In the event of a repetitive test and the player answers 12 out of 15 questions correctly, yielding an 80% success rate and a 100% "Known" status, the strong character would refrain from rewarding the player and instead assign a retest.

5.5.3.4 Conclusions

The examples above demonstrate how the scenario evolves based on the students' modeling. Different variations of these scenarios may arise, further expanding the game's storyline by introducing more locations, objects, and NPCs. When the player is included in the "Known" set but not yet in the "Learned" set during mandatory quizzes, the game expands accordingly, requiring the player to solve additional quizzes and new puzzles. This gradual scenario expansion offers students extra practice in an educational context and boosts their motivation and engagement in the game.

5.6 The Game's Evaluation

5.6.1 Experimental Data

The following section presents the findings of the experiments on the game's real-world usage. Sixty participants, all students enrolled in the MSc program in Informatics at the University of Piraeus, participated in the educational game. Before playing the game, the students had completed one month of classroom lectures on HTML and were given two weeks to play it. On average, the students completed 7.88% of the HTML language quizzes provided. Specifically, 28.33% of users completed a maximum of five quizzes, 58.33% completed between six and ten, and 13.33% completed more than ten. The average difficulty level of the quizzes for the 60 students was 22.29% easy, 28.61% moderate, and 49.10% difficult. The experimental data are detailed in Table 5.3, while Table 5.4 presents a sample of the fuzzy sets indicating students' knowledge level progress.

Table 5.3 Experimental data

Number of interactions–quizzes (q)	The average number of students (%)	The average difficulty levelof the questions of all quizzes		
		Easy (%)	Moderate (%)	Difficult (%)
$q \leq 5$	28.33	20.36	21.52	58.12
$6 \leq q \leq 10$	58.33	22.08	31.53	46.39
$q > 10$	13.33	24.45	32.78	42.78
Total	22.29	28.61	49.10	

Table 5.4 A sample of the changes in the values of fuzzy sets

Users	Values of fuzzy sets per interaction ($\mu_{Un}, \mu_{UK}, \mu_K, \mu_L$)						Total number of questions				Number of quizzes
	Inter. 1	Inter. 2	Inter 3	...	Inter. n-1	Inter. n	Easy	Moderate	Hard	Sum	
1	(1, 0, 0, 0)	(0 , 0, 1, 0)	(0, 0.4, 0.6, 0)	...	(0, 0, 0, 1)	(0, 0, 0, 1)	15	29	60	104	6
2	(1, 0, 0, 0)	(0, 0, 1, 0)	(0, 0.4, 0.6, 0)	...	(0, 0, 0.8, 0.2)	(0, 0, 0, 1)	15	29	45	89	5
3	(1, 0, 0, 0)	(0, 1, 0, 0)	(0, 0, 1, 0)	...	(0, 0, 0.4, 0.6)	(0, 0, 0, 1)	29	37	54	120	8
4	(1, 0, 0, 0)	(0, 0, 1, 0)	(0, 0, 1, 0)	...	(0, 0, 0.8, 0.2)	(0, 0, 0, 1)	15	14	46	75	5
5	(1, 0, 0, 0)	(0, 0, 1, 0)	(0, 0.6, 0.4, 0)	...	(0, 0, 0, 1)	(0, 0, 0, 1)	15	7	38	60	4
6	(1, 0, 0, 0)	(0, 1, 0, 0)	(0, 1 0, 0)	...	(0, 0, 0, 1)	(0, 0, 0, 1)	29	44	77	150	10
7	(1, 0, 0, 0)	(1, 0, 0, 0)	(0.2, 0.8, 0, 0)	...	(0, 0, 0.6, 0.4)	(0, 0, 0.8, 0.2)	52	52	45	149	10
8	(1, 0, 0, 0)	(0.4, 0.6, 0, 0)	(0, 1, 0, 0)	...	(0, 0, 0.6, 0.4)	(0, 0, 0, 1)	44	59	77	180	12
9	(1, 0, 0, 0)	(0, 0, 1, 0)	(0, 0, 1, 0)	...	(0, 0, 0.4, 0.6)	(0, 0, 0.2, 0.8)	29	29	46	104	7
10	(1, 0, 0, 0)	(0, 0.8, 0.2, 0)	(0, 0, 1, 0)	...	(0, 0, 0.6, 0.4)	(0, 0, 0, 1)	15	59	61	135	9

5.6.2 The Evaluation Process

The evaluation phase is critical for any implementation as it demonstrates its success and impact. This section outlines the procedure and findings of evaluating the current work. While there are various evaluation techniques, there is no established standard for a technique or framework for game evaluation. Questionnaires and experiments are the most commonly used evaluation methods, with questionnaires being the prevailing method for game assessment according to Petri and Gresse von Wangenheim [55]. Evaluating games, especially educational games, is a complex task. Virvou points out that the evaluation of educational games should span three axes: education, entertainment, and user-friendliness of the gaming interface [60]. Assessing the context (interface, utility, plot) and educational effectiveness (knowledge acquisition, motivation, engagement) is essential for educational games. Concerning educational effectiveness, it examines whether the game provides motivation and educational benefits to students. However, the impact of educational games can be controversial, as some students may not enjoy playing games or may have difficulties exploring the virtual world, thereby limiting the student's ability to fully benefit from the learning material [68].

It is essential to assess the game's usability and appeal, as these are fundamental aspects [63]. This evaluation is crucial in light of the abundance of modern video games with complex user interfaces, which can detract from the overall gaming experience. Conversely, a digital game featuring educational content may face challenges in maintaining engagement and attractiveness. It is also vital to consider the adaptivity factor. Finally, it is crucial to determine whether the game's educational materials, such as quizzes, are well-suited to the learner's proficiency level and requirements.

Table 5.5 displays the questionnaire used to assess the demonstrated educational game. The survey questionnaire measures the game's educational effectiveness, usability, likability, and adaptivity. It consists of 21 closed-ended questions based on a Likert scale, with responses ranging from "Unsatisfactory" (1) to "Excellent" (5). Six questions assess educational effectiveness, four evaluate usability, eight gauge likability, and three explore the game's adaptivity.

Sixty participants across different age groups engaged with the game. All participants were students enrolled in the postgraduate Informatics program at the University of Piraeus. Before playing the gaming, the students had undergone a one-month classroom instruction on HTML. Subsequently, they had a three-week window to play the game, followed by completing the provided questionnaire. The evaluation results, depicted in Figs. 5.33, 5.34, 5.35, 5.36, and 5.37, reflect positive and encouraging feedback.

Concerning educational effectiveness, 31.11% of the students reported that the game significantly enhanced their understanding of HTML concepts. Additionally, 13.06% mentioned that the educational impact met their expectations, while most students (55.83%) indicated that the game's educational value exceeded their expectations. A small percentage (2.08%) encountered difficulties using the game and suggested various improvements, such as introducing a game map or implementing

Table 5.5 The questionnaire for evaluating the game

	Questions		Answers (circle one for each question) 1: unsatisfactory, 2: Improvement needed, 3: meet expectations, 4: exceeds expectations, 5: excellent				
Usability	Q1	Is the game's goal clear?	1	2	3	4	5
	Q2	Are the game's rules clear?	1	2	3	4	5
	Q3	How easy is it to navigate to the game?	1	2	3	4	5
	Q4	How easy is to play the game?	1	2	3	4	5
Likeability	Q5	How interesting is the game?	1	2	3	4	5
	Q6	How much did you like the graphics/illustrations of the game?	1	2	3	4	5
	Q7	Rate the quality of the game's multimedia elements	1	2	3	4	5
	Q8	How much did you like the particular game?	1	2	3	4	5
	Q9	The game design is attractive?	1	2	3	4	5
	Q10	Do you have fun with the game?	1	2	3	4	5
	Q11	Does gameplay motivate you to participate in it?	1	2	3	4	5
	Q12	Rate your involvement in the game?	1	2	3	4	5
Educational effectiveness	Q13	Is the learning goal clear?	1	2	3	4	5
	Q14	Do you like to use the particular game for learning the knowledge domain about HTML programming?	1	2	3	4	5
	Q15	Would you like to use the particular game as a tool for supporting your learning of other domain concepts?	1	2	3	4	5
	Q16	How interesting is the learning context (quizzes) of the game?	1	2	3	4	5
	Q17	Does the game help you to obtain new knowledge about HTML programming?	1	2	3	4	5
	Q18	Did you feel confident that you were learning, passing through the game?	1	2	3	4	5
Adaptivity	Q19	The game's plot and the continuous quizzes are interesting	1	2	3	4	5
	Q20	Rate the game's quizzes relevance to your knowledge level	1	2	3	4	5
	Q21	Do you think that the game identified each time your knowledge level?	1	2	3	4	5

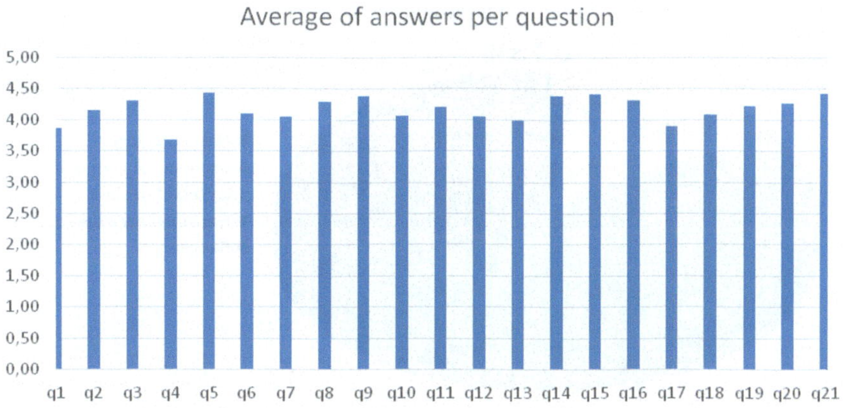

Fig. 5.33 The results of the game's evaluation

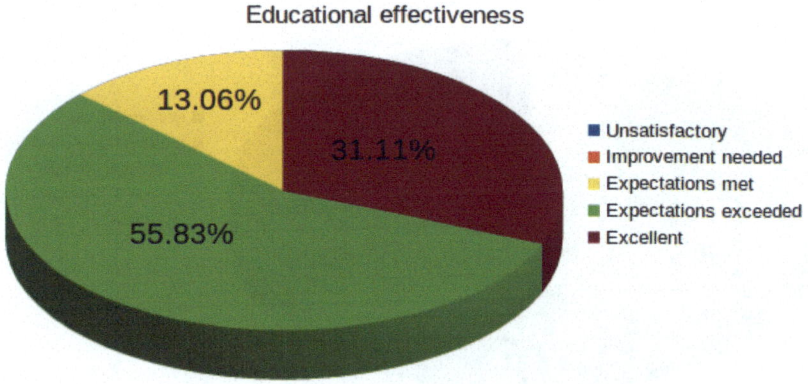

Fig. 5.34 The evaluation results related to educational effectiveness

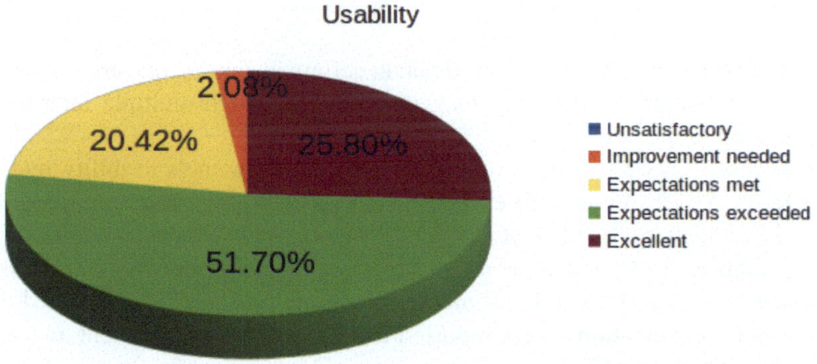

Fig. 5.35 The evaluation results related to usability

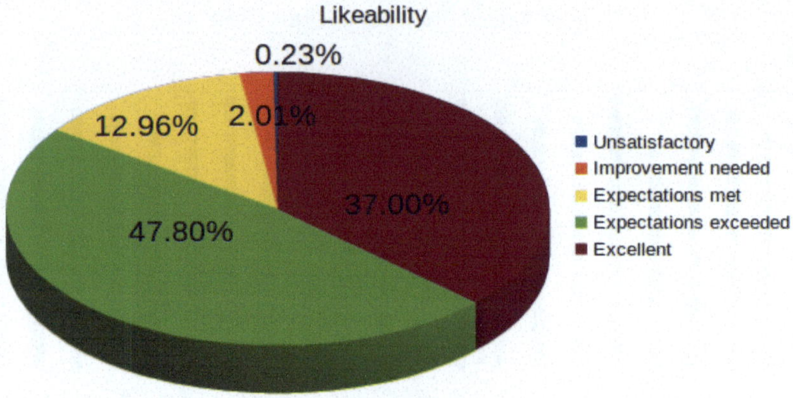

Fig. 5.36 The evaluation results related to likeability

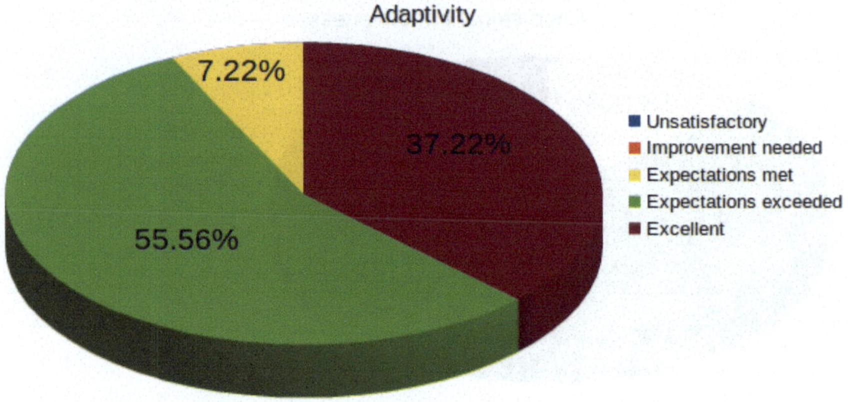

Fig. 5.37 The evaluation results related to adaptivity

easier puzzles to solve. It was observed that in certain instances, players experienced frustration or boredom when they had to navigate through multiple intermediary rooms to reach a specific location.

Furthermore, 20.42% of the students noted that the game's usability met their expectations, 51.7% found it to exceed their expectations, and 25.8% considered the game excellent. Only a few students (0.23%) expressed dissatisfaction with the game's likability. 2.01% mentioned that the game requires improvements or modifications, 12.92% mentioned that it met their expectations, and 47.8% stated that it exceeded their expectations. 37% reported that the game was excellent and served as a motivator to learn HTML.

Finally, the students evaluated the adaptivity of the quizzes, with 7.22% mentioning that the adaptive questions met their expectations, while a significant majority (55.56%) stated that the question selection was better suited than expected based on their knowledge level. Additionally, 37.22% expressed that the adaptability of the questions was excellent and tailored to their understanding of the HTML language.

5.7 Discussion

According to our research, an educational adventure game that offers personalized exercises and scenarios provides a more engaging and beneficial learning experience for the user. Our game, Fuzzy EPATHLO, uses fuzzy logic to assess the student's level of knowledge realistically. Based on this assessment, the game adjusts the difficulty level of the exercises and decides whether to introduce additional challenges, puzzles, and tasks to keep the user engaged.

The literature underscores the importance of adaptive elements in educational games and applications [4, 5, 43, 54, 59]. However, a significant gap exists in the market for educational games that offer advanced adaptive features simultaneously at both educational and gaming levels. Most educational games lack these features or have limited capabilities, focusing solely on the educational or gaming aspects [8, 25, 31, 38, 40]. This gap is where Fuzzy EPATHLO steps in, offering a game that bridges this divide and delivers a truly adaptive and engaging learning experience.

Fuzzy EPATHLO is a novel educational game bridging the connection between the knowledge level and enjoyment gap. It does so by equally focusing on both aspects and providing personalized exercises and tasks based on the learner's current knowledge level. When the player experiences learning difficulties, the game expands its scenario by providing new tasks, characters, puzzles, and exercises. Although many researchers emphasize the importance of adaptive scenarios in a game [1, 9, 32, 71], no educational game adapts its storyline based on the learner's knowledge level as Fuzzy EPATHLO does. This way, the learner's knowledge level forms the basis for a personalized learning experience tailored to their needs. By incorporating this personalized approach, Fuzzy EPATHLO enhances the player's enjoyment of the game and fosters a deeper understanding of the subject matter.

Fuzzy EPATHLO is an advanced educational suite game component that uses fuzzy logic to assess students' knowledge levels. Fuzzy logic is a robust approach that simulates human thinking and helps to make correct decisions even in cases of ambiguity and subjectivity [70]. Many educational applications [2, 11–16, 20, 22, 28, 33, 36, 39, 48, 52] and games [18, 19, 21, 29, 34, 44, 49, 56] make use of this approach. However, Fuzzy EPATHLO goes further and uses fuzzy logic to connect the student's cognitive abilities with the game. The system exploits the potential of fuzzy logic to make decisions about extending the Fuzzy EPATHLO scenario. By extending the scenario, the game motivates the players, increases their engagement, tailors tasks to their weaknesses, and helps them overcome them in an entertaining and immersive environment.

The evaluation of Fuzzy EPATHLO revealed that students enjoyed playing the game. The fuzzy logic technique also helped match the user's knowledge level. The quiz questions were adaptive and effectively catered to the user's needs, allowing them to overcome difficulties. Additionally, the expansion of the game scenario kept users engaged for extended periods. Most users expressed interest in using the game for other educational subjects. The game's innovation in linking its scenario to the student's knowledge level and expanding it under certain conditions has proven highly effective.

5.8 Conclusions

Computer games contain diverse elements such as graphics, animation, sound, and video, contributing to their appeal and widespread popularity. Moreover, their entertainment value makes them beneficial for learning various subject areas. Within this context, narrative-based adventure games are enjoyable as educational tools. Additionally, for an educational game to be effective, it should offer personalized experiences to students to improve learning outcomes.

Fuzzy EPATHLO is an educational adventure game that is part of the EPATHLO Educational Game Suite. Its main objective is to assist students in honing their knowledge of HTML programming within an engaging and interactive environment. In this game, players assume the role of a hero who explores various rooms, collects objects, and engages with NPCs. The game incorporates adaptive features to deliver a personalized educational experience tailored to the individual's current knowledge level. This is achieved using a fuzzy logic algorithm, a mathematical approach dealing with uncertainty and imprecision. The algorithm accurately assesses students' knowledge levels, which can be challenging to quantify precisely. The system adjusts both the educational content and the gameplay experience. In terms of education, it adapts the difficulty level of questions to help students address their weaknesses and enhance their knowledge. Regarding gameplay, the system introduces additional puzzles, rooms, and quizzes to bolster player motivation and provide extra practice opportunities.

The study utilized the questionnaire research method, a commonly employed approach in educational research, to gather student feedback and assess the educational aspects of the game. The questionnaire was structured to determine students' perceptions of the game's usability, appeal, educational effectiveness, and adaptability. The findings regarding the game and its educational elements are positive and encouraging. Students found the game enjoyable, and the application of fuzzy logic techniques significantly matched the users' actual knowledge levels. Additionally, the adaptive nature of the quizzes effectively met the needs of the students, helping them overcome challenges while expanding the game's scenarios and maintaining user engagement. Lastly, most users expressed interest in using the game to practice additional educational subjects, including Mathematics, Physics, and Chemistry.

The study significantly contributes to educational software by introducing an innovative tool. It uses fuzzy logic to develop dynamic student models that accurately reflect their knowledge and learning requirements. These models then adjust game quizzes and modify the game scenario in real time, providing a customized and responsive learning experience. This unique approach allows students to receive additional training without sacrificing the surprise and motivation in the game. The research findings underscore the enormous potential of educational games that combine adventure, fuzzy logic reasoning, and adaptability to enhance learning.

References

1. Aguilar, J., Altamiranda, J., Díaz, F., de Mesa, J.G., Pinto, Á.: Adaptive plot system for serious emerging games based on the ant colony optimization algorithm. In: 2019 XLV Latin American Computing Conference (CLEI), pp. 1–10. IEEE (2019)
2. Al Duhayyim, M., Newbury, P.: Concept-based and fuzzy adaptive e-learning. In: Proceedings of the 3rd International Conference on Information and Education Innovations, pp. 49–56 (2018)
3. Almasri, A., Ahmed, A., Almasri, N., Abu Sultan, Y.S., Mahmoud, A.Y., Zaqout, I.S., Akkila, A.N., Abu-Naser, S.S.: Intelligent tutoring systems survey for the period 2000–2018 (2019)
4. Anania, J.: The Effects of Quality of Instruction on the Cognitive and Affective Learning of Students. The University of Chicago (1981)
5. Anania, J.: The influence of instructional conditions on student learning and achievement. Eval. Educ. **7**(1), 1–92 (1983)
6. Brusilovsky, P., Anderson, J.: Act-r electronic bookshelf: an adaptive system to support learning act-r on the web (1998)
7. Bursilovsky, P., Eklund, J., Schwarz, E.: Web-Based Education for all: a tool for developing adaptive courseware in proceedings of the 7th international world wide web conference (www'7). Comput. Netw. ISDN Syst. **30**(1–7), 291–300 (1998)
8. Capuano, N., King, R.: Adaptive serious games for emergency evacuation training. In: 2015 International Conference on Intelligent Networking and Collaborative Systems, pp. 308–313. IEEE (2015)
9. Carro, R.M., Breda, A.M., Castillo, G., Bajuelos, A.L.: A methodology for developing adaptive educational-game environments. In: Adaptive Hypermedia and Adaptive Web-Based Systems: Second International Conference, AH 2002 Málaga, Spain, 29–31 May 2002. Proceedings, vol. 2, pp. 90–99. Springer (2002)
10. Carton, C.: The history of the adventure video game (2023)
11. Chrysafiadi, K., Virvou, M.: Evaluating the integration of fuzzy logic into the student model of a web-based learning environment. Expert Syst. Appl. **39**(18), 13127–13134 (2012)
12. Chrysafiadi, K., Virvou, M.: Using fuzzy cognitive maps for the domain knowledge representation of an adaptive e-learning system. In: JCKBSE, pp. 257–265 (2012)
13. Chrysafiadi, K., Virvou, M.: Fuzzy logic for dynamic adaptation: Represent and manage changeable user states. In: IISA 2013, pp. 1–6. IEEE (2013)
14. Chrysafiadi, K., Virvou, M.: Fuzzy logic for adaptive instruction in an e-learning environment for computer programming. IEEE Trans. Fuzzy Syst. **23**(1), 164–177 (2014)
15. Chrysafiadi, K., Virvou, M.: Create dynamically adaptive test on the fly using fuzzy logic. In: 2018 9th International Conference on Information, Intelligence, Systems and Applications (IISA), pp. 1–8. IEEE (2018)
16. Chrysafiadi, K., Virvou, M., et al.: Advances in Personalized Web-Based Education. Springer (2015)
17. Chrysafiadi, K., Papadimitriou, S., Virvou, M.: Which is better for learning: a web-based educational application or an educational game? In: 2019 International Symposium on Performance Evaluation of Computer and Telecommunication Systems (SPECTS), pp. 1–6. IEEE (2019)

18. Chrysafiadi, K., Papadimitriou, S., Virvou, M.: Fuzzy states for dynamic adaptation of the plot of an educational game in relation to the learner's progress. In: 2020 11th International Conference on Information, Intelligence, Systems and Applications (IISA), pp. 1–7. IEEE (2020)
19. Chrysafiadi, K., Papadimitriou, S., Virvou, M.: Cognitive-based adaptive scenarios in educational games using fuzzy reasoning. Knowl. Based Syst. **250**, 109111 (2022)
20. Chrysafiadi, K., Virvou, M., Tsihrintzis, G.A.: A fuzzy-based mechanism for automatic personalized assessment in an e-learning system for computer programming. Intell. Decis. Technol. (Preprint), 1–16 (2022)
21. Chrysafiadi, K., Kamitsios, M., Virvou, M.: Fuzzy-based dynamic difficulty adjustment of an educational 3d-game. Multimed. Tools Appl. 1–25 (2023)
22. Chrysafiadi, K.C.: Modeling the learner's cognitive state transitions using fuzzy logic techniques for adaptive learning. Ph.D thesis, University of Piraeus (Greece) (2014)
23. De Bra, P., Ruiter, J.P.: Aha! adaptive hypermedia for all. In: WebNet, pp. 262–268 (2001)
24. De Bra, P.M.: Teaching Hypertext and Hypermedia Through the Web. ERIC (1996)
25. D'Apice, C., Grieco, C., Piscopo, R., Liscio, L.: Dms2015short-2: advanced learning technologies for elearning in the enterprise: Design of an educational adventure game to teach computer security. J. Vis. Lang. Comput. **31**, 260–266 (2015)
26. Fernández-Vara, C.: 29 adventure. In: The Routledge Companion to Video Game Studies (2014)
27. Fernández-Vara, C.: Adventure. In: The Routledge Companion to Video Game Studies, Routledge, pp. 232–240 (2014)
28. Goel, G., Lallé, S., Luengo, V.: Fuzzy logic representation for student modelling: case study on geometry. In: Intelligent Tutoring Systems: 11th International Conference, ITS 2012, Chania, Crete, Greece, 14–18 June 2012. Proceedings, vol. 11, pp. 428–433. Springer (2012)
29. Hamdaoui, N., Idrissi, M.K., Bennani, S.: Learner modeling in educational games based on fuzzy logic and gameplay data. Int. J. Game Based Learn. (IJGBL) **11**(2), 38–60 (2021)
30. Henze, N., Naceur, K., Nejdl, W., Wolpers, M.: Adaptive hyperbooks for constructivist teaching. KI **13**(4), 26–31 (1999)
31. Hocine, N., Gouaïch, A., Cerri, S.A., Mottet, D., Froger, J., Laffont, I.: Adaptation in serious games for upper-limb rehabilitation: an approach to improve training outcomes. User Model. User-Adap. Inter. **25**, 65–98 (2015)
32. Hussaan, A.M., Sehaba, K., Mille, A.: Tailoring serious games with adaptive pedagogical scenarios: A serious game for persons with cognitive disabilities. In: 2011 IEEE 11th International Conference on Advanced Learning Technologies, pp. 486–490. IEEE (2011)
33. Hwang, G.J., Sung, H.Y., Chang, S.C., Huang, X.C.: A fuzzy expert system-based adaptive learning approach to improving students' learning performances by considering affective and cognitive factors. Comput. Educ. Artif. Intell. **1**, 100003 (2020)
34. Ivan, C., Rumondor, P.C., Ricky, M.Y., Yossy, E.H., Budiharto, W.: Help the math town: adaptive multiplayer math-science games using fuzzy logic. Proc. Comput. Sci. **116**, 309–317 (2017)
35. Kamitsios, M., Papadimitriou, S., Chrysafiadi, K., Virvou, M.: Incorporating quizzes into a 3d educational game calling web services. In: EdMedia+ Innovate Learning, Association for the Advancement of Computing in Education (AACE), pp. 1170–1179 (2018)
36. Karaci, A.: Intelligent tutoring system model based on fuzzy logic and constraint-based student model. Neural Comput. Appl. **31**(8), 3619–3628 (2019)
37. Katsionis, G., Virvou, M.: A cognitive theory for affective user modelling in a virtual reality educational game. In: 2004 IEEE International Conference on Systems, Man and Cybernetics (IEEE Cat. No. 04CH37583), vol. 2, pp. 1209–1213. IEEE (2004)
38. Lindberg, R.S., Laine, T.H.: Formative evaluation of an adaptive game for engaging learners of programming concepts in k-12. Int. J. Serious Games **5**(2), 3–24 (2018)
39. Machado, M.A.S., Moreira, T.D.R.G., Gomes, L.F.A.M., Caldeira, A.M., Santos, D.J.: A fuzzy logic application in virtual education. Proc. Comput. Sci. **91**, 19–26 (2016)
40. Mees, M., Jay, T., Habgood, J.: Designing an Adaptive Learner Model for a Mathematics Game. Academic Conferences and Publishing International Limited (2018)

41. Mougiakou, E., Virvou, M.: Based on gdpr privacy in uml: Case of e-learning program. In: 2017 8th International Conference on Information, Intelligence, Systems & Applications (IISA), pp. 1–8. IEEE (2017)
42. Mougiakou, E., Papadimitriou, S., Virvou, M.: Automated decision making and personal data protection in intelligent tutoring systems: Design guidelines. In: Knowledge-Based Software Engineering: 2018: Proceedings of the 12th Joint Conference on Knowledge-Based Software Engineering (JCKBSE 2018) Corfu, Greece, vol. 12, pp. 231–241. Springer (2019)
43. Nwana, H.S.: Intelligent tutoring systems: an overview. Artif. Intell. Rev. **4**(4), 251–277 (1990)
44. Ozdemir, A., Balbal, K.F.: Fuzzy logic based performance analysis of educational mobile game for engineering students. Comput. Appl. Eng. Educ. **28**(6), 1536–1548 (2020)
45. Panagoulias, D.P., Virvou, M., Tsihrintzis, G.A.: Tailored explainability in medical artificial intelligence-empowered applications: Personalisation via the technology acceptance model. In: 2023 IEEE 35th International Conference on Tools with Artificial Intelligence (ICTAI), pp. 486–490. IEEE (2023)
46. Panagoulias, D.P., Virvou, M., Tsihrintzis, G.A.: A novel framework for artificial intelligence explainability via the technology acceptance model and rapid estimate of adult literacy in medicine using machine learning. Expert Syst. Appl. **248**, 123375 (2024)
47. Papadimitriou, S., Chrysafiadi, K., Kamitsios, M., Virvou, M.: Fsp creator: A novel web service api creator of fuzzy students progress profile. In: 2018 9th International Conference on Information, Intelligence, Systems and Applications (IISA), pp. 1–7. IEEE (2018)
48. Papadimitriou, S., Chrysafiadi, K., Virvou, M.: Authoring technological and platform independent learning material and student's progress profile using web services. In: Knowledge-Based Software Engineering: 2018: Proceedings of the 12th Joint Conference on Knowledge-Based Software Engineering (JCKBSE 2018) Corfu, Greece, vol. 12, pp. 242–251. Springer (2019)
49. Papadimitriou, S., Chrysafiadi, K., Virvou, M.: Evaluating the use of fuzzy logic in an educational game for offering adaptation. In: 2019 International Conference on Computer, Information and Telecommunication Systems (CITS), pp. 1–5. IEEE (2019)
50. Papadimitriou, S., Chrysafiadi, K., Virvou, M.: Fuzzeg: Fuzzy logic for adaptive scenarios in an educational adventure game. Multimed. Tools Appl. **78**(22), 32023–32053 (2019). https://doi.org/10.1007/s11042-019-07955-w
51. Papadimitriou, S., Kamitsios, M., Chrysafiadi, K., Virvou, M.: Learn-and-play personalised reasoning from point-and-click to virtual reality mobile educational games. Intell. Decis. Technol. **15**(2), 321–332 (2021)
52. Papadimitriou, S., Chrysafiadi, K., Virvou, M.: Adaptive quizzes using fuzzy genetic algorithm. In: 2023 14th International Conference on Information, Intelligence, Systems & Applications (IISA), IEEE, pp 1–8 (2023)
53. Papanikolaou, K.A., Grigoriadou, M., Kornilakis, H., Magoulas, G.D.: Inspire: an intelligent system for personalized instruction in a remote environment. In: Hypermedia: Openness, Structural Awareness, and Adaptivity: International Workshops OHS-7, SC-3, and AH-3 Aarhus, Denmark, 14–18 Aug 2001. Revised Papers, vol. 3, pp. 215–225. Springer (2002)
54. Paviotti, G., Rossi, P.G., Zarka, D., et al.: Intelligent tutoring systems: an overview. Pensa Multimedia, pp. 1–176 (2012)
55. Petri, G., von Wangenheim, C.G.: How games for computing education are evaluated? A systematic literature review. Comput. Educ. **107**, 68–90 (2017)
56. Pirovano, M., Lanzi, P.L.: Fuzzy tactics: a scripting game that leverages fuzzy logic as an engaging game mechanic. Expert Syst. Appl. **41**(13), 6029–6038 (2014)
57. Politou, E., Alepis, E., Virvou, M., Patsakis, C.: Privacy and Data Protection Challenges in the Distributed Era. Springer International Publishing (2022)
58. Specht, M., Klemke, R.: Ale-adaptive learning environment. In: World Conference on the WWW and Internet, Association for the Advancement of Computing in Education, pp. 1155–1160 (2001)
59. Streicher, A., Smeddinck, J.D.: Personalized and adaptive serious games. In: Entertainment Computing and Serious Games: International GI-Dagstuhl Seminar 15283. Dagstuhl Castle, Germany, 5–10 July 2015, pp. 332–377. Springer, Revised Selected Papers (2016)

60. Virvou, M.: On the evaluation of the combined role of virtual reality games and animated agents in edutainment. Intell. Comput. Graph. **2011**, 79–95 (2012)
61. Virvou, M., Katsionis, G.: Virge: Tutoring english over the web through a game. In: Proceedings 3rd IEEE International Conference on Advanced Technologies, p. 469. IEEE (2003)
62. Virvou, M., Katsionis, G. Web services for an intelligent tutoring system that operates as a virtual reality game. In: SMC'03 Conference Proceedings. 2003 IEEE International Conference on Systems, Man and Cybernetics. Conference Theme-System Security and Assurance (Cat. No. 03CH37483), vol. 1, pp. 378–383. IEEE (2003)
63. Virvou, M., Katsionis, G.: On the usability and likeability of virtual reality games for education: the case of vr-engage. Comput. Educ. **50**(1), 154–178 (2008)
64. Virvou, M., Manos, K.: A simulated student-player in support of the authoring process in a knowledge-based authoring tool for educational games. In: Proceedings 3rd IEEE International Conference on Advanced Technologies, pp. 338–339. IEEE (2003)
65. Virvou, M., Tsihrintzis, G.A.: A novel trust state-chart model for requirements engineering of trustful ai-empowered software. In: 2023 14th International Conference on Information, Intelligence, Systems & Applications (IISA), pp. 1–6. IEEE (2023)
66. Virvou, M., Manos, K., Katsionis, G., Tourtoglou, K.: Vr-integate: a knowledge-based authoring tool for virtual reality educational games. In: Proceedings of ED-MEDIA 2002 World Conferences on Educational Multimedia, Hypermedia and Telecommunications (2002)
67. Virvou, M., Katsionis, G., Manos, K.: On the interaction features of the virtual reality user interface of an educational game. In: EdMedia+ Innovate Learning, Association for the Advancement of Computing in Education (AACE), pp. 2170–2174 (2004)
68. Virvou, M., Katsionis, G., Manos, K.: Combining software games with education: Evaluation of its educational effectiveness. J. Educ. Technol. Soc. **8**(2), 54–65 (2005)
69. Virvou, M., Tsihrintzis, G.A., Tsichrintzi, E.A.: Virtsi: a novel trust dynamics model enhancing artificial intelligence collaboration with human users-insights from a chatgpt evaluation study. Inf. Sci. **675**, 120759 (2024)
70. Zadeh, L.A.: Fuzzy logic=computing with words. IEEE Trans. Fuzzy Syst. **4**(2), 103–111 (1996)
71. Zarraonandía, T., Díaz, P., Aedo, I.: Modeling games for adaptive and personalized learning: the a-grem. In: Learning Designs for Emerging Pedagogies, The Future of Ubiquitous Learning, pp. 217–239 (2016)

Chapter 6
Playing Educational Games in Web–Reusability–Interoperability

Abstract Recently, digital games have gained popularity as a tool for education. By creating an immersive environment, these games generate positive emotions that encourage players to become more actively involved in the learning process. However, every player has individual preferences in terms of graphics and mechanics, given the variety of educational games available. The same educational content needs to be integrated across different categories of games to maximize engagement and educational outcomes. However updating educational content across games can be complex. This chapter describes a mechanism that uses web services to make educational content portable across two types of mobile games (point-and-click games and virtual reality games). This innovative approach ensures that each game category and platform can access the same educational content while exploiting the unique features of each game to inspire and engage students. Evaluation of this system shows that integrating the same educational content across different games positively impacts educational outcomes.

Keywords Educational software personalized educational games · Artificial Intelligence in educational games · Fuzzy logic in education · User experience · AI-empowered software games · User modeling · Human-AI interaction · e-learning games · 21st-century skills · Human-centred Artificial Intelligence

6.1 Introduction

Several types of educational applications exist, ranging from basic online tutorials, tests, and practice software to comprehensive web-based applications and educational games [10, 11, 32, 39, 81]. Educational applications can incorporate multimedia elements such as images, video, and audio, creating an immersive visual environment that engages the student. These applications can be window-based (desktop), browser-based, or mobile, each with its own specific capabilities and limitations [34]. Desktop applications run on standard computers with specific operating systems and typically require more storage due to multimedia elements like images, videos, and

S. Papadimitriou and M. Virvou, *Artificial Intelligence—Based Games as Novel Holistic Educational Environments to Teach 21st Century Skills*, Intelligent Systems Reference Library 93, https://doi.org/10.1007/978-3-031-77464-5_6

audio. Browser-based applications require Internet access for accessibility from any location and are independent of operating systems, needing only a browser for execution. These applications often utilize a database containing data related to learning objectives and student information. Mobile applications provide accessibility from almost any location, allowing students to access learning applications from multiple places [3, 39, 47, 75].

Two popular categories of educational applications are e-learning platforms, such as Moodle [2, 5, 19]), and Intelligent Tutoring Systems (ITSs) [7, 13, 18, 25]. E-learning platforms are typically web-based and require access to the internet and a compatible browser. ITSs utilize intelligent techniques to provide personalized student guidance. They can be desktop or web applications. The primary advantage of a web-based application is its availability and accessibility anytime and anywhere [36]. Educators can also import teaching materials and track student progress using corresponding authoring tools [1, 20, 45, 50, 59, 76].

Educational games, on the other hand, are powerful tools that combine learning with fun to enhance student motivation. They are educational applications embedded within a gaming context and used simultaneously for entertaining and educational purposes. Many researchers advocate integrating educational games into the educational process, as their immersive nature helps to maintain high levels of motivation among players [16, 17, 48, 51]. The potential of educational games to unleash the learner's imagination and facilitate skill acquisition through enjoyable and engaging gameplay is a promising and exciting aspect of modern education.

Many researchers emphasize the significance of educational applications integrating intelligence to customize educational content and guidance according to each learner's needs. These techniques identify students' requirements, learning styles, knowledge levels, errors, and misconceptions, among other features [12, 48, 51, 69, 83]. In many cases, they also aim to perform automatic reasoning to provide help and make the applications more human-like and personalized [14, 71, 73, 77, 78, 82].

Reusability is achieved through authoring tools that enable the creation of adaptable and versatile educational content. Virvou et al. [84] describe VR-INTEGATE, a tool for developing immersive Virtual Reality (VR) educational games. Virvou and Moundridou [80] highlight an algebra tool that provides personalized tutoring. Moundridou and Virvou [44] discuss web-based ITS authoring, which expands educational access and customization. Virvou and Alepis [76] emphasize mobile features for learning on the go. Virvou [73] explores tools that align content with cognitive processes, enhancing system responsiveness, while Virvou [72] presents a domain-independent reasoning mechanism, increasing versatility across various subjects.

However, differences among students exist regarding their perception and processing of information within a particular subject. They also have distinct preferences for educational applications and the games they enjoy. The immersive context of a digital game elicits positive emotions for the player, motivating them to participate actively in the educational experience. However, educational video games vary in visuals and controls. Hence, not all games cause the same feelings for every player.

Players have specific preferences regarding the aesthetics and the types of games they want to play. Each player may have different preferences for game mechanics and categories [8, 54, 61, 63]. Even when a game adapts its educational content and gameplay style and benefits students, it may appeal to only some. Two games can teach the same subject. One may provide better and more personalized learning material, and the other may be more fun. In this case, there is a dilemma about which game the learner will use: the one that provides better education or the one that provides more fun.

Thus, developing games across different categories but with the same teaching domain is critical while maintaining the same educational content. The student's knowledge and progress must remain the same, regardless of the game chosen. In this case, the student knows that whatever game they choose will have the same personalized educational content. The student's choice of game is, therefore, more straightforward, as their focus is on the entertainment value of each game. Integrating the same educational content across different game categories is not just a matter of consistency, but it is crucial for enhancing student engagement and improving learning outcomes. Failing to do so may lead to a game that does not align with the student's knowledge, ultimately failing to motivate and improve the player's cognitive level. Therefore, a significant challenge is ensuring that the educational material is accessible from any game the student plays.

Designing educational content for all games (or educational applications) can be challenging and time-consuming. Integrating educational theory and adaptive exercises into multiple games requires significant effort. When updating the educational section of a game, it is crucial to ensure that other games are also updated accordingly. This can add complexity and cost to software development. Ensuring that different games provide students with an equivalent educational experience and results necessitates consistent underlying business logic and algorithms. Having to rewrite the same algorithm for various development platforms targeting different games increases the time and effort required by the programmer. An incorrect implementation of the same algorithm in one of the games leads to inconsistencies in the educational material compared to other games. Delivering inappropriately adapted educational material in a game is not only a technical issue but can significantly impact the student's learning experience, emphasizing the developers' crucial role in game development.

6.2 Software Engineering Aspects and Related Work

6.2.1 Functionality of Educational Games: A Review of the Literature

Educational games have become a transformative tool in education, merging entertainment with pedagogy to create engaging learning environments. These games utilize various technologies, including Artificial Intelligence (AI), Virtual Reality (VR), and data analytics, to provide personalized and interactive learning experiences across multiple disciplines.

Player Agency and Engagement

Player agency is an important component of educational games, enhancing engagement and motivation. Joshi et al. [29] explore how giving players control over their actions in educational games leads to increased investment in the learning process. This autonomy helps students develop critical thinking and decision-making skills, which are essential for effective learning.

Immersive and AI Technologies

The integration of immersive technologies and AI has revolutionized the landscape of educational games. Rapaka et al. [55] discuss how VR and AI create interactive learning environments that adapt to individual learner needs. These technologies allow for realistic simulations and personalized feedback, enhancing the educational impact of games.

Evolution and Research in Serious Games

Özyurt et al. [46] provide a topic modeling perspective on the evolution of educational serious games, highlighting their growing complexity and educational value. This evolution includes incorporating narrative elements, real-world problem-solving, and collaborative tasks that enhance learning outcomes.

Language Learning and Assessment

Educational games play a significant role in language learning. Khaleh Sar [57] examines how digital games facilitate language acquisition in EFL (English as Foreign Language) contexts by providing immersive and interactive environments. These games help learners practice language skills in a context-rich setting, enhancing their language proficiency.

Physical Activity and Holistic Learning

Games that promote physical activity are gaining attention for their holistic educational approach. San Blas et al. [60] describe a platform that uses interactive games to encourage physical movement and learning, addressing both cognitive and physical development. This dual focus supports healthier and more engaged learners.

Learning Outcomes and Effectiveness

Ren et al. [56] conduct a meta-analysis demonstrating educational games' positive impact on learning outcomes across various subjects. These games engage students through interactive experiences, making learning more effective and enjoyable compared to traditional methods.

Accessibility and Inclusivity

Hassan [26] emphasizes the importance of accessibility in educational games, ensuring they are inclusive for learners with disabilities. By incorporating inclusive design principles, educational games can provide equitable learning opportunities for all students, regardless of their abilities.

Design Principles and Cognitive Skills

Effective educational game design balances educational content with engaging gameplay mechanics. Shang and Zhou [62] discuss the principles of scientific game design, emphasizing the need to align games with curriculum objectives while maintaining high levels of engagement. Chou et al. [9] highlight how cognitive-based game mechanisms promote critical thinking and problem-solving skills.

Data-Driven Personalization

Lu et al. [41] explore how big data can inform the design of educational games, allowing for personalized learning experiences. Data analytics provide insights into learner behavior, enabling developers to tailor content and game mechanics to individual needs, thus enhancing educational impact.

Interdisciplinary Applications

Educational games are applied across various disciplines, from computing to environmental education. Zimbrão et al. [89] discuss the use of digital games to teach computing concepts, while Kristic et al. [35] highlight holographic technology's role in raising environmental awareness through games.

Cultural and Ethical Learning

Games also serve as platforms for exploring cultural and ethical issues. Hutson and Fulcher [28] describe a VR game that addresses cultural heritage repatriation ethics, providing a nuanced exploration of complex topics through interactive gameplay.

Game Design and Development Methodologies

Archuby et al. [4] introduce the DIJS methodology for designing and developing educational serious games, emphasizing iterative design and collaboration among educators, designers, and learners. This approach ensures that games are both pedagogically sound and engaging.

Educational Game Genres and Effectiveness

López-Fernández et al. [40] compare the effectiveness of different educational game genres in computer science education, highlighting the strengths and weaknesses of various approaches. Understanding genre effectiveness helps educators choose the right type of game for specific learning objectives.

Gamification and Motivation

Gamification elements, such as rewards and challenges, are essential to maintain motivation and engagement in educational games. Tyni et al. [70] explore how game reward types influence player motivation, providing insights into designing more effective educational experiences.

Computational Thinking and Development

Sun et al. [64] conduct a meta-analytic review, demonstrating how educational games promote the development of students' computational thinking. These games challenge students to apply logic, creativity, and analytical skills, preparing them for the demands of the digital age.

Heuristics and Language Learning

Marques and Miranda [43] provide heuristics to evaluate optimal experiences in educational games designed for learning Japanese as a second language. Their work underscores the importance of tailored educational game design in language acquisition.

Educational Game Design for Lifestyle and Health

Azizi-Soleiman et al. [6] discuss the development of educational-therapeutic games aimed at improving children's healthy lifestyles. By integrating health education with engaging gameplay, these games encourage positive behavioral changes in young learners.

Exploring Gamification and Digital Economy

Lampropoulos [37] highlights the role of educational technology and gamification in cognitive and social-emotional development, focusing on extended reality and learning analytics. This work demonstrates how gamification strategies can be used to prepare students for the digital economy.

Predictive Modeling in Educational Games

Hooshyar et al. [27] explore predictive modeling in educational computer games to anticipate student performance. By using AI and data analytics, educators can provide timely interventions and personalized learning paths.

Educational Game Development in Production Engineering

Machado et al. [42] developed a database of educational games applicable to production engineering, illustrating the importance of aligning game content with industry-specific skills and knowledge.

AI and Student Performance Prediction

Gomez et al. [24] utilize explainable AI to enhance real-time student performance prediction in serious games, providing educators with valuable insights into student learning patterns and needs.

Reward Mechanisms and Game Effectiveness

Wang [86] examines the effects of reward mechanisms in online educational games on learners' motivation, providing insights into designing effective motivational strategies in game-based learning.

Teaching-Learning Support Through Digital Games

De Oliveira Zimbrão et al. [89] highlight the essential aspects of digital educational games in supporting teaching-learning processes, emphasizing the need for thoughtful integration of digital content into educational contexts.

Scientific Studies in Mathematics Education

Düsünen and Ata [21] conduct a bibliometric analysis of educational games used in mathematics education, providing insights into trends and gaps in current research.

Educational Games for Computational Thinking

Tikva et al. [66] evaluate the scaffolded aMazeD game, designed to develop computational thinking skills, highlighting its effectiveness in promoting logical reasoning and problem-solving.

Virtual Reality for Ethics Education

Hutson and Fulcher [28] describe a VR game that explores the ethics of cultural heritage repatriation, providing an interactive platform for students to engage with complex ethical dilemmas.

Educational games offer a unique blend of entertainment and education, making learning more engaging and effective. By employing advanced technologies and innovative design principles, these games provide dynamic learning environments that cater to diverse learner needs. As the field continues to evolve, educational games will play an increasingly important role in shaping the future of education.

6.2.2 Programming Educational Games

Programming the educational part of all games (or educational applications) can be complex and time-consuming. It takes substantial time to incorporate theory and adaptive exercises into numerous games. The burden of ensuring that other games are also aware of these modifications can increase the complexity and cost of software development. However, with the right technological solution, the relief from the burden of repetitive tasks can be significant. Employing the same underlying business logic and algorithms is essential to ensure that two distinct games offer an equal educational experience and produce similar results for the learner. Rewriting the same algorithm for different development platforms that target various games leads to increased time and effort for the programmer. Moreover, an inaccurate algorithm implementation in one game results in inconsistency in the educational content compared to the other games. Providing inappropriately adapted educational material in a game is not just a technical issue, but it can have a significant negative impact on the learning benefit for the student, underscoring the importance of the developer's role in game development.

The above issue necessitates a technological solution that decreases the development workload and integrates learning materials into any game in a standardized format with minimal human intervention. The key to overcoming this obstacle lies in leveraging web services technology [38, 53, 58]. Web services comprise a collection of functions acting as a unified entity, accessible to external applications [31, 50]. They promote interoperability [22, 87, 88] between applications and streamline development by utilizing the same programming language. Due to these significant advantages, web services technology is widely adopted across various applications. This efficient solution can instill confidence in developers and educational technology researchers alike, affirming that they are moving in the right direction.

In addition, an educational application with web services enables external programs to access various educational components, including theory, tutorials, exercises, assignments, student models, and personalized content features tailored to each student's profile. Allowing students to access the learning materials when needed can significantly enhance their learning experience. Participating in specific game categories can help students better understand the curriculum or actively engage in the learning process. Additionally, some students prefer using a computer, while others opt for a mobile device. Therefore, delivering the same educational resources across different game types and platforms is crucial to ensure a high-quality learning experience for a diverse student population. However, there are limited examples of educational software applications utilizing web services in the current literature. For instance, Kabassi and Virvou developed a multi-agent learning environment using web services to assist users in managing their file repositories [30]. This environment allows students to use the system's features and capabilities regardless of their computing device. However, their work focuses not on the benefits of web services but on an application's ability to access its data online. They illustrate the scenario where a user in a laboratory may need simultaneous access to their model from a different computer. It is worth noting that their work relates to an educational application that is not a game.

Thyagharajan and Nayak developed a web-based tutoring system using web services and an educational application that leveraged the web services of the former to utilize its adaptive features [65]. Again, this work is not an educational game but a web-based application. The authors discuss the advantages of sharing educational material across different applications, and their application stands out as the sole user of their web services to demonstrate the value of sharing educational content. Regarding educational games, Virvou and Katsionis employed web services technology in a VR adventure game that constructs the student's profile based on their affective characteristics [33]. The innovation in this work lies in the student's modeling occurring on a central server, and each player utilizes the game (client) to connect to this server. In addition, the tutor can access each student's profile and monitor their progress. However, this is the only game that utilizes web services technology to showcase the transferability of the student profile across games and the students' benefits in choosing their preferred game type.

This chapter introduces the EPATHLO Student Profile Creator, a vital component of the Educational Game Suite. The EPATHLO Student Profile Creator is an abstract learning mechanism that effectively addresses the challenge of delivering personalized education through a centralized platform. It ensures that the educational material is tailored to the specific needs of each student, thereby enhancing the learning experience and improving the overall quality of education. The mechanism's flexibility and adaptability to different games allow for seamless integration and compatibility with various educational games, providing a robust and efficient solution to the complex problem of educational game development.

The following sections describe the EPATHLO Student Profile Creator, used by several educational components of the Educational Game Suite. The EPATHLO Student Profile Creator uses web services to provide the ability to transfer the same learning content to different types of games. Thus, any external game can communicate with the suite and access the same educational material and student models. In this chapter, we present the mobile version of Fuzzy EPATHLO, the leading game integrated into the EPATHLO Educational Game Suite, and Surviving Business Man, an external 3D VR game that takes advantage of the capabilities of the EPATHLO Inter-Game Module. This component facilitates communication and data sharing between different games. The two games are not and do not need to be aware of student modeling mechanisms and personalized educational content generation.

The games operate independently from the educational activities, aiming to create an engaging environment that fosters students' participation and enthusiasm while seamlessly incorporating educational content. This innovative approach allows the educational content and mechanism to be easily integrated into any game. As a result, regardless of the game category or platform, each game leverages its unique characteristics to motivate learners and enhance their engagement in the learning process.

6.3 The Description of the Abstract Learning Mechanism

The EPATHLO Student Profile Creator, as depicted in Fig. 6.1, is built on an abstract learning mechanism that mirrors a typical ITS architecture. However, it stands out with its unique capabilities, such as calculating a student's knowledge level and generating personalized exercises. The domain model encompasses courses, their corresponding theory and exercises stored in the database, and a tool to evaluate student answers. The user model incorporates a fuzzy logic mechanism that computes students' knowledge levels based on their exercise performance. Both the exercises and the student's answers are stored in the database. The virtual tutor model interacts with the domain and user models to select suitable questions tailored to the student's knowledge level.

One of the key features that sets the EPATHLO Student Profile Creator apart from a typical ITS is that it does not communicate directly with the user interface model but with a component called the Game Type Selector. This allows the student-player to select the game they want to play, and the selected game then provides its UI for interaction. The communication between the game and the EPATHLO Student Profile Creator is facilitated via web services, ensuring both have internet access. The EPATHLO Student Profile Creator provides the game with learning material,

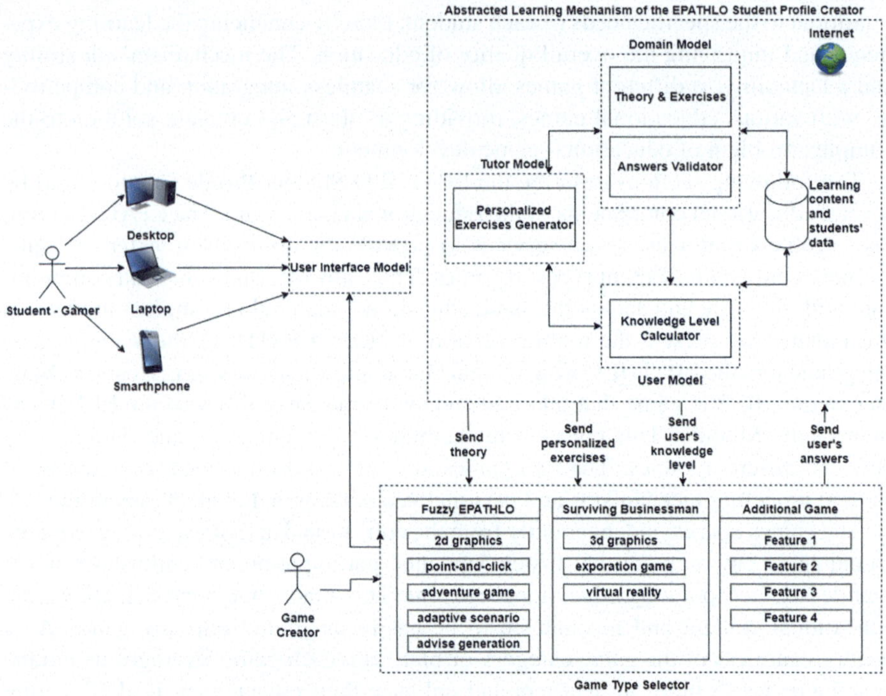

Fig. 6.1 The architecture of EPATHLO student profile creator

personalized exercises, and the student's current knowledge level. The game then displays and uses the information as designed and implemented by the developer. It also collects the student's answers to an exercise through its user interface and sends them to the EPATHLO Student Profile Creator. The latter checks their correctness and reassesses the student's knowledge level. When the game requests a new exercise, the EPATHLO Student Profile Creator creates one, considering the student's knowledge level, and delivers it to the game. Each game can also request the student's current knowledge level from the EPATHLO Student Profile Creator and add its additional adaptive features. For example, Fuzzy EPATHLO examines the knowledge level determined after completing a quiz and extends its scenario to motivate the player to continue playing and practicing.

Therefore, according to the figure, the whole educational part is within the abstract learning mechanism, and no game is involved in the educational functionalities of the mechanism. Thus, any game developer can emphasize the explicit distinct motivational features of the game and use them to keep the player's interest at a high level. For example, Fuzzy EPATHLO uses two-dimensional graphics, is point-and-click, and is perfect for adventure game enthusiasts. On the other hand, The Surviving Businessman is an exploration game that uses 3D graphics and VR technology. Both games communicate with the abstract learning mechanism and receive the same personalized educational content.

In addition, each game can incorporate additional educational features and enhance the learning experience. For example, Fuzzy EPATHLO modifies the storyline according to the student's level of knowledge or gives recommendations on studying specific curriculum sections when the student encounters difficulties. In other words, the tool for generating hints, and when this tool judges that it is necessary to give advice, it is part of Fuzzy EPATHLO and not of EPATHLO Student Profile Creator. In this way, Fuzzy EPATHLO enhances the educational process by using the capabilities of EPATHLO Student Profile Creator and adding new educational features. When the student chooses to study the parts of the syllabus recommended by Fuzzy EPATHLO, the latter retrieves the material from the EPATHLO Student Profile Creator and presents it on the student's screen. This educational approach not only enhances the learning process but also inspires students to continue their educational journey.

Following the information above, an external application, regardless of its implementation platform, can utilize the methods offered by the EPATHLO Student Profile Creator API (Table 6.1). These applications can be web-based, windowed, educational, or even game-like. EPATHLO Student Profile Creator empowers the application to integrate educational resources and exercises in the form of quizzes within its framework. The API operates using RESTful web services and encompasses methods for user registration, user authentication, quiz creation for specific courses, submission of a student's answer to the latest quiz question, and retrieval of information about the student's profile. The student's profile and data are stored in a database accessible only through the EPATHLO Student Profile Creator backend system. The interoperation between third-party applications and the API offers the benefit of keeping each application utterly independent from the business logic executed by the abovementioned methods. Therefore, the sole requirement for an external application is to invoke the specified methods as per the EPATHLO Student Profile Creator specifications. Table 6.1 outlines the available services and provides illustrative examples of the data produced by each service in JSON format.

The game triggers the specified method by sending an HTTP request to the relevant URI, retrieving the educational tool's response in JSON format, and processing it according to the requirements. JSON is a user-friendly data exchange format that is easily readable and writable by humans and readily parsed and generated by machines [23]. The system offers support for a range of functionalities through the provision of corresponding web services:

Students can create an account in the system, which stores information about them and generates a suitable model. After signing up, students are authenticated and can take quizzes and practice their skills. Whenever a student completes a quiz, their student model gets updated. The RESTful API supplies the fuzzy reasoning component with the student's answers. This fuzzy reasoner then processes the answers and generates inferences about the student's level of proficiency using the fuzzy logic technique outlined in Chap. 5.

Table 6.1 The web services provided by EPATHLO Student Profile Creator

Service description	URI	JSON response example
Register student	/user/register	{"code": 200, "message": "Account created successfully.", "user_id": 20}
Authenticate student	/user/authenticate	{"code": 200, "message":"Successful authentication.", "user_id": 18}
Retrieve learning content (course, chapter, or lesson)	/chapter/view/14	{"code":200, "chapter_id": 14, "name": "HTML", "content": "HTML (HyperText Markup Language) is the most fundamental building..."}
Create a quiz about a course, chapter, or lesson.	/quiz/chapter/create/14	{"code": 200, "message": "Successful quiz creation.", "quiz_id": 727}
Retrieve the current question of a quiz	/quiz/view/727	{"code": 200, "quiz_id": 727, "current_question": {"question_id": 13760, "description": "How does the colour attribute of HTML input elements work?", "answers": [{"answer_id": 35275, "description": "Changes colour of the text."},...]}}
Submit an answer regarding a current quiz question	/quiz/answer/727	{"code": 200, "message": "Correct answer.", "correct": 1}
Retrieve the student's knowledge regarding a course, chapter, or lesson.	/chapter/model/14	{"code": 200, "chapterid": 14, "userid": 18, "knowledge_level": {"knowledge_level_id": 2, "description": "Unsatisfactorily known"}}

6.4 Examples of Operation in Different Games of the EPATHLO Suite

The EPATHLO Student Profile Creator utilizes fuzzy logic to evaluate students' levels of knowledge. When a student finishes an assignment, the system assesses their performance and assigns a verbal knowledge level, such as "Unknown," "Unsatisfactorily Known," "Known," or "Learned." The students' models and assignments are preserved if they pause the game and resume later or switch to another game using the same educational tool to study the same topic.

Two educational games for learning HTML use the EPATHLO Student Profile Creator. The first game, Fuzzy EPATHLO, is the primary game in the EPATHLO Educational Game Suite. The player's main objective in this game is to help the trapped hero escape from a locked building. By clicking or tapping at various points on the game screen, the player directs the hero through different rooms, gathers items, and interacts with non-player characters (NPCs). The NPCs assign quests to the hero, which involve finding objects or solving quizzes related to the HTML language.

Surviving Businessman is a mobile VR game developed by Margaritis Kamitsios for research purposes [31]. In this game, players seek to discover four hidden keys, unlock corresponding business doors, and gain access to private corporations. The search for these keys occurs in a maze and a castle. The maze features additional locked gateways that can only be accessed by obtaining the matching keys. At different points in the game, players are presented with HTML-related quizzes by the game's agents. The player's performance in the quiz determines the type of prize they receive. Excelling in the quiz earns the player a key to open one of the available gateways. However, if the player's performance is satisfactory but not perfect, the game expands to include more areas to explore and additional locked gates. A poor performance in the quiz requires the player to participate in a new quiz.

The web application utilizes the EPATHLO Student Profile Creator's API to create an environment focused on studying rather than gaming. It allows students to learn the theory of HTML language and engage in exercises to enhance their knowledge. This application is designed for students who prefer not to play games and are solely interested in learning and testing their knowledge. Although not a primary focus of the current work, this example illustrates how the student model and educational material can be integrated into software beyond educational games.

The initial pair of examples demonstrate how the student's progress in education is maintained while shifting from one game to another. Ultimately, the third instance succinctly explains how the educational content can be transferred to a web application.

6.4.1 Example of Student-Player Model Initilization Between Educational Games

The first player starts playing Fuzzy EPATHLO by registering or authenticating their user profile to ensure the system can verify that a profile already exists in the database. The game provides these functions for the student by utilizing the relevant API methods. After completing the registration process, the player starts exploring the rooms within the building, collecting items, interacting with characters, and solving puzzles. Eventually, they converse with an NPC wearing a tie. The NPC offers to help the hero escape if he can answer quiz questions with an acceptable success rate (Fig. 6.2). Since the player has not participated in other HTML-related quizzes, their knowledge level is marked as "Unknown." As a result, the initial quiz

Fig. 6.2 The first player converses with the man wearing a tie

consists of easy questions tailored to the player's proficiency. After achieving a high success rate, the system categorizes the player's knowledge level as "Learned."

Later, the player continues progressing through the game, solving more puzzles. They reach a storage room where an NPC presents them with a new quiz. The player answers with a satisfactory success rate. The system's educational component updates the student's profile, indicating a "Known" knowledge level due to some challenges in understanding the HTML topic. After solving more puzzles, the player encounters a witch who presents them with additional HTML questions (Fig. 6.3). The quiz consists of questions of varying difficulties, suitable for students with a "Known" knowledge level. The player answers some questions and stops playing before completing the quiz (Fig. 6.4).

Afterward, the player utilizes their mobile device to engage in Surviving Businessman. They navigate through a three-dimensional world to find the keys necessary to unlock the gates of hidden, closed-off businesses. They encounter a character who quizzes them on HTML concepts during their journey. At this point, the game incorporates educational elements by presenting quiz questions. These questions are the same ones the player encountered in the previous game, Fuzzy EPATHLO, but they still need to be completed. Consequently, the player answers the remaining questions (Fig. 6.5). However, despite their efforts, they receive a low score, and their knowledge level is classified as "Unsatisfactorily known."

Afterward, they stop playing Surviving Businessman and return to playing Fuzzy EPATHLO. They return to the witch's room, where she gives them another quiz. This time, the difficulty level is "Unsatisfactorily Known," so the quiz includes a mix of

Fig. 6.3 The first player talks to the witch for the first time

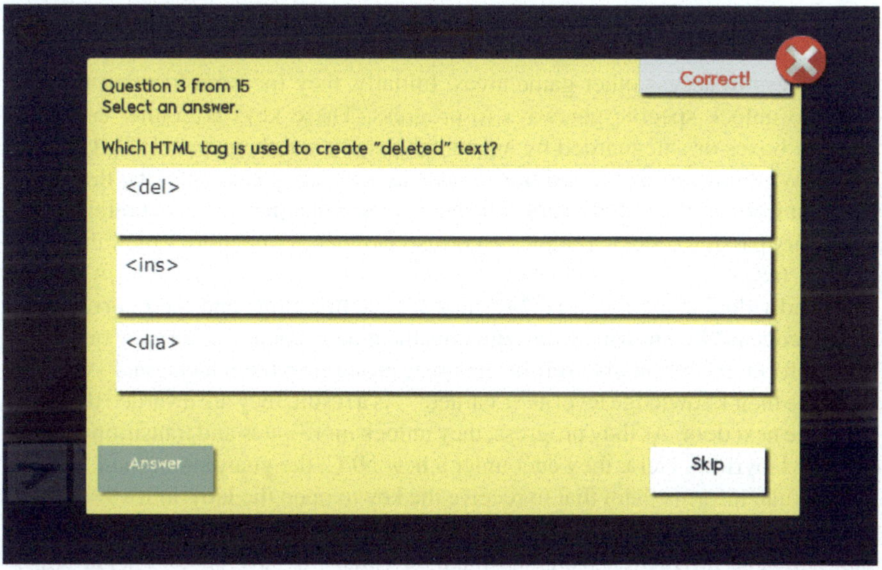

Fig. 6.4 The first player quits the quiz before finishing it

easy and moderate questions. They answer every question flawlessly and advance to the next level, "Learned." The suite's educational component refreshes the student's profile and adjusts the difficulty of the questions based on their performance.

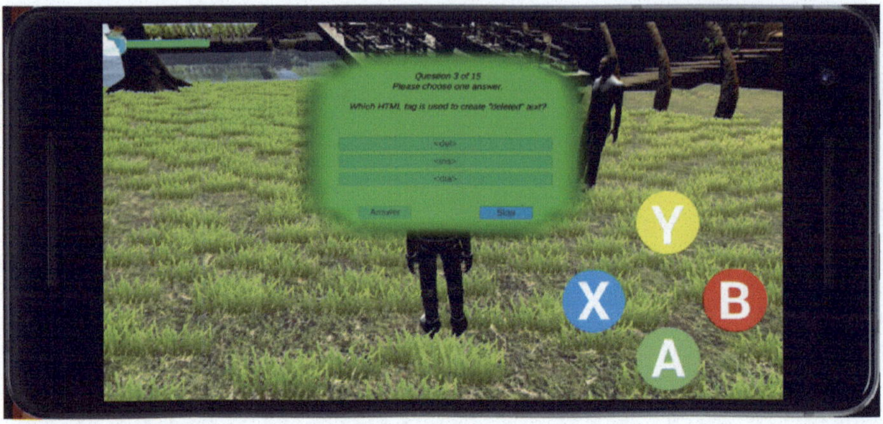

Fig. 6.5 The first player continues the abandoned quiz in The Surviving Businessman

6.4.2 Example of Student Model Concerning Quizzes Between Games

Another player utilizes their mobile phone to engage in Surviving Businessman. They navigate the 3D environment, dodge enemies, unlock boxes, and acquire a life elixir or keys to access other game areas. Initially, they focus on locating the keys needed to unlock specific gateways to progress. These keys are either concealed in hidden boxes or safeguarded by agents who challenge them with HTML-related quizzes. While searching for additional gateway keys, they encounter the first agent. The agent informs them that the player must answer the questions accurately with a satisfactory passing score to obtain the corresponding key and unlock the subsequent gate (Fig. 6.6).

The individual is taking an HTML quiz for the first time, and their current level of knowledge is "Unknown." Consequently, the quiz is composed of easy questions. They perform exceptionally well in the quiz, prompting the educational system to categorize their knowledge level as "Learned." As a result, they are awarded the key to unlock the next door. As they progress, they unlock more gates and transition to a new area - the labyrinth. Here, they encounter a new NPC, the guardian of the labyrinth. The guardian informs them that to receive the key to open the labyrinth's gates, they must satisfactorily answer questions. Failing the quiz would require exploration of a new zone and interaction with additional guardians. In such cases, the participant would be permitted to engage in review exercises to allow the system to reassess their knowledge level (Fig. 6.7).

The player initially engages with the quiz but discontinues it before completing it. They transition from playing Surviving Businessman to immersing themselves in the educational experience offered by Fuzzy EPATHLO. As they navigate through the various rooms of the building, they collect items, solve riddles, and engage in

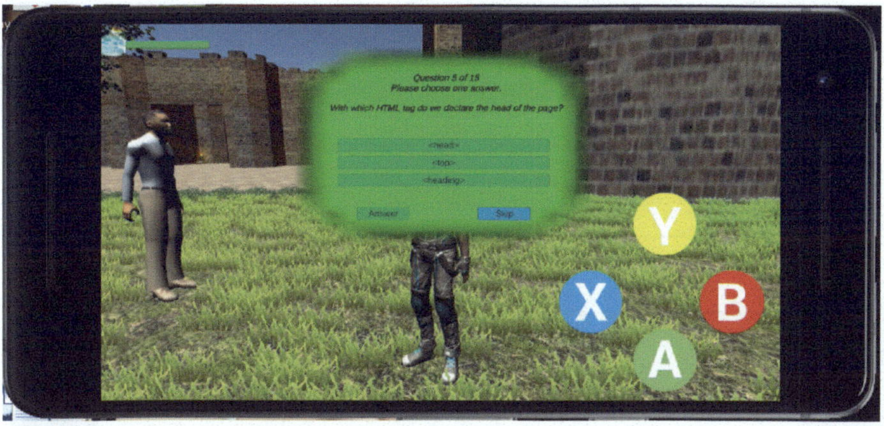

Fig. 6.6 The first agent assigns the second player a quiz of HTML questions

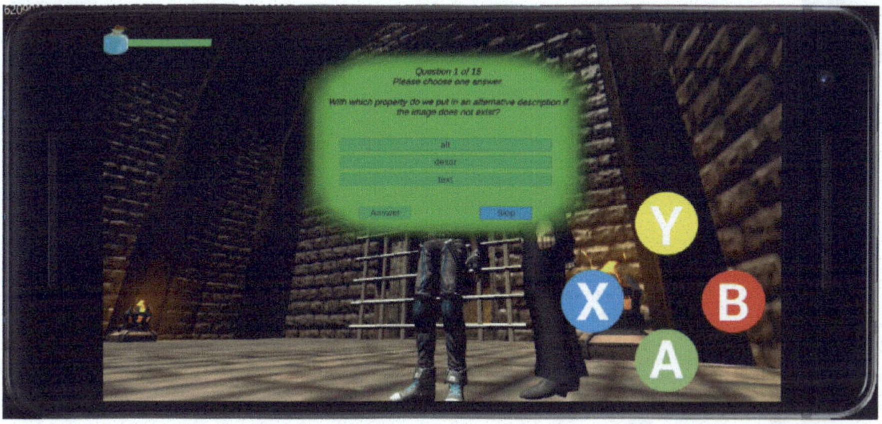

Fig. 6.7 The guardian of the labyrinth challenges the second player with another quiz containing HTML questions

conversations with NPCs. During their exploration, they encounter the NPC dressed in a tie, who offers to assist them in escaping the building if they can answer his questions satisfactorily. The questions presented are identical to the ones the player had left unanswered while playing Surviving Businessman.

Consequently, the player must respond to the remaining questions (Fig. 6.8). Following the player's satisfactory responses to the agent's questions, the system evaluates their knowledge level and categorizes it within the fuzzy set "Known." Subsequently, the game retrieves the updated knowledge level from the educational tool and utilizes it implicitly to introduce additional rooms, objects, NPCs, and riddles.

The player enters the kitchen of the building and encounters an NPC, a cook, who appears after the game story has advanced (Fig. 6.9). The chef presents the

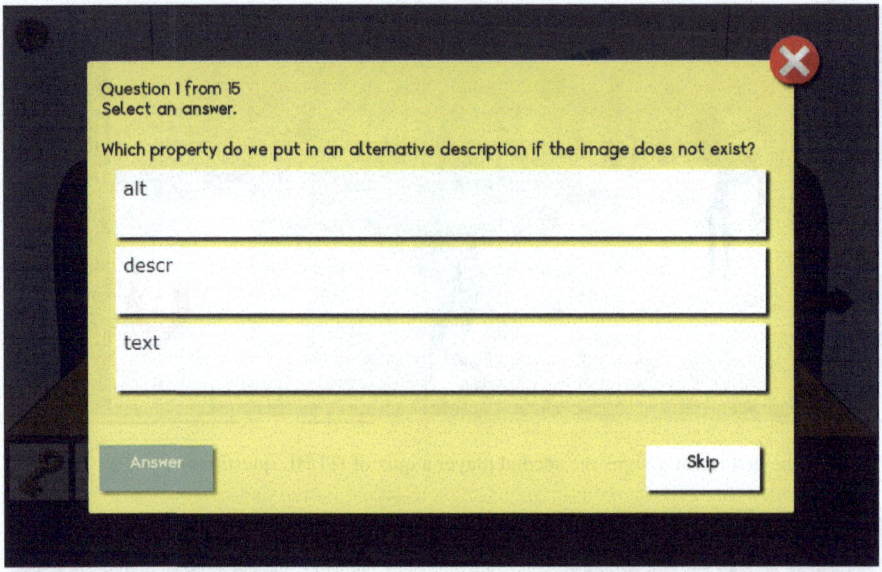

Fig. 6.8 The second player continues the abandoned quiz in Fuzzy EPATHLO

Fig. 6.9 The second player meets the chef while playing Fuzzy EPATHLO

player with a new quiz featuring questions tailored to their "Known" knowledge level, mixing in moderate and challenging difficulty levels. Once again, the player excels, outperforming their weaker areas, and their knowledge level advances to the "Learned" category.

Later, they pause their game of Fuzzy EPATHLO and opt to play Surviving Businessman again. They reencounter the labyrinth guardian, who tests them with a new, more challenging HTML quiz tailored to their expertise. Despite the increased difficulty, they achieve a perfect score, maintaining their "Learner" status. As a result, the guardian awards them a key to unlock the labyrinth entrances and progress further into the game environment.

6.4.3 Example of Student Model Extracted from Educational Games to Educational Platforms

In addition to implementing Fuzzy EPATHLO, our current work involves creating an original integrated e-learning environment designed to teach various subjects, including the HTML programming language (Fig. 6.10). The platform offers courses not only in computer science but also in other domains, such as content for students seeking a driving license, providing a wide range of educational opportunities. The platform utilizes web services to integrate educational material and quiz-format exercises related to HTML and manage student profiles. Student profiles reflect their knowledge levels using fuzzy sets as described in Chap. 5. The application uses this information from EPATHLO Student Profile Creator to determine which

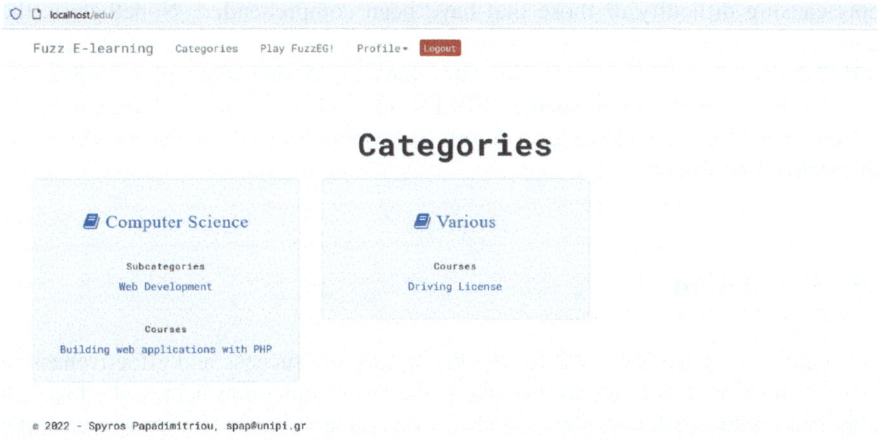

Fig. 6.10 The e-learning platform that communicates with the suite

learning material to present to each student. Furthermore, the application leverages EPATHLO Student Profile Creator to identify concepts that students have insufficient knowledge of and provides the necessary theory for them to study. As a result, the application enhances the capabilities of EPATHLO Student Profile Creator by implementing additional adaptive rules.

The first time a student engages with the application, the system communicates with EPATHLO Student Profile Creator to verify if the user already has a profile. If no profile is found, the student can proceed with the registration process, and EPATHLO Student Profile Creator will then create their user profile. If there is an existing profile, the application will authenticate the student and retrieve the corresponding profile information through interaction with the EPATHLO Student Profile Creator.

For instance, Mary and John are students interacting with the web-based educational environment for the first time. Mary is identified as a new user, so she registers herself in the system, and a corresponding profile is created for her. Since she is considered a beginner, the system presents her with the fundamental concepts of the teaching material. On the other hand, John already has an existing profile, so the web application retrieves his student profile from the EPATHLO Student Profile Creator and customizes the learning material based on his level of knowledge. John's profile is already stored in the EPATHLO Student Profile Creator database due to his prior interaction with Fuzzy EPATHLO.

The web application dynamically organizes the learning materials in both scenarios, providing personalized tutoring. It leverages the student model information obtained from EPATHLO Student Profile Creator to create models of the two students' objectives, preferences, and knowledge, utilizing them in their interactions. For instance, it associates an icon and a label with each concept in the domain to indicate the student's current level of understanding, clearly highlighting any concepts causing difficulty or those that have been comprehended. Nonetheless, the adaptive mechanisms of the web application are beyond the scope of the current work. Nevertheless, this example illustrates that any external application with access to a portable student model, such as EPATHLO Student Profile Creator, can apply additional intelligent techniques, offering the student an alternative or expanded personalized experience.

6.5 Evaluation

The evaluation phase is pivotal for demonstrating the success and effectiveness of the work presented. It is focused on the pedagogical outcomes achieved when students can choose different games to learn the same subject. However, evaluating educational games is a complex process involving various aspects. Specifically, the game's usability, appeal, and educational effectiveness must be evaluated to determine whether the game is engaging and beneficial for students in an educational context [79, 85]. This study's evaluation method is based on Virvou's framework for assessing edutainment [74], emphasizing the importance of conducting empirical

Table 6.2 The questionnaire

	Questions	Answers (circle one for each question) 1: Not at all 2: Slightly satisfied 3: Moderately 4: Very 5: Completely					
Usability and Likeability	Q1	How clear is the game's goal?	1	2	3	4	5
	Q2	How interesting is the game?	1	2	3	4	5
	Q3	How much do you like the game?	1	2	3	4	5
	Q4	How easy is it to navigate the game?	1	2	3	4	5
	Q5	How much fun do you have playing the game?	1	2	3	4	5
	Q6	Rate your involvement in the game.	1	2	3	4	5
Educational effectiveness	Q7	How much does the gameplay motivate you to participate in the learning process?	1	2	3	4	5
	Q8	How much do you like to use the game to learn the knowledge domain of HTML language?	1	2	3	4	5
	Q9	How much do you like to use the game to learn another knowledge domain concept?	1	2	3	4	5
	Q10	How much did the game help you obtain new knowledge about the knowledge domain of HTML programming language?	1	2	3	4	5
	Q11	How much does the gameplay increase your engagement in the learning process?	1	2	3	4	5
	Q12	How confident do you feel about your learning results through the game?	1	2	3	4	5

evaluations on edutainment software's appeal, usability, educational objectives, and pedagogical goals. Further details regarding the evaluation approach and its outcomes will be provided.

6.5.1 *The Method*

To evaluate the games' effectiveness, we examined the feedback and perspectives of student-players from three distinct groups who utilized the games to learn the HTML programming language. A survey measured user satisfaction with the user interface, game story, gameplay, and educational impact. The survey (Table 6.2) comprised twelve closed-ended questions – six focused on game usability and appeal, and the remaining six on the educational effectiveness of the game. The questions were rated on a Likert scale, spanning from "Not at all (1)" to "Absolutely (5)." After engaging with the games to acquaint themselves with the HTML scripting language, student-players from all three groups completed the survey. The initial group engaged with Fuzzy EPATHLO, while the second group interacted with Surviving Businessman. The third group could switch between the two games, continuing their learning journey from where they left off, using the same educational resources. Both games encompass identical learning content and teaching methods.

Furthermore, an actual condition t-test was conducted to analyze the three groups' educational outcomes and student achievement based on their performance. A comparative analysis of the mean scores of student performance across the three groups was then carried out to determine whether selecting the game environment or the category representing the same learning content improves educational outcomes.

6.5.2 The Testbed

Three distinct student groups participated in the assessment. Each group consisted of 35 students enrolled in the Computer Science graduate program at the University of Piraeus. Group 1 engaged with Fuzzy EPATHLO, while Group 2 interacted with The Surviving Businessman. Both games encompassed identical educational content and utilized similar learning tools. Lastly, Group 3 had the flexibility to switch freely between the two games, continuing the educational session from where they had left off, using the same educational material based on their preference for the gaming environment. The students underwent HTML lessons for one month and two weeks of game practice. Subsequently, they completed a corresponding questionnaire, and a comparison of their performance results was conducted.

6.5.3 Evaluation Results

Based on the survey results (Figs. 6.11 and 6.12), there is no significant difference in the responses of students from Group 1 and Group 2. Both groups expressed equal satisfaction with the learning material's usability, likability, and educational effectiveness. However, students in Group 3 showed higher satisfaction with the gaming environment and gameplay and demonstrated more interest and motivation to participate in the educational experience. Consequently, the educational effectiveness of students in Group 3 was found to be greater than that of Groups 1 and 2. Therefore, the newly introduced feature, which allows embedding the same learning material in any genre of game or platform, is beneficial for engaging students in the learning process. This approach leverages the game context and the features students prefer to enhance the learning experience.

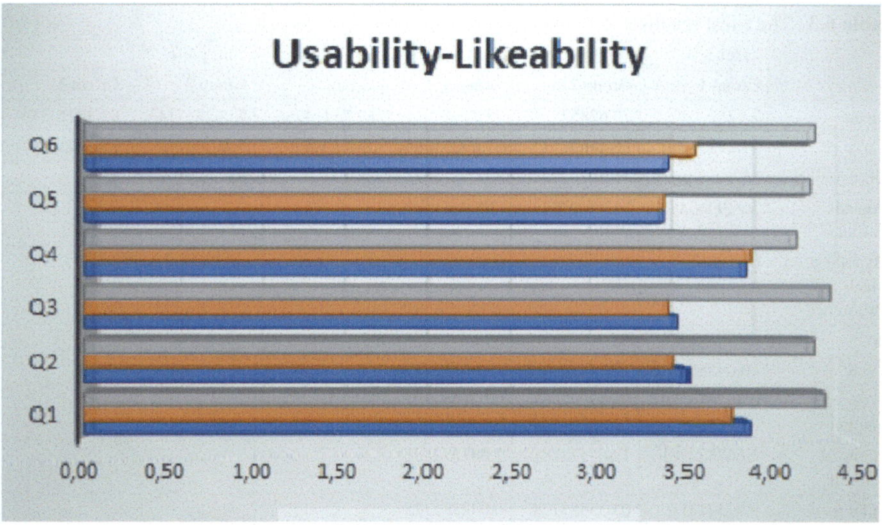

Fig. 6.11 The questionnaire's results about usability-likeability

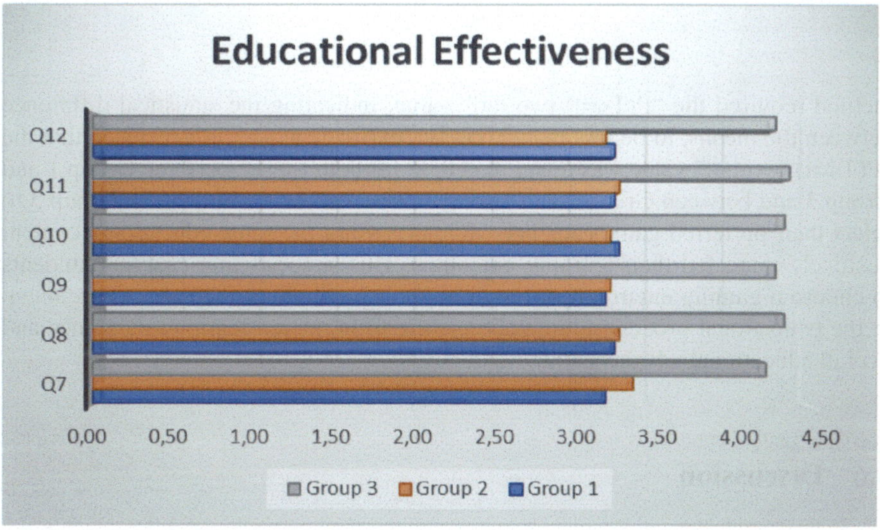

Fig. 6.12 The questionnaire's results about educational effectiveness

Table 6.3 presents a comparison of the mean performance of students in three different groups. The mean performance for Groups 1, 2, and 3 is 73.2%, 73.63%, and 82.26% respectively. There were no significant differences in the mean scores between Groups 1 and 2, but there was a noticeable difference in the mean score of Group 3. To ensure the statistical significance of the different mean scores of Group 3 compared to the other two groups, the Independent-sample T-test was conducted. This

Table 6.3 The t-test results

	1st t-test		2nd t-test		3rd t-test	
	Group 1	Group 2	Group 1	Group 3	Group 2	Group 3
Mean	73,2	73,62857	73,2	82,25714286	73,62857143	82,25714286
Variance	373,2235294	360,7697	373,2235294	251,1966387	360,7697479	251,1966387
Observations	35	35	35	35	35	35
Pooled Variance	366,9966387	312,210084	305,9831933			
Hypothesized Mean Difference	0	0	0			
df	68	68	68			
t Stat	−0,093586067	−2,144306091	−2,063522158			
P(T< = t) one-tail	0,462856589	0,017792864	0,021441387			
t Critical one-tail	1,667572281	1,667572281	1,667572281			
P(T< = t) two-tail	0,925713179	0,035585728	0,042882775			
t Critical two-tail	1,995468931	1,995468931	1,995468931			

method required the "P(T<=t) two-tail" value, indicating the statistical difference between the means, to be less than 0.05. The experimental results revealed that the "P(T<=t) two-tail" value was lower than 0.05 for both t-tests: between Group 1 and Group 3 and between Group 2 and Group 3. Thus, allowing students in Group 3 to select their preferred game category while accessing the same educational content positively impacted their learning outcomes. This suggests that enabling students to choose a gaming environment based on their preferences increases engagement in the educational process, ultimately leading to improved learning outcomes and overall educational advancement.

6.6 Discussion

The chapter discusses the absence of educational applications that employ web services technology to provide online accessibility to the student model, particularly in the form of games. Although several works emphasize the importance of using web services to allow students to use their applications on any computer and still access their model, along with their personalized content, only a few take advantage of them [30, 33, 65]. Hence, further efforts are required to create an independent tool that builds the student model and offers customized material while allowing external applications to access it [49]. There is a need for such an educational gaming tool that can improve both learning and overall gaming experiences for students.

The literature on games highlights that players have varying preferences regarding the game or device they want to use for playing or learning [8, 15, 54, 61, 63, 67, 68]. However, there has yet to be concrete work to address the gap of providing each student with the freedom to choose the game they want to play while still having access to the same educational content. The current work uses an abstraction learning mechanism integrated into the EPATHLO Educational Game suite to address the issue. This mechanism enables students to choose the game that aligns with their game preferences, resolving the problem. This approach empowers the players with the freedom of choice, leading to greater engagement and improved learning outcomes.

While many games are available for teaching the same subject, such as programming or mathematics, most incorporate no or limited adaptive elements. Hence, they provide different educational material, which is inconsistent with the student's knowledge level. Each game retains no or different information regarding the student's model. Therefore, when the player selects another educational game, an essential problem emerges in student modeling that needs addressing. Games must remember the student model as calculated each time, and the educational content must be the same across all games. The inconsistent provision of educational material by each game can cause a dilemma for the student-player, as they have to choose between the game with more exciting features or better-suited educational material.

The abstraction learning mechanism integrated into the EPATHLO Educational Game Suite successfully addresses the issue of delivering inconsistent and different educational material to each game. The mechanism entails the student modeling process by calculating the students' knowledge level and providing customized educational content consistent with their current knowledge. Thus, students can select games that cater only to their gaming preferences [31, 52]. All games that communicate with the abstracted learning mechanism provide the same customized content tailored to the student's knowledge level. The player can choose games based on criteria such as the unique features characterizing each game. Overall, this mechanism helps relieve students of choosing between games based on educational material. It ensures they receive consistent and personalized educational material across all games, ultimately leading to a more effective and enjoyable gaming and learning experience.

According to the evaluation findings on the game choice feature, it significantly improves their enjoyment when students can choose the game they want to play [52]. Also, it contributes towards enhancing their knowledge and understanding of the subject matter. This feature allows students to engage more actively in learning by choosing a game that aligns with their interests and preferences, making the entire experience more personalized and enjoyable. Overall, the game choice feature has been a valuable addition to educational games, as it promotes learning and makes it a fun and engaging experience for students.

References

1. Ahn, J.Y., Mun, G.S., Han, K.S., Choi, S.H.: An online authoring tool for creating activity-based learning objects. Educ. Inf. Technol. **22**, 3005–3015 (2017)
2. Al-Ajlan, A., Zedan, H.: Why moodle. In: 2008 12th IEEE International Workshop on Future Trends of Distributed Computing Systems, pp. 58–64. IEEE (2008)
3. Alepis, E., Virvou, M., Alepis, E., Virvou, M.: Mobile versus desktop educational applications. In: Object-Oriented User Interfaces for Personalized Mobile Learning, pp. 65–72 (2014)
4. Archuby, F., Sanz, C., Manresa-Yee, C.: Dijs: methodology for the design and development of digital educational serious games. IEEE Trans. Games **15**(2), 273–284 (2023)
5. Athaya, H., Nadir, R.D.A., Indra Sensuse, D., Kautsarina, K., Suryono, R.R.: Moodle implementation for e-learning: a systematic review. In: Proceedings of the 6th International Conference on Sustainable Information Engineering and Technology, pp. 106–112 (2021)
6. Azizi-Soleiman, F., Heidari-Beni, M., Hemati, Z.: Kelishadi R (2023) Designing and developing an educational-therapeutic game for improving healthy lifestyle in children and adolescents. Int. J. Comput. Games Technol. **1**, 9897496 (2023)
7. Burns, H.L., Capps, C.G.: Foundations of intelligent tutoring systems: An introduction. In: Foundations of Intelligent Tutoring Systems, pp. 1–19. Psychology Press (2013)
8. Chesham, A., Wyss, P., Müri, R.M., Mosimann, U.P., Nef, T., et al.: What older people like to play: genre preferences and acceptance of casual games. JMIR Serious Games **5**(2), e7025 (2017)
9. Chou, Y.S., Hou, H.T., Chang, K.E., Su, C.L.: Designing cognitive-based game mechanisms for mobile educational games to promote cognitive thinking: an analysis of flow state and game-based learning behavioral patterns. Interact. Learn. Environ. **31**(5), 3285–3302 (2023)
10. Chrysafiadi, K., Virvou, M.: Evaluating the integration of fuzzy logic into the student model of a web-based learning environment. Exp. Syst. Appl. **39**(18), 13127–13134 (2012)
11. Chrysafiadi, K., Virvou, M.: Dynamically personalized e-training in computer programming and the language c. IEEE Trans. Educ. **56**(4), 385–392 (2013)
12. Chrysafiadi, K., Virvou, M.: Student modeling approaches: a literature review for the last decade. Exp. Syst. Appl. **40**(11), 4715–4729 (2013)
13. Chrysafiadi, K., Virvou, M.: Fuzzy logic for adaptive instruction in an e-learning environment for computer programming. IEEE Trans. Fuzzy Syst. **23**(1), 164–177 (2014)
14. Chrysafiadi, K., Virvou, M.: Enhancing automatic reasoning of human errors in an operating system using fuzzy logic. In: 2019 10th International Conference on Information, Intelligence, Systems and Applications (IISA), pp. 1–4. IEEE (2019)
15. Chrysafiadi, K., Papadimitriou, S., Virvou, M.: Which is better for learning: a web-based educational application or an educational game? In: 2019 International Symposium on Performance Evaluation of Computer and Telecommunication Systems (SPECTS), pp. 1–6. IEEE (2019)
16. Chrysafiadi, K., Papadimitriou, S., Virvou, M.: Fuzzy states for dynamic adaptation of the plot of an educational game in relation to the learner's progress. In: 2020 11th International Conference on Information, Intelligence, Systems and Applications (IISA), pp. 1–7. IEEE (2020)
17. Chrysafiadi, K., Papadimitriou, S., Virvou, M.: Cognitive-based adaptive scenarios in educational games using fuzzy reasoning. Knowl. Based Syst. **250**, 109111 (2022)
18. Chrysafiadi, K., Virvou, M., Tsihrintzis, G.A., Hatzilygeroudis, I.: Evaluating the user's experience, adaptivity and learning outcomes of a fuzzy-based intelligent tutoring system for computer programming for academic students in greece. Educ. Inf. Technol. **28**(6), 6453–6483 (2023)
19. Costa, C., Alvelos, H., Teixeira, L.: The use of moodle e-learning platform: a study in a portuguese university. Proc. Technol. **5**, 334–343 (2012)
20. Dermeval, D., Paiva, R., Bittencourt, I.I., Vassileva, J., Borges, D.: Authoring tools for designing intelligent tutoring systems: a systematic review of the literature. Int. J. Artif. Intell. Educ. **28**, 336–384 (2018)

21. Düsünen, K., Ata, R.: Bibliometric analysis of scientific studies on the use of educational digital games in mathematics education. In: Uden, L., Liberona, D., (eds.) Learning Technology for Education Challenges-12th International Workshop, LTEC 2024, Kaohsiung, Taiwan, July 29–August 1, 2024, Proceedings, Springer, Communications in, vol. 2082, pp. 40–50, (2024). https://doi.org/10.1007/978-3-031-61678-5_4
22. Ford, T.C., Colombi, J.M., Graham, S.R., Jacques, D.R.: (2007) A survey on interoperability measurement. Gateways **2**(3)
23. Friesen, J.: Java XML and JSON. Springer (2016)
24. Gomez, M.J., Armada Sánchez, Á., Albaladejo-González, M., García Clemente, F.J., Ruipérez-Valiente, J.A.: Utilizing explainable ai to enhance real-time student performance prediction in educational serious games. In: Proceedings of the 39th ACM/SIGAPP Symposium on Applied Computing, pp. 93–94 (2024)
25. Graesser, A., Conley, M., Olney, A.: Intelligent Tutorıng Systems, APA Educational Psychology Handbook (2012)
26. Hassan, L.: Accessibility of educational games and game-based approaches to people with learning and physical disabilities: a systematic literature review (2024)
27. Hooshyar, D., El Mawas, N., Milrad, M., Yang, Y.: Modeling learners to early predict their performance in educational computer games. IEEE Access **11**, 20399–20417 (2023)
28. Hutson, J., Fulcher, B.: A virtual reality educational game for the ethics of cultural heritage repatriation. Games Cult. **18**(6), 759–782 (2023)
29. Joshi, A., Mousas, C., Tokey, S.S., Kao, D.: Exploring player agency in educational video games. Entertain. Comput. **52**, 100754 (2025)
30. Kabassi, K., Virvou, M.: Using web services for personalised web-based learning. J. Educ. Technol. Soc. **6**(3), 61–71 (2003)
31. Kamitsios, M., Papadimitriou, S., Chrysafiadi, K., Virvou, M.: Incorporating quizzes into a 3d educational game calling web services. In: EdMedia+ Innovate Learning, Association for the Advancement of Computing in Education (AACE), pp. 1170–1179 (2018)
32. Katsionis, G., Virvou, M.: A cognitive theory for affective user modelling in a virtual reality educational game. In: 2004 IEEE International Conference on Systems, Man and Cybernetics (IEEE Cat. No. 04CH37583), vol. 2, pp. 1209–1213. IEEE (2004)
33. Katsionis, G., Virvou, M.: Personalised e-learning through an educational virtual reality game using web services. Multimed. Tools Appl. **39**, 47–71 (2008)
34. Khan, I., Hollebeek, L.D., Fatma, M., Islam, J.U., Rather, R.A., Shahid, S., Sigurdsson, V.: Mobile app vs. desktop browser platforms: the relationships among customer engagement, experience, relationship quality and loyalty intention. J. Mark. Manag. **39**(3–4), 275–297 (2023)
35. Kristić, M., Dilberović, A.K., Car, Ž., Žubrinić, K.: Research instrument for analyzing user interactions and accessibility of holographic technology in educational games for environmental awareness. In: 2024 47th MIPRO ICT and Electronics Convention (MIPRO), pp. 1967–1973. IEEE (2024)
36. Kularbphettong, K., Kedsiributt, P., Roonrakwit, P.: Developing an adaptive web-based intelligent tutoring system using mastery learning technique. Proc. Soc. Behav. Sci. **191**, 686–691 (2015)
37. Lampropoulos, G.: The role of educational technology, artificial intelligence, and gamification in improving education and students' cognitive and social-emotional development: development and evaluation of educational applications and serious games using extended reality and learning analytics. Ph.D thesis, International Hellenic University, Greece (2023). http://hdl.handle.net/10442/hedi/54240
38. Lim, B., Wen, H.J.: Web services: an analysis of the technology, its benefits, and implementation difficulties. Inf. Syst. Manag. **20**(2) (2003)
39. Liu, X., Shao, X.: Modern mobile learning technologies in online piano education: online educational course design and impact on learning. Interact. Learn. Environ. **32**(4), 1279–1290 (2024)
40. López-Fernández, D., Gordillo, A., Lara-Cabrera, R., Alegre, J.: Comparing effectiveness of educational video games of different genres in computer science education. Entertain. Comput. **47**, 100588 (2023)

41. Lu, Y., Lan, J., Li, K.: Enhancing ai education through educational games: evidence from a meta-analysis study. In: 2024 13th International Conference on Educational and Information Technology (ICEIT), pp. 72–77. IEEE (2024)
42. Machado, B.A., Bernardi de Souza, F., de Souza Rodrigues, J.: Development of a database of educational games applicable to production engineering. Int. J. Emerg. Technol. Learn. **19**(4) (2024)
43. Marques, F.P.R., de Miranda, L.C.: Heuristics to support the evaluation of optimal experience in educational games for learning japanese as a second language. J. Interact. Syst. **14**(1), 494–517 (2023)
44. Moundridou, M., Virvou, M.: Authoring intelligent tutoring systems over the world wide web. In: Proceedings First International IEEE Symposium Intelligent Systems, vol. 1, pp. 160–165. IEEE (2002)
45. Moundridou, M., Virvou, M.: Analysis and design of a web-based authoring tool generating intelligent tutoring systems. Comput. Educ. **40**(2), 157–181 (2003)
46. Ozyurt, H., Ozyurt, O., Mishra, D.: Exploring the evolution of educational serious games research: a topic modeling perspective. IEEE Access (2024)
47. Papadakis, S., Kalogiannakis, M.: Mobile educational applications for children: what educators and parents need to know. Int. J. Mobile Learn. Organ. **11**(3), 256–277 (2017)
48. Papadimitriou, S., Virvou, M.: Adaptivity in scenarios in an educational adventure game. In: 2017 8th International Conference on Information, Intelligence, Systems & Applications (IISA), pp. 1–6. IEEE (2017)
49. Papadimitriou, S., Chrysafiadi, K., Kamitsios, M., Virvou, M.: Fsp creator: a novel web service api creator of fuzzy students progress profile. In: 2018 9th International Conference on Information, Intelligence, Systems and Applications (IISA), pp. 1–7. IEEE (2018)
50. Papadimitriou, S., Chrysafiadi, K., Virvou, M.: Authoring technological and platform independent learning material and student's progress profile using web services. In: Knowledge-Based Software Engineering: 2018: Proceedings of the 12th Joint Conference on Knowledge-Based Software Engineering (JCKBSE 2018) Corfu, Greece, vol. 12, pp. 242–251. Springer (2019)
51. Papadimitriou, S., Chrysafiadi, K., Virvou, M.: Fuzzeg: Fuzzy logic for adaptive scenarios in an educational adventure game. Multimed. Tools Appl. **78**(22), 32023–32053 (2019). https://doi.org/10.1007/s11042-019-07955-w
52. Papadimitriou, S., Kamitsios, M., Chrysafiadi, K., Virvou, M.: Learn-and-play personalised reasoning from point-and-click to virtual reality mobile educational games. Intell. Decis. Technol. **15**(2), 321–332 (2021)
53. Papazoglou, M.: Web Services: Principles and Technology. Pearson Education (2008)
54. Peever, N., Johnson, D., Gardner, J.: Personality & video game genre preferences. In: Proceedings of the 8th Australasian Conference on Interactive Entertainment: Playing the System, pp. 1–3 (2012)
55. Rapaka, A., Dharmadhikari, S., Kasat, K., Mohan, C.R., Chouhan, K., Gupta, M.: Revolutionizing learning-a journey into educational games with immersive and ai technologies. Entertain. Comput. **52**, 100809 (2025)
56. Ren, J., Xu, W., Liu, Z.: The impact of educational games on learning outcomes: Evidence from a meta-analysis. Int. J. Game Based Learn. (IJGBL) **14**(1), 1–25 (2024)
57. Rezadoust Siah Khaleh Sar, H.: The role of clandestine assessment in iranian efl context: a teacher's perceptions about the impact of educational digital-based games on teaching and learning english. Educ. Inf. Technol. 1–25 (2023)
58. Richardson, L., Ruby, S.: RESTful web services. O'Reilly Media, Inc. (2008)
59. Roldán-Álvarez, D., Martín, E., Haya, P.A., García-Herranz, M., Rodríguez-González, M.: Dedos: an authoring toolkit to create educational multimedia activities for multiple devices. IEEE Trans. Learn. Technol. **11**(4), 493–505 (2018)
60. San Blas, H.S., Mendes, A.S., de la Iglesia, D.H., Silva, L.A., González, G.V.: A multiagent platform for promoting physical activity and learning through interactive educational games using the depth camera recognition system. Entertain. Comput. **49**, 100629 (2024)

61. Scharkow, M., Festl, R., Vogelgesang, J., Quandt, T.: Beyond the "core-gamerâŁž: genre preferences and gratifications in computer games. Comput. Hum. Behav. **44**, 293–298 (2015)
62. Shang, J., Zhou, J.: How to design scientific, happy and effective educational games: research on educational game design from the perspective of learning sciences. In: International Conference on Blended Learning, pp. 21–38. Springer (2024)
63. Sherry, J.L., Lucas, K., Greenberg, B.S., Holmstrom, A.: Child development and genre preference: research for educational game design. Cyberpsychology Behav. Soc. Netw. **16**(5), 335–339 (2013)
64. Sun, L., Guo, Z., Hu, L.: Educational games promote the development of students' computational thinking: a meta-analytic review. Interact. Learn. Environ. **31**(6), 3476–3490 (2023)
65. Thyagharajan, K., Nayak, R.: Adaptive content creation for personalized e-learning using web services. J. Appl. Sci. Res. **3**(9), 828–836 (2007)
66. Tikva, C., Mousiou, M., Tambouris, E.: Educational games for computational thinking: Evaluation of the scaffolded amazed game. In: Sifaleras A, Lin F (eds) Generative Intelligence and Intelligent Tutoring Systems-20th International Conference, ITS 2024, Thessaloniki, Greece, 10–13 June 2024, Proceedings, Part II, Springer, Lecture Notes in Computer Science, vol. 14799, pp. 96–103 (2024). https://doi.org/10.1007/978-3-031-63031-6_9
67. Tondello, G.F., Nacke, L.E.: Player characteristics and video game preferences. In: Proceedings of the Annual Symposium on Computer-Human Interaction in Play, pp. 365–378 (2019)
68. Tondello, G.F., Arrambide, K., Ribeiro, G., Cen, A.J.l., Nacke, L.E.: "I don't fit into a single type": a trait model and scale of game playing preferences. In: Human-Computer Interaction–INTERACT 2019: 17th IFIP TC 13 International Conference, Paphos, Cyprus, 2–6 Sept 2019, Proceedings, Part II, vol. 17, pp. 375–395. Springer (2019)
69. Tsiriga, V., Virvou, M.: A framework for the initialization of student models in web-based intelligent tutoring systems. User Model. User Adapt. Interact. **14**, 289–316 (2004)
70. Tyni, J., Turunen, A., Kahila, J., Bednarik, R., Tedre, M.: Reward types in popular recreational and educational mobile games. IEEE Access **11**, 1166–1174 (2022)
71. Virvou, M.: Automatic reasoning and help about human errors in using an operating system. Interact. Comput. **11**(5), 545–573 (1999)
72. Virvou, M.: A domain-independent reasoning mechanism for an its authoring tool. In: EdMedia+ Innovate Learning, Association for the Advancement of Computing in Education (AACE), pp. 1762–1764 (2000)
73. Virvou, M.: Modelling the knowledge and reasoning of users in a knowledge-based authoring tool. Int. J. Bus. Perform. Manag. **5**(2–3), 399–412 (2003)
74. Virvou, M.: On the evaluation of the combined role of virtual reality games and animated agents in edutainment. Intell. Comput. Graph. **2011**, 79–95 (2012)
75. Virvou, M., Alepis, E.: Mobile versus desktop facilities for an e-learning system: users' perspective. In: 2004 IEEE International Conference on Systems, Man and Cybernetics (IEEE Cat. No. 04CH37583), vol. 1, pp. 48–52. IEEE (2004)
76. Virvou, M., Alepis, E.: Mobile educational features in authoring tools for personalised tutoring. Comput. Educ. **44**(1), 53–68 (2005)
77. Virvou, M., Kabassi, K.: Reasoning about users' actions in a graphical user interface. Hum. Comput. Interact. **17**(4), 369–398 (2002)
78. Virvou, M., Kabassi, K.: Evaluating an intelligent graphical user interface by comparison with human experts. Knowl. Based Syst. **17**(1), 31–37 (2004)
79. Virvou, M., Katsionis, G.: On the usability and likeability of virtual reality games for education: the case of vr-engage. Comput. Educ. **50**(1), 154–178 (2008)
80. Virvou, M., Moundridou, M.: An authoring tool for algebra-related domains. In: HCI (2), pp. 647–651. Citeseer (1999)
81. Virvou, M., Papadimitriou, S.: Use of digital 'guess who board game'in teaching english. In: IISA 2014, The 5th International Conference on Information, Intelligence, Systems and Applications, pp. 439–443. IEEE (2014)

82. Virvou, M., Stavrianou, A.: User modelling in a gui. In: Proceedings of HCI International (the 8th International Conference on Human-Computer Interaction) on Human-Computer Interaction: Ergonomics and User Interfaces, vol. I, pp. 261–265 (1999)
83. Virvou, M., Maras, D., Tsiriga, V.: Student modelling in an intelligent tutoring system for the passive voice of english language. J. Educ. Technol. Soc. **3**(4), 139–150 (2000)
84. Virvou, M., Manos, K., Katsionis, G., Tourtoglou, K.: Vr-integate: A knowledge-based authoring tool for virtual reality educational games. In: Proceedings of ED-MEDIA 2002 World Conferences on Educational Multimedia, Hypermedia and Telecommunications (2002)
85. Virvou, M., Katsionis, G., Manos, K.: Combining software games with education: evaluation of its educational effectiveness. J. Educ. Technol. Soc. **8**(2), 54–65 (2005)
86. Wang, M.: Effects of touch-type online educational games on learners' learning motivations. Int. J. Emerg. Technol. Learn. (Online) **18**(6), 4 (2023)
87. Wegner, P.: Interoperability. ACM Comput. Surv. (CSUR) **28**(1), 285–287 (1996)
88. Zeng, M.L.: Interoperability. KO. Knowl. Organ. **46**(2), 122–146 (2019)
89. Zimbrão, R.D.O., Oliveira, A.M.D., Oliveira, E.W., Valle, P.H.D.: Essential aspects in digital educational games to support the teaching-learning of computing contents. In: Proceedings of the 20th Brazilian Symposium on Information Systems, pp. 1–10 (2024)

Chapter 7
Authoring Intelligent Educational Games with Fuzzy Reasoning

Abstract Educators can utilize authoring tools, namely software designed to create and update educational content without requiring technical programming expertise. These tools also enable educators to track their students' learning progress. However, most authoring tools accompanying Intelligent Tutoring Systems (ITSs) do not allow interference with the AI mechanisms operating in the background. This chapter introduces an innovative authoring tool that not only empowers educators to monitor students' progress and create courses but also allows them to intervene with the AI mechanism and modify its functions. The ITS in this work utilizes a fuzzy mechanism, which processes imprecise or uncertain information to enhance adaptability and personalize the learning experience. The tutor can introduce additional fuzzy variables to assess the student's level of knowledge, such as question frequency, percentage of incorrect responses, and the time taken to answer a question beyond the expected duration. This additional customization feature is expected to enable the tutor to tailor the intelligent mechanisms and draw even more precise conclusions about the student's academic performance or the material to be provided. As a result, students can benefit from a progressively personalized learning experience and improve their learning abilities.

Keywords Educational software personalized educational games · Artificial Intelligence in educational games · Fuzzy logic in education · User experience · AI-empowered software games · User modeling · Human-AI interaction · e-learning games · 21st-century skills · Human-centred Artificial Intelligence · Game authoring tools

7.1 Introduction

When using educational applications, there can be various challenges related to the learning process. Initially, a tutor needs the help of a programmer or someone with programming expertise to create lessons. However, collaborating with a developer and integrating educational content requires significant time and effort. Each time a lesson needs an update, the developer must be available to work with the tutor,

S. Papadimitriou and M. Virvou, *Artificial Intelligence—Based Games as Novel Holistic Educational Environments to Teach 21st Century Skills*, Intelligent Systems Reference Library 93, https://doi.org/10.1007/978-3-031-77464-5_7

understand the requirements, make the necessary technical adjustments, and conduct testing to ensure everything works properly. Additionally, manual course development can create proprietary code that may need significant technical improvements for compatibility. As a result, using previous courses as templates for new material may require additional effort, as the resulting code will need further modifications.

Authoring tools effectively resolve the previously mentioned issues. They have become integral in developing educational content, as they offer a range of functionalities that enhance learning experiences across various domains. The abundance of authoring tools caters to various fields, domains, and technologies due to their potential. They are software programs that empower educators to develop digital courses and share them in multiple formats [11, 17, 25, 41, 45, 46, 78].

Authoring tools are available for Virtual or Augmented Reality (AR) and are designed for both windowed and web-based environments [10, 16, 70]. They can be standalone or built-in. When using a standalone authoring tool, the educator creates content within it, exports it, and then integrates it into an educational application [28]. The exported file must adhere to the format supported by the application. On the other hand, built-in authoring tools are integrated within an educational application [1, 14, 31], eliminating compatibility issues during export and import. Rather than manually coding a course, an authoring tool allows instructors to utilize user-friendly interfaces, including drag-and-drop features, to construct a course without the need for programming expertise. Upon creating the course content, it is uploaded to the preferred learning platform. The digital learning content produced using authoring tools encompasses e-learning courses, video lectures, assignments, assessments, and simulations [49]. Generally, authoring tools notably reduce the time required to update learning materials, improving retention rates by allowing lessons to be more precisely tailored to students' needs. Moreover, instructors can replicate existing learning content, make necessary adjustments, and release new content more efficiently than manually editing course materials.

The need for such tools is particularly evident when sophisticated automatic reasoning is required for user modeling and personalization purposes [2, 62, 63, 68, 69, 75]. These tools enable educators to create interactive and personalized learning environments without extensive programming knowledge [35]. Authoring tools for Intelligent Tutoring Systems (ITSs) leverage intelligent techniques, such as adaptive learning paths, personalized feedback, and cognitive modeling, to create educational experiences that dynamically adjust to individual learner needs and enhance the effectiveness of instruction. They often incorporate Artificial Intelligence (AI) and knowledge-based techniques [60] to analyze student performance data, allowing them to tailor educational content and provide targeted interventions, thereby improving learning outcomes through real-time personalization and adaptation. An authoring tool may provide a generalization mechanism that facilitates easy development [65]. Authoring tools specific to ITSs aim to establish contexts for the streamlined development of tutoring systems that can intelligently adapt to individual learners [39, 66, 67, 72].

For example, Virvou et al. [76] discuss VR-INTEGATE, a knowledge-based authoring tool for developing Virtual Reality (VR) educational games, which immerses students in interactive learning scenarios that enhance engagement and retention. A very different case in education combined with entertainment is the INMA tool, a knowledge-based authoring tool designed for music education, focusing on enhancing music perception and understanding through intelligent educational techniques [77]. On another domain, Virvou and Moundridou [71] highlight an authoring tool designed explicitly for algebra-related subjects, allowing for the development of ITSs that adapt to individual learner needs, thus facilitating personalized learning paths and immediate feedback. Moundridou and Virvou [36] explore authoring ITS over the World Wide Web (WWW), enabling remote access and personalized instruction, which expands educational accessibility and supports adaptive learning experiences.

Furthermore, Virvou and Alepis [67] emphasize integrating mobile educational features in authoring tools, providing learners with the convenience of accessing personalized tutoring on the go and supporting continuous learning through mobile devices. The research by Virvou [65] on modeling knowledge and reasoning within authoring tools underscores the development of systems that align content with learners' cognitive processes, thereby enhancing the responsiveness and effectiveness of educational systems. Lastly, Virvou [64] introduces a domain-independent reasoning mechanism for ITS authoring tools and gaming environments, which increases versatility and applicability across various educational domains, reducing development time and effort while promoting scalability. Otherwise, developing cognitive models in specific ITSs can be very sophisticated and costly, such as in [24].

Although some authoring tools offer limited functionalities, such as creating only quizzes, in most cases, it is more accurate to depict an authoring tool as a comprehensive suite of tools that encompass the majority of tasks required by educational designers. The remaining challenge that researchers must tackle concerns the tutor's interaction with the intelligent mechanisms of an ITS. These mechanisms are responsible for evaluating the student's cognitive state and determining the personalized material to provide [61]. Refining these mechanisms can result in more thorough and precise assessments, ultimately enhancing learning outcomes. Consequently, the tutor must be able to directly adjust these mechanisms without relying on a programmer or an AI expert for assistance.

Considering the information provided, the EPATHLO Author is a tool for creating content that enables educators to input data relevant to any chosen domain. This tool is fundamental to the EPATHLO Educational Game Suite, which works alongside the EPATHLO Student Profile Creator. The EPATHLO Student Profile Creator is an essential component that generates student models independent of platforms or technologies, utilizing web services technology. This approach is innovative for two main reasons. Firstly, it produces a student profile that seamlessly integrates with any educational application, regardless of its technological requirements or programming language. This allows for easy access and utilization of student data by different applications. Secondly, the system leverages web services technology, enabling swift recognition and usage of the student profile by various applications. This technology improves the system's efficiency and ensures uniformity in the data used across different applications.

Furthermore, the EPATHLO Author allows for creating and managing educational content through a user-friendly interface accessible on both computers and mobile phones. This enables the easy development of new courses and updating existing ones while maintaining a high degree of reusability. Additionally, external applications like Fuzzy EPATHLO [48] or Surviving Businessman [23] can access supplementary courses, providing students with the opportunity to apply their learnings to various subjects beyond HTML programming. The unique feature of the EPATHLO Author lies in its ability to allow instructors to modify the fuzzy inference mechanisms. This can be accomplished by using simple linguistic rules [80] by adjusting mathematical functions and their parameters. Notably, the former option does not require any AI knowledge, while the latter requires a basic understanding of fuzzy logic. In both cases, the involvement of a programmer is unnecessary.

7.2 A Review of Authoring Tools

The academic literature has explored various authoring tools for intelligent learning systems. Moundridou and Virvou [37–40, 72] made significant contributions to the field by developing a web-based authoring tool designed to assist with learning algebraic equations through an intelligent system. This tool, which is user-friendly and accessible from any device with an internet connection, initially requires a human teacher to describe a specific domain, aids the teacher in creating exercises, monitors the student's progress, and offers tailored feedback. Notably, the tool's diagnostic component includes intelligent features for identifying and diagnosing student errors. Furthermore, it handles teaching material flexibly and personally, thereby significantly advancing the field of ITSs.

Cabada et al. introduced EDUCA, a web authoring tool designed for building adaptive and ITSs using a Kohonen network [6]. Olsen et al. enhanced the Cognitive Tutor Authoring Tools with an authoring tool extension to facilitate the development of collaborative ITSs through multiple synchronized tutor engines [45]. Escudero and Fuentes developed an authoring tool to create courses for ITSs. The authoring tool can export the created courses to another ITS, thereby reusing the domain model represented by the teaching course [12]. These tools are designed to be adaptable to various teaching and learning scenarios, providing educators with the necessary flexibility to update the educational content.

Virvou and Alepis developed Mobile Author, a mobile authoring tool designed for creating ITSs [67]. With Mobile Author, tutors can use a computer or a mobile phone to produce ITSs and distribute them to their students. These tutoring systems can evaluate students' performance, update databases that track students' progress, and offer customized advice to individual students. The same researchers also designed a web authoring tool that enables teachers to create tutoring characters for student interaction [66]. With this tool, tutors can parameterize every aspect of how each character communicates, including their voice's tone, speed, and volume. The authoring tool ensures that the characters reflect the teaching style of the human instructor and can address emotional issues for students who may lack access to in-person interaction.

Herbert et al. developed the Generalised, Rapid Authoring Tool (GRAT) [20] to simplify the creation of ITSs. This tool features a unified web-based wizard-style graphical user interface and programming-by-demonstration methods, making authoring ITS logic without extensive technical knowledge easier. They used GRAT to develop network cabling courses and console device configuration tasks. Another authoring tool, Situated Pedagogical Authoring (SitPed) [7], is specifically designed to train soft skills, such as counseling. It emphasizes the importance of considering the student's perspective and provides tools for creating task lists, authoring assessment knowledge, and composing tutor messages.

Ramírez-Noriega and Martínez-Ramírez argue that the domain model is the cornerstone of the ITS because it accommodates the knowledge to be imparted and evaluated [51]. Bearing this in mind, they have designed a framework for creating authoring tools that automatically generate a suitable domain model. The resulting domain model forms the basis for the student and tutor modules, encompassing learned topics and knowledge inference.

Virvou and Papadimitriou have created a digital version of the "Guess Who" board game as an ITS for language learning [73]. With its unique authoring tool [74], this system allows educators to monitor student progress and introduce additional educational material. It also enhances the learning experience with new game elements like avatars and levels, making language learning more engaging and effective.

Papadimitriou, Chrysafiadi, and Virvou have developed an authoring tool that allows teachers to oversee student performance, create and modify educational material, and facilitate the integration of its content into third-party applications [47]. This unique mechanism, featuring student modeling and personalized exercise generation functions, enables any educational application to access the learning content, regardless of its technological characteristics or programming language. The system's adaptability is further demonstrated by the fact that a student's progress profile, generated using two different educational applications that utilize the same web services accompanying the authoring tool, will be recognized by both applications. This means that if a student's knowledge level changes during their interaction with one application, the other application will also acknowledge the updated knowledge level of the student, thereby ensuring a practical, personalized learning experience.

Virvou et al. presented the concept of VR-INTEGATE, which stands for Virtual Reality-Intelligent Game Authoring Tuition Environment [76]. VR-INTEGATE is a comprehensive authoring tool tailored for educators aiming to develop ITSs that are integrated with VR games. By leveraging VR-INTEGATE and ITSs, educators can gain access to sophisticated diagnostic evaluations of students' interactions and performances within educational contexts.

Augmented Reality Authoring Tools

AR authoring tools have transformed the way that educators and creators develop interactive educational content. Blattgerste [4] explores the design space of AR authoring tools, emphasizing their utility in procedural training contexts. The study highlights the need for intuitive interfaces that allow users to create and modify AR experiences efficiently. Silva et al. [58] further expand on this by developing design principles for AR authoring tools based on teacher feedback, underscoring the importance of user-centric design to enhance educational outcomes. Ez-zaouia et al. [14] discuss the democratization of AR in education, emphasizing the need for accessible tools that empower educators to integrate AR into their teaching without technical barriers.

Bräker et al. [5] derive design principles for no-code AR authoring tools, empowering users to develop AR solutions without programming knowledge. This democratization of AR content creation fosters innovation and broadens the scope of AR applications in education and industry.

Kim et al. [27] discuss RealityBrush, an AR authoring system that captures the kinetic properties of everyday objects, enabling users to create dynamic AR experiences. This approach leverages the natural interactions of physical objects to enhance digital content, providing a seamless blend of the real and virtual worlds.

Kim et al. [26] present RealityBrush, an AR authoring system that leverages the kinetic properties of everyday objects to create dynamic AR experiences. This approach allows users to incorporate real-world interactions into their digital creations, enhancing the realism and engagement of AR content.

Virtual Reality Authoring tools

VR authoring tools have significantly impacted e-learning and training by facilitating the creation of immersive experiences. Horst et al. [21] explore the reuse of pattern-based VR content across different authoring toolkits, highlighting the efficiency and flexibility of modular design. This approach allows educators to adapt and expand VR content to suit various learning scenarios. Müser et al. [42] evaluate VR authoring tools in educational settings, revealing their potential to enhance teaching by providing students with engaging, interactive environments.

Müser et al. [42] evaluate VR authoring tools designed for teaching, highlighting their potential to enhance learning through interactive simulations and virtual environments. Their study emphasizes the importance of usability and accessibility in VR tools to ensure they meet educators' and learners' needs.

De Oliveira et al. [43] explore the acceptance and use of VR authoring tools through the UTAUT2 model, which analyzes user perceptions and behavioral intentions. Their findings indicate that perceived ease of use, usefulness, and social influence significantly impact the adoption of VR authoring tools, emphasizing the need for intuitive interfaces and robust support systems.

Quintana et al. [50] focus on authoring tools for procedural modeling in VR-based rehabilitation exercises, showcasing the potential of VR in healthcare education. These tools enable the creation of tailored rehabilitation programs, allowing for personalized and adaptive therapeutic experiences.

User Interaction and Playfulness

Improving user interaction in immersive environments is a key focus of current research. Schäfer [55] examines novel hand gesture authoring tools that enhance interactions in VR, offering users more natural and intuitive ways to engage with virtual content. Liapis et al. [30] discuss designing for playfulness in human-AI authoring tools, suggesting that integrating playful elements can foster creativity and engagement, making content creation more enjoyable and effective.

Collaborative Authoring Tools

Collaboration is crucial to modern content creation, particularly in game development. Hossain and Zaman [22] introduce NCCollab, a collaborative behavior tree authoring tool that supports teamwork in game development, allowing multiple users to contribute to the creative process.

3D Content Authoring and Tangible Interaction

Creating 3D content is a complex task that authoring tools aim to simplify. Yu [79] focuses on 3D sketching tools that enable users to design intricate models interactively, enhancing creativity and precision in content creation. Lam et al. [29] explore tangible interaction techniques with authoring capabilities for kitchen design, demonstrating how physical interactions can be seamlessly integrated into digital content creation, providing users with a more immersive design experience.

Evaluating and Improving Authoring Tools

Evaluating authoring tools is essential to ensure they meet user needs and accessibility standards. Salmen and Schroeder [54] propose explorable authoring requirements for evaluating these tools, offering a framework to assess their usability and effectiveness. Baldiris et al. [3] evaluate authoring tools under ATAG and WCAG recommendations, emphasizing the importance of accessibility in ensuring that tools are usable by a diverse range of users, including those with disabilities.

Interactive Narratives and Cultural Heritage

Authoring tools play a significant role in developing interactive digital narratives (IDN) and preserving cultural heritage. Massidda et al. [33] review IDN authoring tools, highlighting their use in creating hybrid experiences that engage audiences and preserve cultural stories. These tools allow creators to craft immersive narratives that blend historical facts with interactive elements, enhancing cultural understanding and appreciation.

Gómez-Albarrán et al. [15] explore authoring tools for creating interactive fiction using conventional web technologies. Their work emphasizes the flexibility and accessibility of web-based tools in developing engaging storytelling experiences, making them suitable for various educational and entertainment applications.

Green et al. [18] explore UX principles for these tools, emphasizing the need for intuitive design that facilitates seamless user interaction. By aligning tool functionality with user needs, developers can create narratives that captivate and engage audiences, fostering deeper connections with the content.

Zarei et al. [81] focus on designing enactment-scaffolded narrative authoring tools for elementary school children. Their research highlights the potential of narrative tools to enhance educational outcomes by providing scaffolded support that guides children through the creative process. By fostering creativity and storytelling skills, these tools contribute to developing critical thinking and literacy in young learners.

Adaptive Learning and Mulsemedia

Adaptive learning environments benefit significantly from smart authoring tools. Olney et al. [44] introduces a special issue on creating and improving adaptive learning through smart authoring tools, highlighting the importance of adaptive content that responds to individual learner needs and preferences.

De Mattos et al. [34] advocate for developing mulsemedia authoring tools that integrate multiple sensory modalities into digital experiences. These tools aim to enhance user engagement and learning by providing a rich, multisensory environment that goes beyond traditional multimedia content.

Coelho et al. [9] evaluate authoring tools designed for creating 360-degree multisensory videos, focusing on the effectiveness of different interfaces. Their research highlights the importance of intuitive and user-friendly interfaces in enabling users to produce engaging multisensory content. These tools enhance the viewer's experience by supporting various sensory modalities, making the content more immersive and impactful.

Evaluation and Design Guidelines

The evaluation of authoring tools is critical to their success and adoption. Shiels and Tanner [57] investigate the perceived affordances and constraints of learning design authoring tools among South African high school teachers, identifying key factors that influence their effectiveness in educational settings.

Chamusca et al. [8] evaluate design guidelines for developing intuitive VR authoring tools using NVIDIA Omniverse as a case study. Their findings highlight the importance of user-centered design and iterative testing in creating tools that meet the needs of diverse user groups.

Accessibility and Interaction Design

Ensuring accessibility in authoring tools is vital for inclusivity. Roig et al. [52] provide recommendations for improving the accessibility of word processors and interaction design, emphasizing the need for tools that cater to users with varying abilities.

Ez-zaouia et al. [13] explore the design space of educational AR authoring tools, focusing on creating accessible and user-friendly solutions that facilitate the integration of AR into educational practices.

Comparative Analysis and Resources

Shibolet and Lombardo [56] offer resources for a comparative analysis of IDN authoring tools, providing insights into their features and capabilities. This analysis aids developers and educators in selecting the most appropriate tools for their specific needs and objectives.

Data Visualization and Cognitive Interaction

Rubab et al. [53] examine interaction techniques in data visualization authoring tools, analyzing their effectiveness from cognitive and goal-oriented perspectives. The study underscores the importance of designing tools that cater to human cognitive processes, enabling users to create clear and impactful visualizations. These tools support decision-making and data interpretation, making them invaluable in fields that rely heavily on data analysis.

AI and Deep Authoring for Multimedia

Takács and Vincze [59] introduce a deep authoring AI toolset for creating immersive multimedia experiences. By utilizing AI, these tools simplify the content creation process, allowing users to generate complex multimedia projects easily. AI-driven features such as automated content generation and optimization play a crucial role in expanding the creative possibilities for users.

XR Applications and Educational Workshops

Masneri et al. [32] discuss using workshop-authoring tools for XR applications in education. Their study highlights the potential of XR technologies to transform educational environments by providing immersive and interactive learning experiences. These tools enable educators to create XR content that enhances student engagement and learning outcomes.

Procedural Content Creation for Gaming

Halperin [19] examines procedural tools used to develop Marvel's Spider-Man: Miles Morales for PlayStation 5. These tools facilitate the creation of complex game environments and mechanics by automating certain aspects of content generation. By streamlining the development process, procedural tools allow developers to focus on creativity and innovation in game design.

As shown above, a wide range of authoring tools are available for educational software, but the options are more limited when it comes to ITSs. Nonetheless, their primary function remains consistent –enabling tutors to create and modify educational content and monitor student progress. It may be necessary for an ITS to modify its AI mechanisms to enhance the personalized learning experience and improve educational outcomes. However, this process requires the involvement of a programmer and an AI expert, and it is both time-consuming and expensive. The EPATHLO Author addresses this challenge by allowing educators to customize the intelligent knowledge assessment mechanism, thus assisting the tutoring system in delivering a more personalized experience for each student. This tool can potentially revolutionize the field of ITSs by making personalization more accessible and cost-effective.

7.3 The Architecture of the EPATHLO Author

The tutor uses a web browser to interact with EPATHLO Author through a user-friendly interface. The authoring tool consists of three main modules: the learning content module, the fuzzy content module, and the statistics content module (Fig. 7.1). All modules interact with a database to retrieve, add, or update data. The database stores information about the teaching domains, related exercises, student data, and the parameter values needed to operate the fuzzy logic algorithm.

7.3.1 The Learning Content Module

The content module for learning is in charge of updating the teaching domain-related content. The tutor uses this module to incorporate theory and exercises. A simplified entity-relationship (ER) diagram illustrating the structure of the educational content is shown in Fig. 7.2.

In the ER diagram, categories serve as entities to which courses can be related. For instance, categories like "Computer Science," "High School Courses," or "Foreign Language Learning" can exist. These categories can have subcategories, which can, in turn, have their own subcategories. The potential depth of these category levels is limitless in theory. For example, "Computer Science" could be further divided into "Software Development" and "Algorithms."

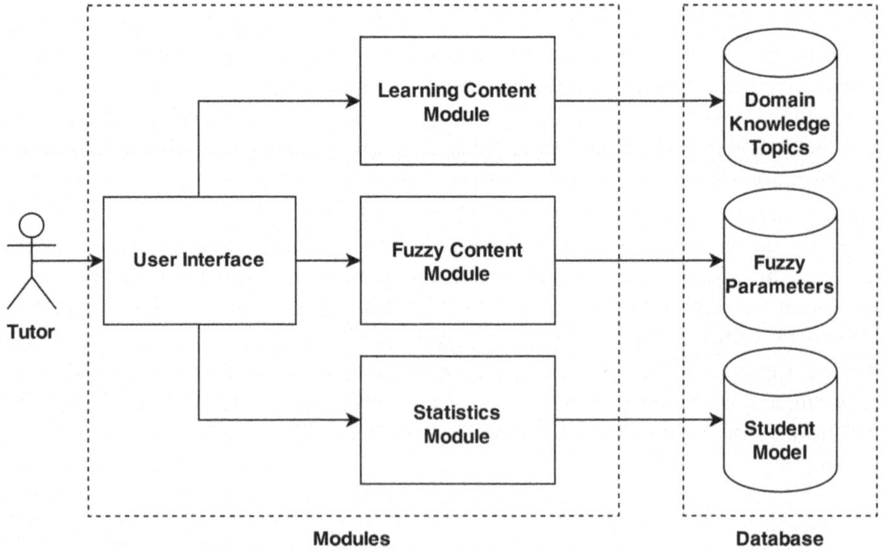

Fig. 7.1 The architecture of EPATHLO author

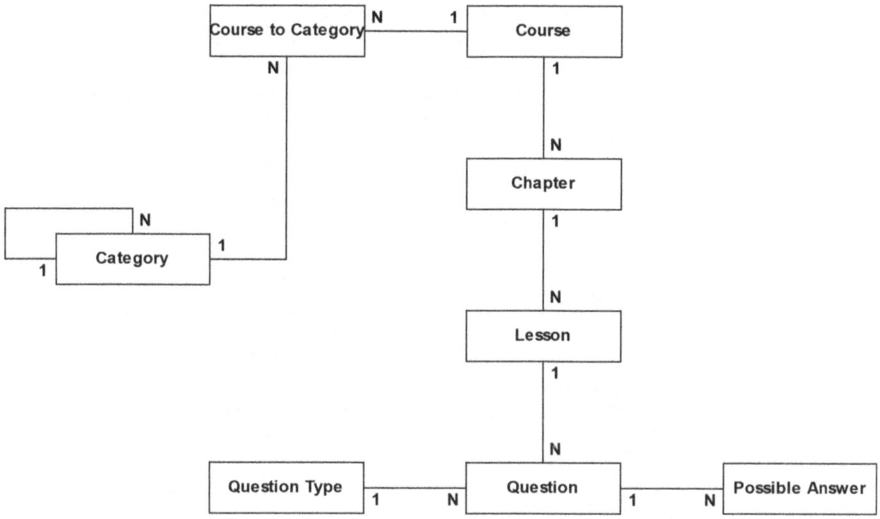

Fig. 7.2 The ER diagram of the educational content

The "Foreign Language Learning" category can be subdivided into "English Language Learning" and "Greek Language Learning." A course can be linked to multiple categories and subcategories. For instance, the course "Introduction to HTML5" might be categorized under 'Computer Science' and its "Web Development" subcategory. Each course comprises specific chapters, and each chapter includes multiple lessons.

The details presented above are related to the theoretical aspect of the educational material. Educators can formulate suitable questions and potential answers for the exercises in each lesson. Questions can be categorized as single-choice (e.g., true or false), multiple-choice, fill-in-the-blank, or sorting. Furthermore, each question is linked to a specific difficulty level: easy, moderate, or difficult. There are no limitations on the number of possible answers for each question. The layout illustrated in Fig. 7.3 offers an example of how educational content can be organized.

Within this diagram are two primary categories: "Computer Science" and "Web Development," which is a subset of the former. The course "Building Web Applications with PHP" falls within these categories and consists of three chapters: "HTML," "CSS," and "PHP," each of which contains multiple lessons. For example, the "HTML" chapter covers topics such as "HTML Tags" and "Headings." The "Headings" section includes various related questions, such as "Select the correct HTML element for the largest heading," where the student must choose from options like "<h6>," "<heading>," or "<h1>." The diagram does not include additional lessons, questions, or corresponding answers to maintain simplicity.

The EPATHLO Author offers a user interface for tutors to import educational material using the Learning Content Module, as illustrated in the screenshots below.

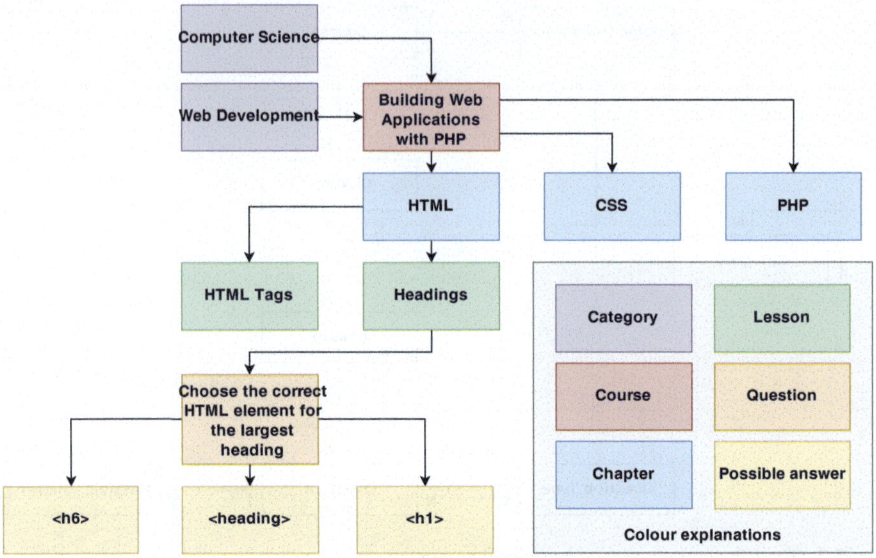

Fig. 7.3 A sample of the structure of the educational content

Figure 7.4 displays the login form users can submit their credentials. The system identifies the user as a tutor or administrator upon submission. Administrators have unrestricted access to all data, including user, course, assignment, and student information. Conversely, tutors are only granted access to their respective courses, exercises, and the enrolled students. Figure 7.5 showcases the EPATHLO Author dashboard that tutors encounter upon successful login. The dashboard features a left sidebar presenting all of the tutor's options. The options in the main content area to the right of the sidebar are represented as cards, each providing additional explanations about its functionality.

In the "Categories" section, the tutor can include, adjust, or delete course categories and their subcategories (Fig. 7.6). For example, Fig. 7.7 illustrates the subcategory "Web Development" within the overarching "Computer Science" category.

In the "Courses" section, the tutor can create or update their courses. Figure 7.8 showcases two courses the tutor created. It is worth noting that there are no limitations on the subjects the tutor can create; for instance, one of the courses is "Building Web Applications with PHP," which falls under the "Computer Science" and "Web Development" categories. In contrast, the course "Driving License" belongs to the category "Various," which includes courses of diverse domains. For each course, the tutor can view the corresponding chapters. In the "Chapters" section, the tutor can add, edit, or delete chapters.

In addition, there is an option to arrange the sequence of the chapters within a course. Furthermore, tutors can incorporate various types of rich content, such as text, images, audio, and videos, to complement each chapter (Fig. 7.9). Moreover,

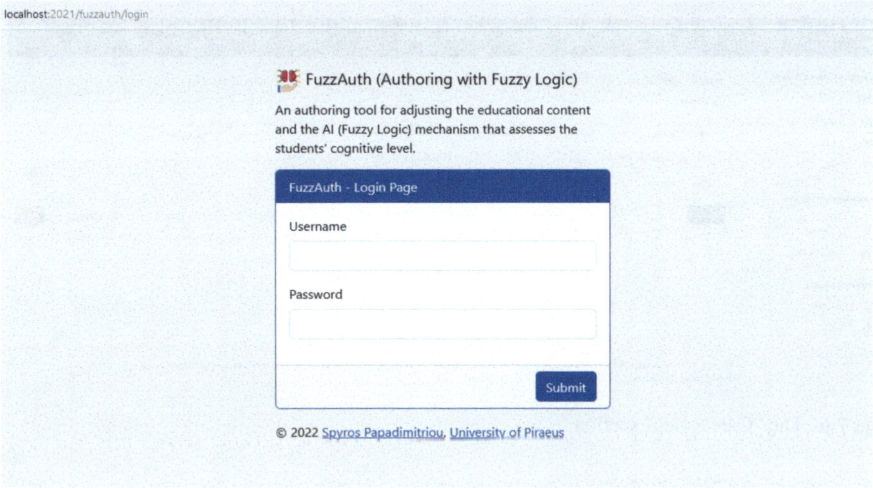

Fig. 7.4 The login page of the EPATHLO author

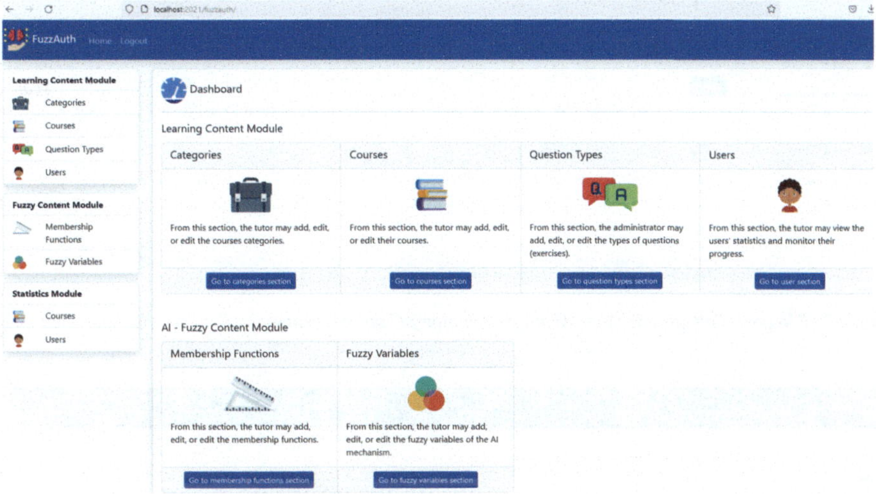

Fig. 7.5 The dashboard of EPATHLO author

each chapter is broken down into lessons, featuring its own rich content for student access and study. Educators can include questions for each lesson and align them with specific exercise types, including single-choice, multiple-choice, sequence, and fill-the-gaps (Fig. 7.10). Educators can provide a question description and optionally upload an image when adding a question. Subsequently, they can create possible answers for each question, with each answer having a description and an indicator of whether it is true or false. Educators can also specify the order in which the possible

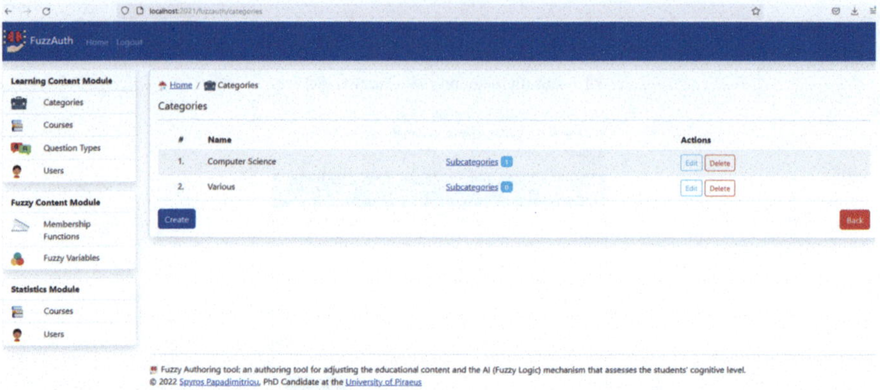

Fig. 7.6 The "Categories" section

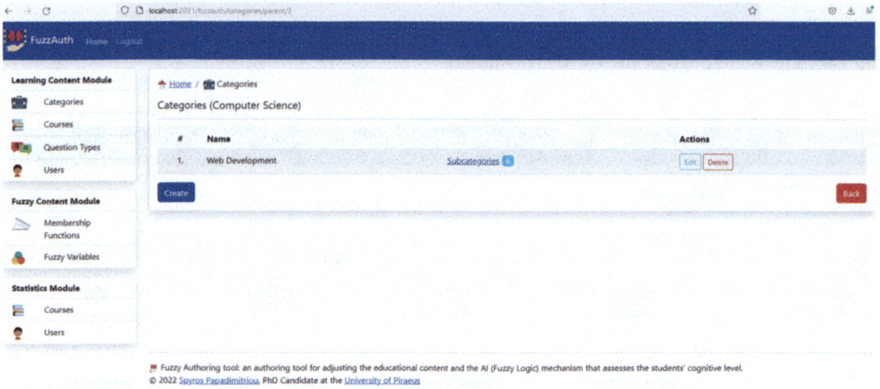

Fig. 7.7 "Web development" is a "Computer Science" subcategory

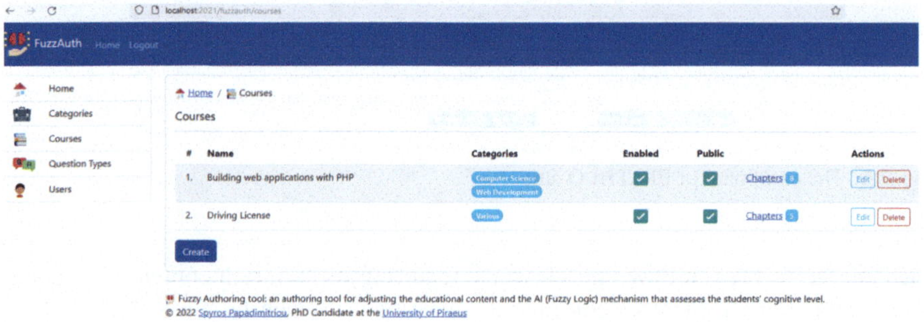

Fig. 7.8 The "Courses" section

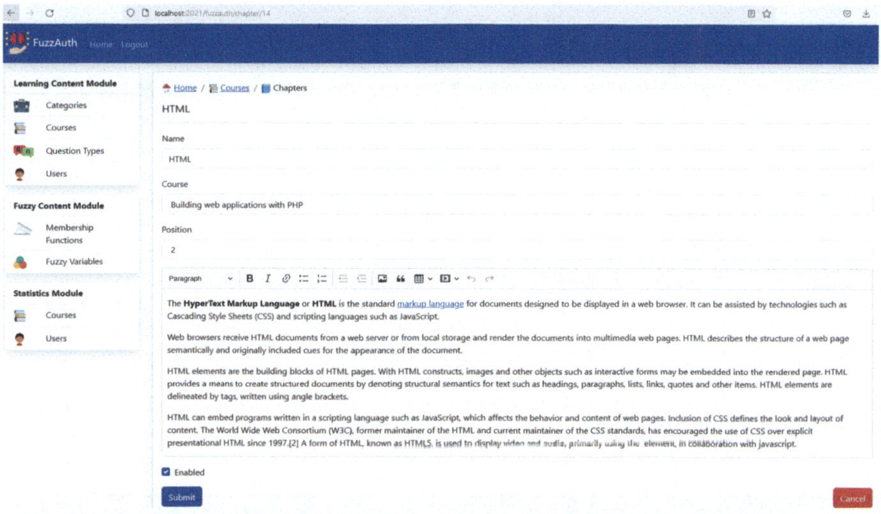

Fig. 7.9 The "Edit Chapter" section

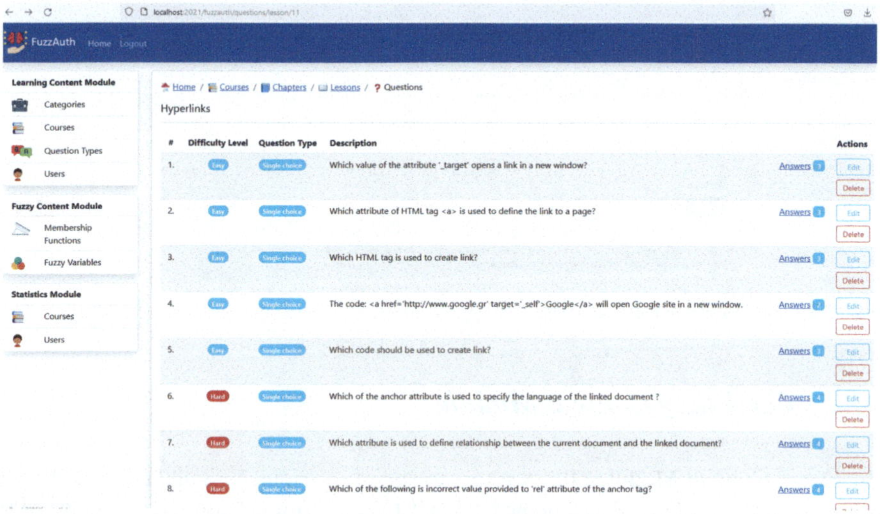

Fig. 7.10 The "Questions" section

answers are displayed. It is worth noting that while an external application retrieves the answers in the order defined by the educator, it can present them differently if necessary. As depicted in Fig. 7.11, there is an example of a list of possible answers for the question "Choose the correct HTML element for the largest heading," which is related to the lesson "Headings" in the "HTML" chapter. Lastly, Fig. 7.12 illustrates the edit page for an answer.

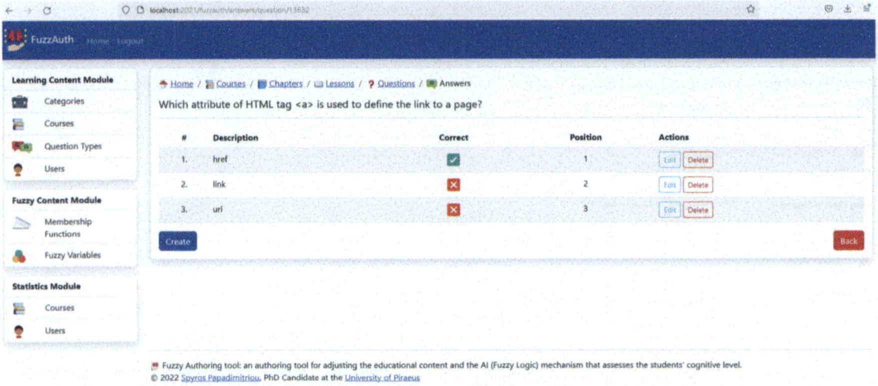

Fig. 7.11 The “Answers” section

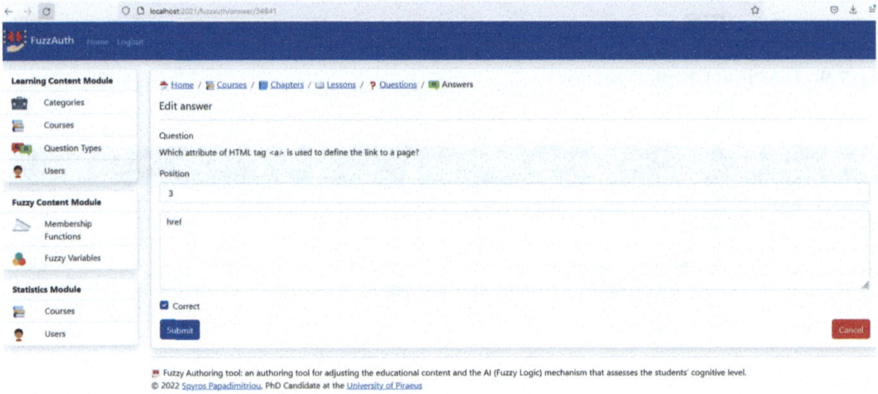

Fig. 7.12 The “Edit Answer” section

7.3.2 *The Fuzzy Content Module*

The Fuzzy Content Module allows tutors to specify various parameters of the fuzzy inference system (FIS) utilized in the EPATHLO Educational Game Suite. It provides two options for setting up the AI mechanism-one designed for AI experts and another for tutors who lack expertise in the field.

If the tutor is well-versed in AI and fuzzy logic concepts, they can define membership functions (MFs) to evaluate the student’s level of understanding. Alternatively, they can opt to utilize the default parameters provided by the platform. Initially, the tutor will choose the function type from a predefined list. EPATHLO Author supports the following functions: triangular, trapezoidal, and Gaussian. The triangular function computes fuzzy membership values using the following equation:

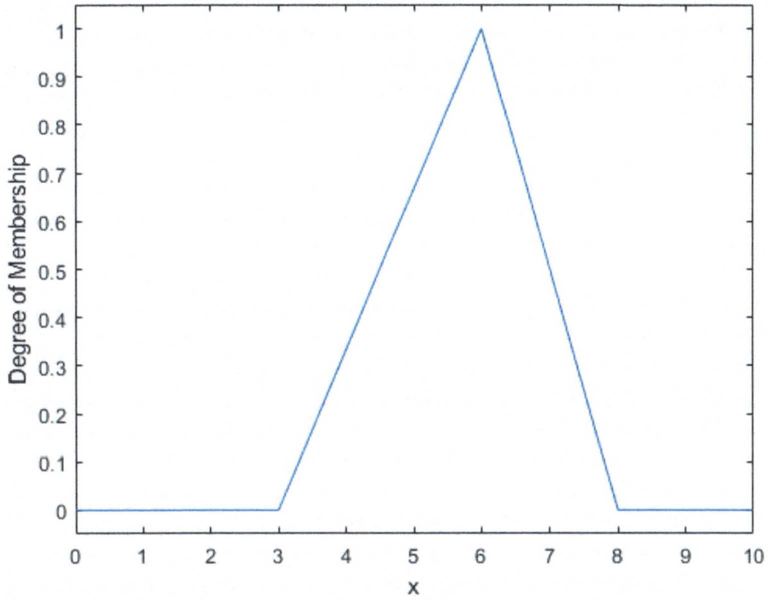

Fig. 7.13 The plot of a triangular membership function

$$f(x, a, b, c) = \left\{ \begin{array}{ll} 0, & x < a \\ \frac{x-a}{b-a}, & a \leq x \leq b \\ \frac{c-x}{c-b}, & b \leq x \leq c \\ 0, & c \leq x \end{array} \right\} \tag{7.1}$$

Or, more compactly:

$$f(x, a, b, c) = max\left(min\left(\frac{x-a}{b-a}, \frac{c-x}{c-b}\right), 0\right) \tag{7.2}$$

Hence, x represents the input variable of the membership function, while a, b, and c are the parameters specified by the tutor. For instance, the plot in Fig. 7.13 demonstrates the triangular membership function with the parameters a = 3, b = 6, and c = 8.

Similarly, the trapezoidal function produces fuzzy participation values obtained by the following equation:

$$f(x, a, b, c, d) = max\left(min\left(\frac{x-a}{b-a}, 1, \frac{d-x}{d-c}\right), 0\right) \tag{7.3}$$

When x is the input variable of the membership function, the tutor specifies the parameters a, b, and c. The parameters b and c represent the shoulders of the

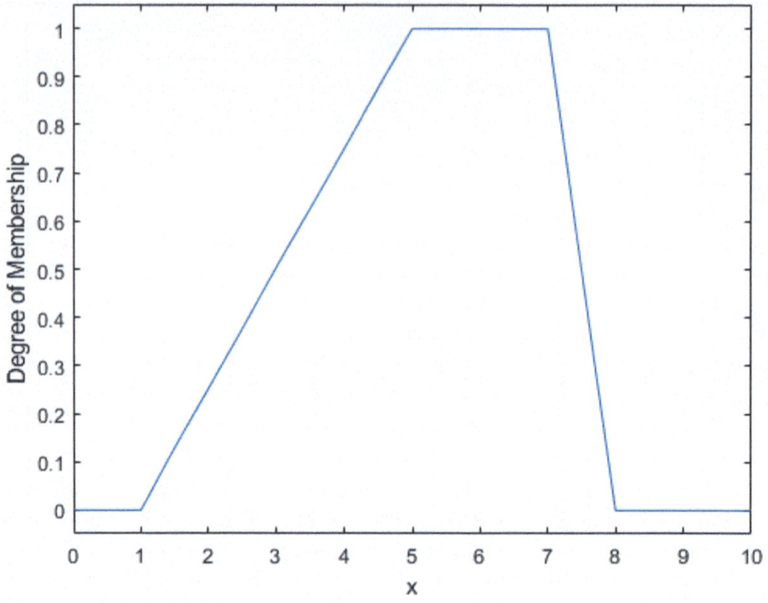

Fig. 7.14 The plot of a trapezoidal membership function

membership function, while a and d represent the function's legs. The specific values of b and c determine the shape of the membership function.

- When c is greater than b, the resulting membership function has a trapezoidal shape.
- When b equals c, the resulting membership function is analogous to a triangular membership function with parameters [a b d].
- When c is less than b, the resulting membership function is triangular with a maximum value of less than 1.

Figure 7.14 illustrates the plot of the trapezoidal membership function with the following parameters: a = 1, b = 7, c = 7, and d = 8.

Finally, the Gaussian function returns fuzzy membership values calculated by the following equation:

$$f(x, \sigma, c) = e^{\frac{-(x-c)^2}{2\sigma^2}} \tag{7.4}$$

Parameter σ is the standard deviation, while parameter c is the mean. A Gaussian membership function differs from a Gaussian probability distribution, as it always has a maximum value of 1. Figure 7.15 illustrates the plot of the Gaussian membership function with the following parameters: $\sigma = 2$, c = 5.

The authoring tool provides detailed information about the meaning of each parameter for every equation and enables the tutor to input their preferred values. Following this, the tutor can navigate to the "Fuzzy Variables" section to review the preset fuzzy variables used in the suite, including "Knowledge Level," "Frequency

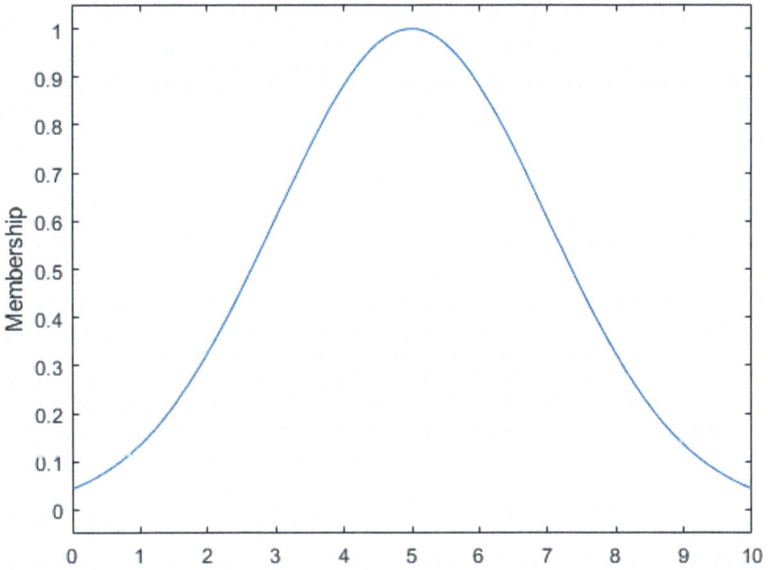

Fig. 7.15 The plot of a Gaussian membership function

of Errors," "Frequency of Asking Help," "Response Time," "Quiz Performance," and "Reduction Rate." The suite's Fuzzy Logic Student Modeling component utilizes the "Knowledge Level" variable. The assessment of Fuzzy EPATHLO and the demonstration of the benefits of adjusting the scenario based on the estimated student's knowledge level were carried out using this data. By selecting the "Knowledge Level" fuzzy variable, the tutor can observe the fuzzy sets employed to characterize the student's knowledge level, such as "Unknown," "Unsatisfactorily Known," "Known," and "Learned." For each fuzzy set, the tutor can define the type of membership function and its associated parameters. This process can also be applied to the other fuzzy variables and their respective fuzzy sets.

The default parameters for each trapezoidal membership function of the four fuzzy sets are:

- Unknown: a = 0, b = 0, c = 45, d = 50.
- Unsatisfactorily Known: a = 45, b = 50, c = 65, d = 70.
- Known: a = 65, b = 70, c = 85, d = 90.
- Learned: a = 85, b = 90, c = 100, d = 100.

The tutor can update the existing membership functions to assess the student's level of understanding more accurately. Afterward, by selecting a test option, they can input the quiz score and view the characterization of the student's knowledge level as determined by the membership functions they have specified. This allows the tutor to thoroughly test if the AI system's outcomes correspond with their estimations. Additionally, the tutor can restore the default membership functions in the system.

While not an expert in AI, the tutor can evaluate the student's level of knowledge using simple linguistic rules. The default determination of the knowledge level is primarily based on the student's quiz score. Nevertheless, the tutor can incorporate additional factors influencing the knowledge level assessment. As discussed in the "Fuzzy Variables" section, these factors encompass the error rate, frequency of help requests, and response time. The "Frequency of Errors" denotes the percentage of errors the student makes on a question. The "Frequency of Asking for Help" represents the percentage of times the student requests help with a question. The "Response Time" reflects the extra time the student takes to complete a quiz, considering that the average completion time is typically around 10 minutes. Finally, the "Reduction Rate" indicates the percentage by which the student's success score on a quiz is diminished, thereby impacting their knowledge level ranking.

For instance, if the estimated reduction rate is 10% and the student scored 90% on the quiz, the score would change to 90 - 90 * 0.1 = 81. As outlined in Sect. 5.4, with a success rate of 90%, the student's knowledge level is categorized as "Learned" at 100%. However, with a reduced success rate of 81%, the knowledge level drops to 100% "Known." Consequently, any modification to the student modeling AI mechanism immediately impacts the selection of questions for subsequent quizzes and the adaptive mechanisms of the games connected to the EPATHLO Educational Game Suite.

Once again, it is essential to emphasize that evaluating Fuzzy EPATHLO and the benefits derived from its goal to enhance student engagement and simultaneously improve their knowledge revolves around the predefined mechanism of student modeling, which considers their quiz success rates. The EPATHLO Author administrator can input new fuzzy variables and corresponding membership functions, which are readily accessible and visible to the tutor. Additionally, the administrator can incorporate more function types, such as piecewise linear or singleton. However, integrating the appropriate source code into the EPATHLO Fuzzy Student Modeling component is necessary for this scenario's new fuzzy variables or membership function types. Once again, it is evident that the EPATHLO Educational Game Suite leverages web services technology to ensure that the results of student modeling are immediately accessible from any external application without necessitating further adjustments by their creators.

The screenshots below showcase the user interface of EPATHLO Author, which enables tutors to manage the student modeling AI mechanism using the Fuzzy Content Module. Figure 7.16 displays the page for modifying membership function types, while Fig. 7.17 presents the predefined fuzzy variables within EPATHLO Author. Administrators have the option to include additional fuzzy variables. In addition, Fig. 7.18 shows the page containing the fuzzy sets associated with the "Knowledge Level" fuzzy variable. Tutors can use this page to adjust the parameters of the membership functions linked to each fuzzy set.

In Fig. 7.19, you can see the page where the instructor can observe all the regulations used in the fuzzy inference process. Figure 7.20 displays the interface that allows a non-specialist tutor to construct linguistic rules.

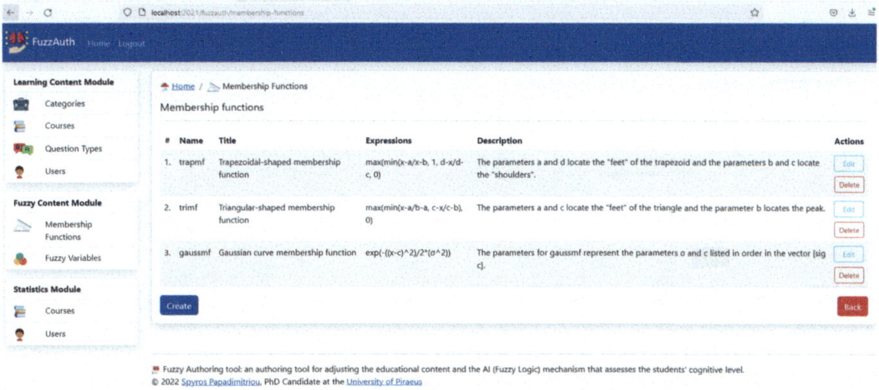

Fig. 7.16 The types of membership functions EPATHLO Author supports

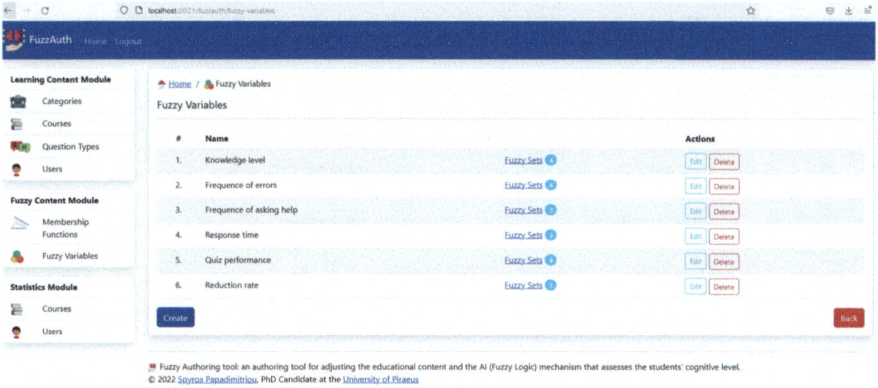

Fig. 7.17 The predefined Fuzzy Variables that EPATHLO Author supports

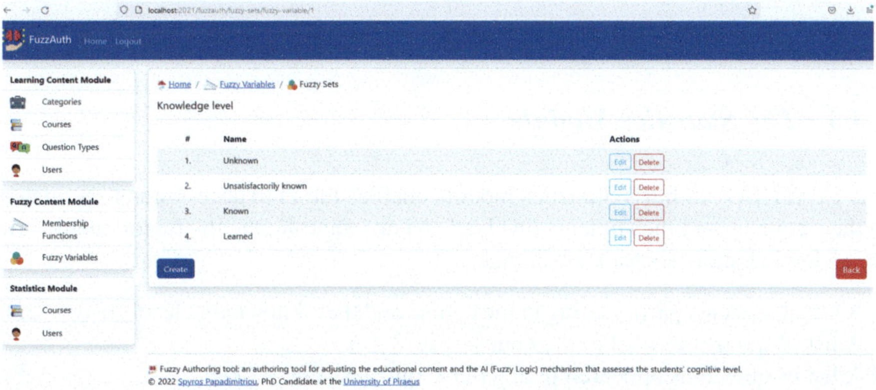

Fig. 7.18 The fuzzy sets related to the fuzzy variable "Knowledge Level"

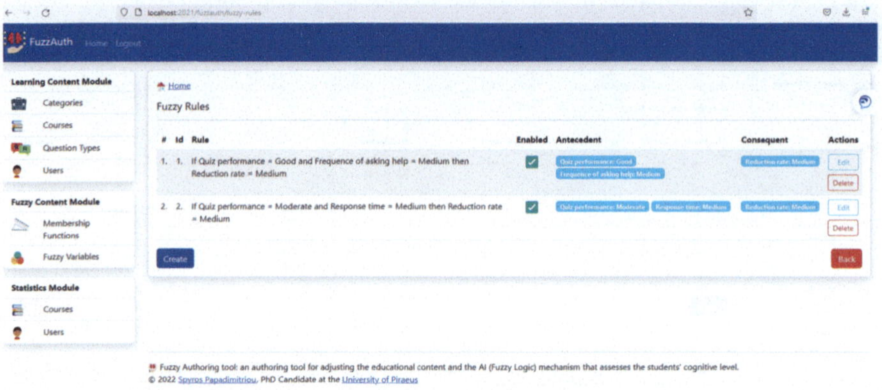

Fig. 7.19 The tutor can view the fuzzy inference system's rules

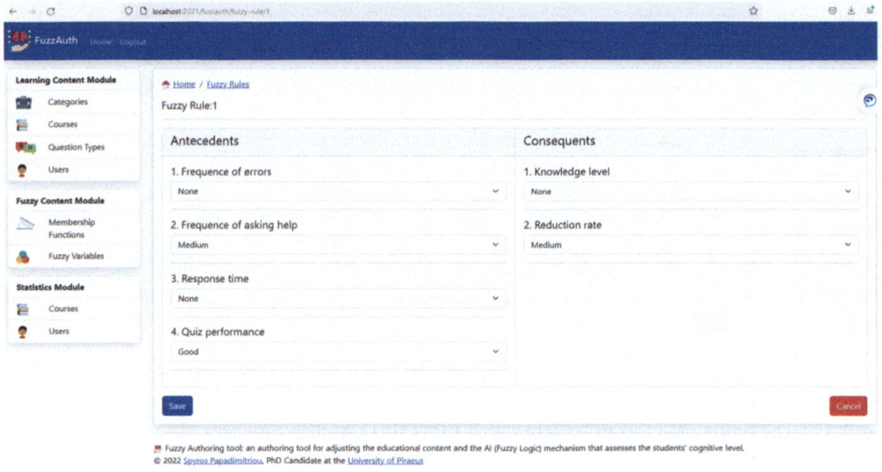

Fig. 7.20 The tutor without AI knowledge can create their own fuzzy rules

7.3.3 The Statistics Module

The EPATHLO Author's Statistics module enables tutors to access statistics related to their courses and the students involved. As an example, the instructor can review the following statistics for each course:

- A list of students participating in the course and their knowledge levels (Fig. 7.21).
- A list of quizzes related to the course (Fig. 7.22).
- A list of questions appeared in all course quizzes (Fig. 7.23).
- The tutor can view the answers given by the students and the corresponding percentages for each question (Fig. 7.24). Furthermore, the EPATHLO Author presents the percentages of the questions categorized by their difficulty level.

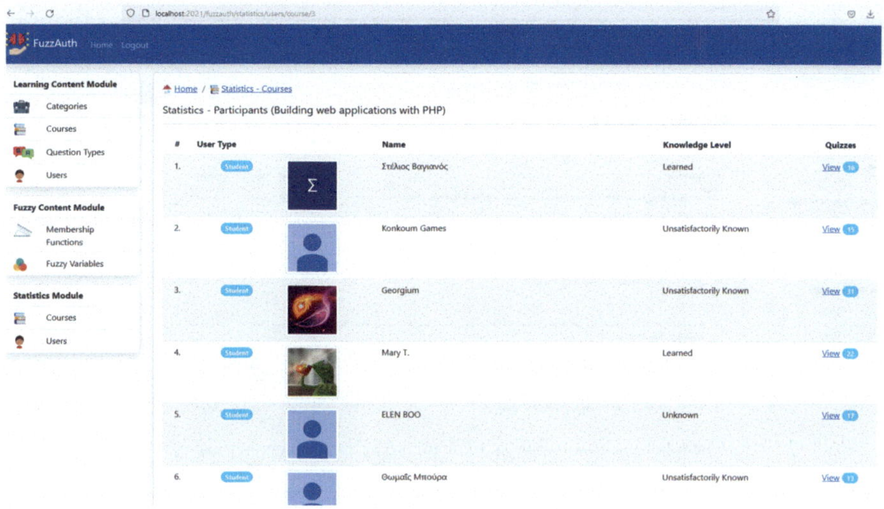

Fig. 7.21 The list of participants in the course "Building Web Applications with PHP"

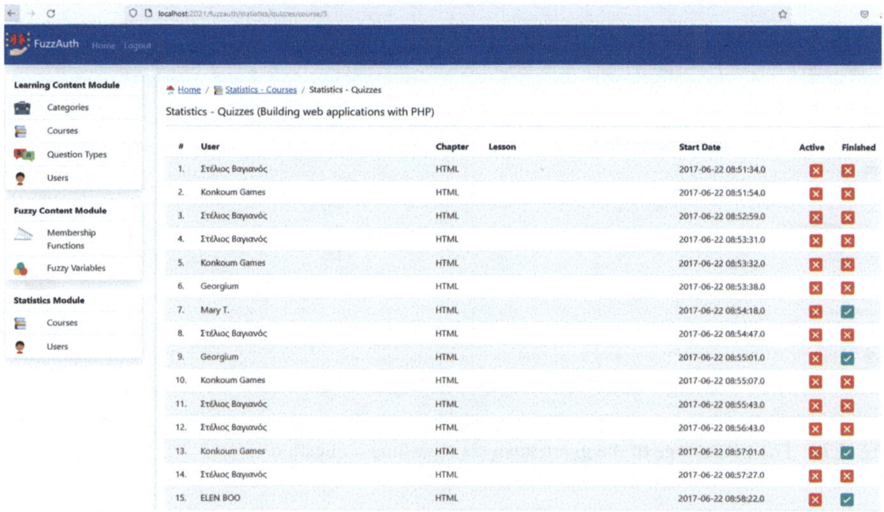

Fig. 7.22 The list of the quizzes of the course "Building Web Applications with PHP"

All the above information types also apply to a chapter or lesson (Fig. 7.25).

In addition, the instructor has access to individual student data for each course. This includes details such as the quizzes they attempted, completion status, their responses, and corresponding success rates (Figs. 7.26 and 7.27). Furthermore, EPATHLO Author categorizes questions based on their difficulty level, allowing the

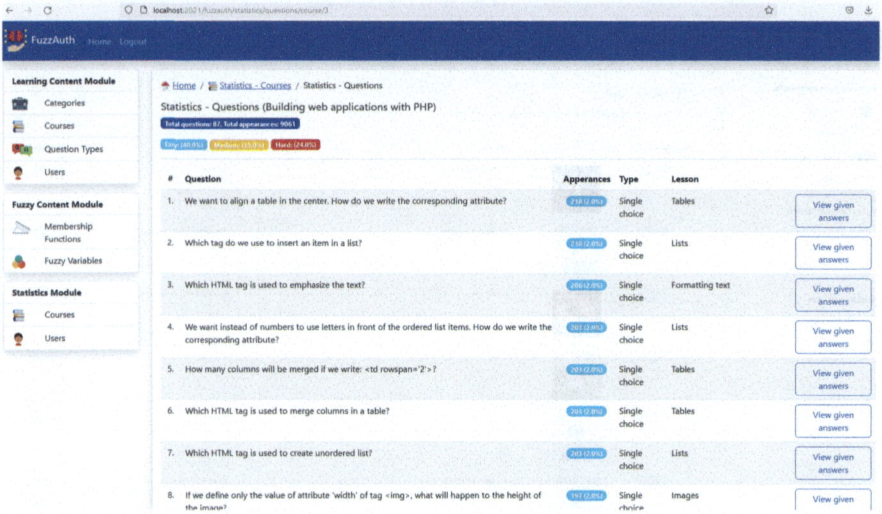

Fig. 7.23 The list of all questions that appeared in the quizzes of the course "Building Web Applications with PHP"

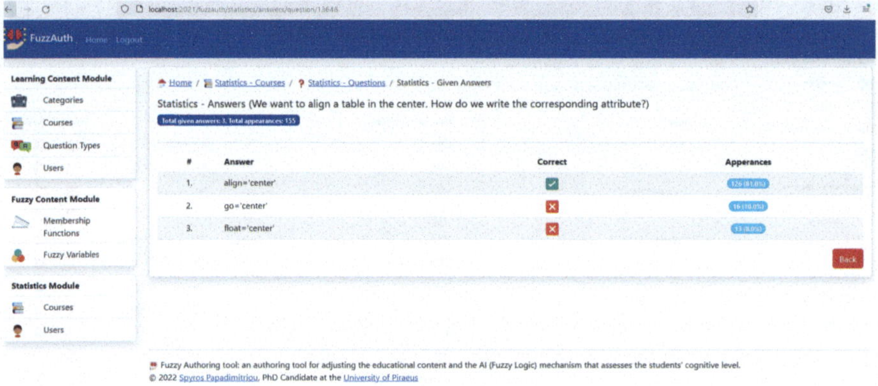

Fig. 7.24 The percentage of the given answers regarding a question

instructor to understand the challenges students have encountered or are currently facing across the quizzes associated with the course.

The tutor can evaluate the student's comprehension after each quiz, enabling them to monitor their learning progress. By analyzing this data in conjunction with the quiz questions, the tutor can evaluate the efficacy of the AI question selection process and determine if the questions were appropriately tailored to the student's knowledge level. If the tutor identifies any questions that may not align with the intended difficulty level, they can make necessary adjustments. Furthermore, they can add or remove questions to ensure comprehensive curriculum coverage and

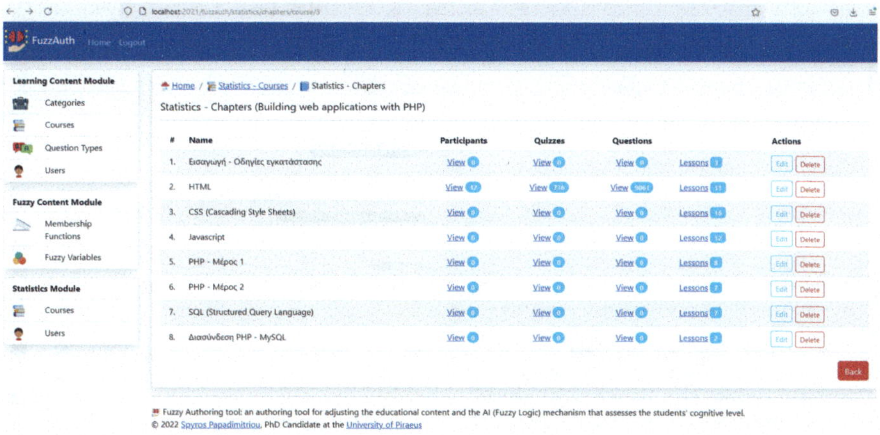

Fig. 7.25 Statistics regarding the chapters of the course "Building Web applications with PHP"

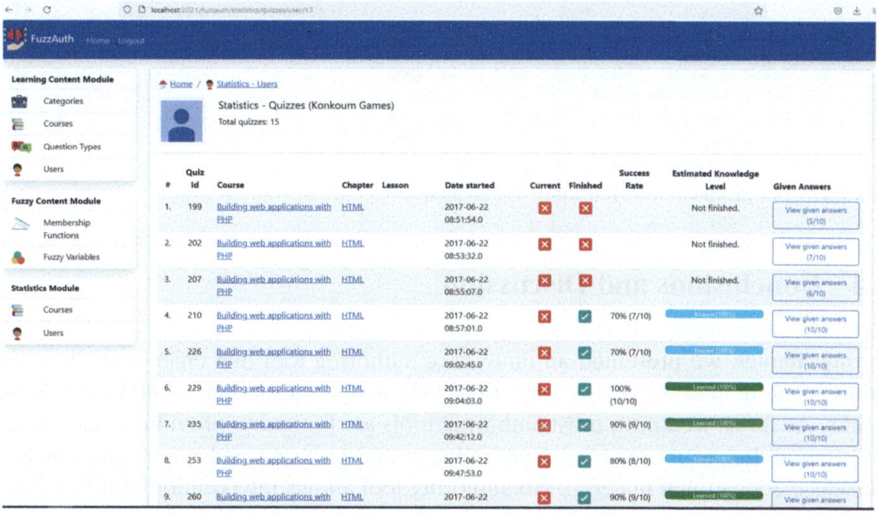

Fig. 7.26 Statistics regarding a specific student

support students to acquire the desired knowledge. These considerations also pertain to individual chapters and lessons.

The figures below illustrate the tutor's access to the features provided by EPATHLO Author.

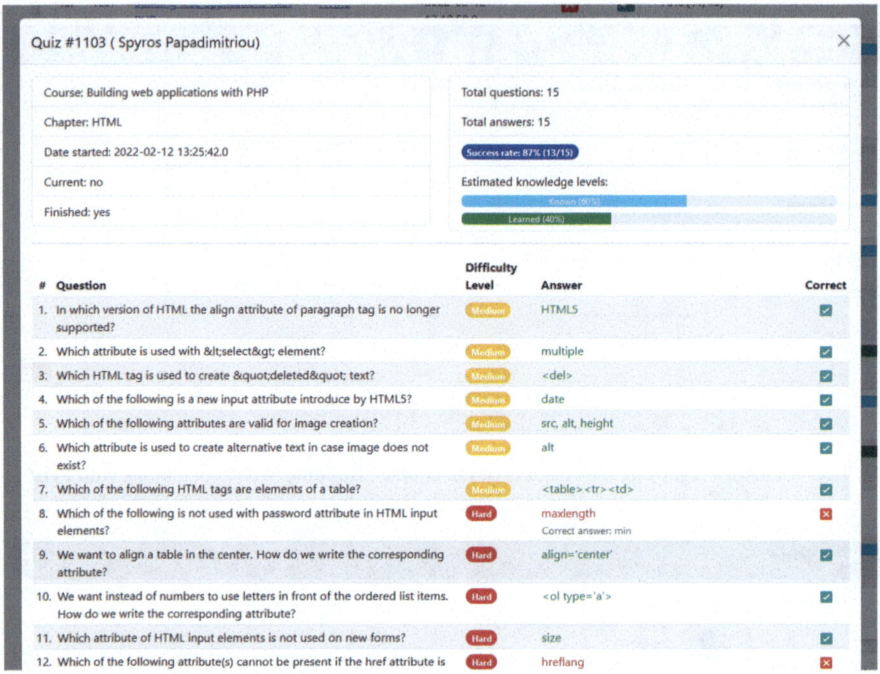

Fig. 7.27 The given answers related to a specific quiz

7.4 Conclusions and Discussion

In this chapter, we presented an innovative authoring tool that enables the teacher to intervene in how the AI mechanism used to assess the student's knowledge level works. As far as we know, most authoring tools are limited to providing the teacher with the ability to monitor students' educational progress and easily and quickly introduce educational material. No authoring tool allows intervention in the student modeling functions to produce appropriate personalized materials.

EPATHLO Author provides two ways of configuring how the learner's knowledge level is calculated. One targets teachers with no AI knowledge and uses simple verbal rules. The second is aimed at teachers who know AI, specifically how an FIS works. They can use additional fuzzy variables to create the corresponding fuzzy set and membership functions for each.

As we move forward, the full implementation of EPATHLO Author opens up exciting possibilities. The upcoming real-world evaluation by tutors will help us confirm that the ability to modify AI techniques in an intelligent teaching system indeed serves the tutor, eliminates the need for programmer and AI expert intervention, and, most importantly, enhances the user's personalized learning experience.

References

1. Ahn, J.Y., Mun, G.S., Han, K.S., Choi, S.H.: An online authoring tool for creating activity-based learning objects. Educ. Inf. Technol. **22**, 3005–3015 (2017)
2. Alepis, E., Kabassi, K., Virvou, M.: Personalized museum exploration by mobile devices. In: Auer, M.E., Tsiatsos, T. (eds,) Interactive Mobile Communication Technologies and Learning - Proceedings of the 11th IMCL Conference, 30 November–1 December 2017, Mediterranean Palace Hotel, Thessaloniki, Greece, Advances in Intelligent Systems and Computing, vol. 725, pp. 353–360. Springer (2017). https://doi.org/10.1007/978-3-319-75175-7_36
3. Baldiris, S., Vargas, D., Garzón, J., Ávila-Garzón, C., Burgos, D.: Evaluation of authoring tools under atag and wcag recommendations. Univ. Access Inf. Soc. **22**(3), 919–930 (2023)
4. Blattgerste, J.: The design space of augmented reality authoring tools and its exploration for the procedural training context. Ph.D thesis, Bielefeld University (2023)
5. Bräker, J., Hertel, J., Semmann, M.: Empowering users to create augmented reality-based solutions–deriving design principles for no-code ar authoring tools (2023)
6. Cabada, R.Z., Estrada, M.L.B., García, C.A.R.: Educa: a web 2.0 authoring tool for developing adaptive and intelligent tutoring systems using a kohonen network. Exp. Syst. Appl. 38(8):9522–9529 (2011)
7. Chad Lane, H., Core, M.G., Hays, M.J., Auerbach, D., Rosenberg, M.: Situated pedagogical authoring: authoring intelligent tutors from a student's perspective. In: Artificial Intelligence in Education: 17th International Conference, AIED 2015, Madrid, Spain, 22–26 June 2015. Proceedings, vol. 17, pp. 195–204. Springer (2015)
8. Chamusca, I.L., Santos, F.V.D.J., Ferreira, C.V., Murari, T.B., Junior, A.L.A., Winkler, I.: Evaluation of design guidelines for the development of intuitive virtual reality authoring tools: a case study with nvidia omniverse. In: 2022 IEEE International Symposium on Mixed and Augmented Reality Adjunct (ISMAR-Adjunct), pp. 357–362. IEEE (2022)
9. Coelho, H., Melo, M., Barbosa, L., Martins, J., Teixeira, M.S., Bessa, M.: Authoring tools for creating 360 multisensory videos-evaluation of different interfaces. Expert. Syst. **38**(5), e12418 (2021)
10. Coelho, H., Monteiro, P., Gonçalves, G., Melo, M., Bessa, M.: Authoring tools for virtual reality experiences: a systematic review. Multimed. Appl. **81**(19), 28037–28060 (2022)
11. Dermeval, D., Paiva, R., Bittencourt, I.I., Vassileva, J., Borges, D.: Authoring tools for designing intelligent tutoring systems: a systematic review of the literature. Int. J. Artif. Intell. Educ. **28**, 336–384 (2018)
12. Escudero, H., Fuentes, R.: Exchanging courses between different intelligent tutoring systems: a generic course generation authoring tool. Knowl. Based Syst. **23**(8), 864–874 (2010)
13. Ez-Zaouia, M., Marfisi-Schottman, I., Oueslati, M., Mercier, C., Karoui, A., George, S.: A design space of educational authoring tools for augmented reality. In: International Conference on Games and Learning Alliance, pp. 258–268. Springer (2022)
14. Ez-Zaouia, M., Marfisi-Schottman, I., Mercier, C.: Authoring tools: the road to democratizing augmented reality for education. In: Proceedings of the 15th International Conference on Computer Supported Education, vol. 1, pp. 115–127 (2023)
15. Gómez-Albarrán, M., Sarasa-Cabezuelo, A., Sierra-Rodríguez, J.L., Temprado-Battad, B.: Authoring and playing interactive fiction with conventional web technologies. Multimed. Tools Appl. **81**(11), 14705–14747 (2022)
16. Gordillo, A., Barra, E., Gallego, D., Quemada, J.: An online e-learning authoring tool to create interactive multi-device learning objects using e-infrastructure resources. In: 2013 IEEE Frontiers in Education Conference (FIE), IEEE, pp 1914–1920 (2013)
17. Gordillo, A., Barra, E., Quemada, J.: An easy to use open source authoring tool to create effective and reusable learning objects. Comput. Appl. Eng. Educ. **25**(2), 188–199 (2017)
18. Green, D., Hargood, C., Charles, F.: Use of tools: Ux principles for interactive narrative authoring tools. J. Comput. Cult. Heritage (JOCCH) **14**(3), 1–25 (2021)
19. Halperin, X.: Marvel's spider-man: Miles morales procedural tools for playstation 5 content authoring. In: SIGGRAPH Asia 2021 Technical Communications, pp. 1–4 (2021)

20. Herbert, B., Billinghurst, M., Weerasinghe, A., Ens, B., Wigley, G.: A generalized, rapid authoring tool for intelligent tutoring systems. In: Proceedings of the 30th Australian Conference on Computer-Human Interaction, pp. 368–373 (2018)
21. Horst, R., Gerstmeier, S., Naraghi-Taghi-Off, R., Wagner, J., Rau, L., Dörner, R.: Virtual reality content creation based on self-contained components in the e-learning domain: Re-using pattern-based vr content in different authoring toolkits. Multimed. Tools Appl. **83**(15), 46557–46594 (2024)
22. Hossain, M.Y., Zaman, L.: Nccollab: collaborative behavior tree authoring in game development. Multimed. Appl. **82**(3), 4671–4708 (2023)
23. Kamitsios, M., Papadimitriou, S., Chrysafiadi, K., Virvou, M.: Incorporating quizzes into a 3d educational game calling web services. In: EdMedia+ Innovate Learning, Association for the Advancement of Computing in Education (AACE), pp. 1170–1179 (2018)
24. Katsionis, G., Virvou, M.: A cognitive theory for affective user modelling in a virtual reality educational game. In: 2004 IEEE International Conference on Systems, Man and Cybernetics (IEEE Cat. No. 04CH37583), vol. 2, pp. 1209–1213. IEEE (2004)
25. Khademi, M., Haghshenas, M., Kabir, H.: A review on authoring tools. In: Proceedings of the 5th International Conference on Distance Learning and Education, IPCSIT, vol. 12, pp. 40–44 (2011)
26. Kim, H., Hong, S., Kim, J., Jang, T., Woo, W., Heo, S., Lee, B.: Realitybrush: an ar authoring system that captures and utilizes kinetic properties of everyday objects. Multimed. Tools Appl. **80**, 31135–31158 (2021)
27. Kim, H., Hong, S., Kim, J., Jang, T., Woo, W., Heo, S., Lee, B.: Correction to: Realitybrush: an ar authoring system that captures and utilizes kinetic properties of everyday objects. Multimed. Tools Appl. **81**(22), 32703–32703 (2022)
28. Kwon, S.O., Kim, J.O., Ju, S.Y., Jeong, J.S., Yoo, K.H.: Design and implementation of educational content authoring tool for smart devices. J. Korea Contents Assoc. **13**(12), 1–8 (2013)
29. Lam, M.C., Muhammad Nizam, S.S., Arshad, H., Suwadi, N.A.: Tangible interaction technique with authoring capability for kitchen design. Multimed. Tools Appl. **82**(19), 30125–30150 (2023)
30. Liapis, A., Guckelsberger, C., Zhu, J., Harteveld, C., Kriglstein, S., Denisova, A., Gow, J., Preuss, M.: Designing for playfulness in human-ai authoring tools. In: Proceedings of the 18th International Conference on the Foundations of Digital Games, pp. 1–4 (2023)
31. Marne, B., Labat, J.M.: Model and authoring tool to help teachers adapt serious games to their educational contexts. Int. J. Learn. Technol. **9**(2), 161–180 (2014)
32. Masneri, S., Dominguez, A., Mangina, E.: Workshop-authoring tools for xr applications in education. In: 2021 7th International Conference of the Immersive Learning Research Network (iLRN), pp. 1–2. IEEE (2021)
33. Massidda, M., Travaglini, L., Pescarin, S., et al.: Interactive digital narrative authoring tools and hybrid experiences in cultural heritage: an integrated review. In: GCH 2023-Eurographics Workshop on Graphics and Cultural Heritage (2023)
34. Mattos, D.P.D., Muchaluat-Saade, D.C., Ghinea, G.: Beyond multimedia authoring: On the need for mulsemedia authoring tools. ACM Comput. Surv. (CSUR) **54**(7), 1–31 (2021)
35. Mehm, F., Konert, J., Göbel, S., Steinmetz, R.: An authoring tool for adaptive digital educational games. In: 21st Century Learning for 21st Century Skills: 7th European Conference of Technology Enhanced Learning, EC-TEL 2012, Saarbrücken, Germany, 18–21 Sept 2012. Proceedings, vol. 7, pp. 236–249. Springer (2012)
36. Moundridou, M., Virvou, M.: Authoring intelligent tutoring systems over the world wide web. In: Proceedings First International IEEE Symposium Intelligent Systems, vol. 1, pp. 160–165. IEEE (2002)
37. Moundridou, M., Virvou, M.: An empirical study for the development of a web-based authoring tool for intelligent tutoring systems. In: The 4th International Workshop on Computer Science and Information Technologies (CSIT), September, pp. 18–20 (2002)
38. Moundridou, M., Virvou, M.: Wear: a web-based authoring tool for building intelligent tutoring systems. In: Proceedings of the 2nd Helenic Conference on AI, Companion volume, Thessaloniki, Greece, Citeseer, pp. 203–214 (2002)

39. Moundridou, M., Virvou, M.: Analysis and design of a web-based authoring tool generating intelligent tutoring systems. Comput. Educ. **40**(2), 157–181 (2003)
40. Moundridou, M., Virvou, M.: Usability evaluation of a web-based authoring tool for building intelligent tutoring systems. In: Proceedings of the 10th International (2003). Accessed http://citeseerx.ist.psu.edu/viewdoc/download
41. Mousavinasab, E., Zarifsanaiey, N., Niakan Kalhori, R.: S, Rakhshan M, Keikha L, Ghazi Saeedi M,: Intelligent tutoring systems: a systematic review of characteristics, applications, and evaluation methods. Interact. Learn. Environ. **29**(1), 142–163 (2021)
42. Müser, S., Maiero, J., Meyer, J., Hinkenjann, A.: Authoring tools for teaching in vr–an evaluation study. In: 2023 IEEE International Conference on Advanced Learning Technologies (ICALT), pp. 258–260. IEEE (2023)
43. Pires de Oliveira, V., de Jesus Macedo, R., Vinicius de Freitas, F., Machado, A., Murari, T., Winkler, I.: Virtual reality authoring tools acceptance and use: an exploratory study with the utaut2 model. In: Proceedings of the 25th Symposium on Virtual and Augmented Reality, pp. 299–303 (2023)
44. Olney, A.M., Gilbert, S.B., Rivers, K.: Preface to the special issue on creating and improving adaptive learning: smart authoring tools and processes (2022)
45. Olsen, J.K., Belenky, D.M., Aleven, V., Rummel, N., Sewall, J., Ringenberg, M.: Authoring tools for collaborative intelligent tutoring system environments. In: Intelligent Tutoring Systems: 12th International Conference, ITS 2014, Honolulu, HI, USA, 5–9 June 2014. Proceedings, vol. 12, pp. 523–528. Springer (2014)
46. Orsborn, G.: Use of rapid e-learning authoring tools in higher education. Whitireia Nurs. Health J. **24**, 47–51 (2017)
47. Papadimitriou, S., Chrysafiadi, K., Virvou, M.: Authoring technological and platform independent learning material and student's progress profile using web services. In: Knowledge-Based Software Engineering: 2018: Proceedings of the 12th Joint Conference on Knowledge-Based Software Engineering (JCKBSE 2018) Corfu, Greece, vol. 12, pp. 242–251. Springer (2019)
48. Papadimitriou, S., Chrysafiadi, K., Virvou, M.: Fuzzeg: Fuzzy logic for adaptive scenarios in an educational adventure game. Multimed. Tools Appl. **78**(22), 32023–32053 (2019). https://doi.org/10.1007/s11042-019-07955-w
49. Quah, C.Y., Ng, K.H.: A systematic literature review on digital storytelling authoring tool in education: January 2010 to January 2020. Int. J. Hum. Comput. Interact. **38**(9), 851–867 (2022)
50. Quintana, D., Rodríguez, A., Boada, I.: Authoring tools for procedural modeling of virtual reality-based rehabilitation exercises. IEEE Access **10**, 131567–131578 (2022)
51. Ramírez-Noriega, A., Martínez-Ramírez, Y., García, J.E.S., Bojórquez, E.M., Francisco Figueroa Pérez, J., Mendivil-Torres, J., Miranda, S.: Towards the automatic construction of an intelligent tutoring system: Domain module. In: New Knowledge in Information Systems and Technologies, vol. 1, pp. 293–302 . Springer (2019)
52. Roig, J., Ribera, M., Alcaraz, R.: Authoring tools are critical to the accessibility of the documents: Recommendations for word processors, including interaction design. In: Proceedings of the 10th International Conference on Software Development and Technologies for Enhancing Accessibility and Fighting Info-exclusion, pp. 160–165 (2022)
53. Rubab, S., Tang, J., Wu, Y.: Examining interaction techniques in data visualization authoring tools from the perspective of goals and human cognition: a survey. J. Vis. **24**, 397–418 (2021)
54. Salmen, F., Schroeder, U.: Evaluating authoring tools with the explorable authoring requirements (2024). arXiv:2403.17714
55. Schäfer, A.: Improving essential interactions for immersive virtual environments with novel hand gesture authoring tools. Ph.D thesis, Rheinland-Pfälzische Technische Universität Kaiserslautern-Landau (2023)
56. Shibolet, Y., Lombardo, V.: Resources for comparative analysis of idn authoring tools. In: International Conference on Interactive Digital Storytelling, pp. 513–528. Springer (2022)
57. Shiels, J., Tanner, M.: The perceived affordances and usage constraints of learning design authoring tools by south african high school teachers (2022)

58. Silva, M., Roberto, R., Radu, I., Cavalcante, P., Schneider, B., Teichrieb, V.: Development of design principles for ar authoring tools for education based on teacher's perspectives. IEEE Trans. Learn, Technol (2023)
59. Takacs, B., Vincze, Z.: Deep authoring-an ai tool set for creating immersive multimedia experiences. Multimed. Tools Appl. **80**(20), 31105–31134 (2021)
60. Tsihrintzis, G.A., Virvou, M., Saruwatari, T.: Special collection of extended selected papers on "Novel Research Results Presented at the 14th International Joint Conference on Knowledge-Based Software Engineering (jckbse2022), 22–24 Aug 2022, Larnaca, Cyprus" (2022). https://easyconferences.eu/jckbse2022/
61. Utterberg Modén, M., Tallvid, M., Lundin, J., Lindström, B.: Intelligent tutoring systems: why teachers abandoned a technology aimed at automating teaching processes (2021)
62. Virvou, M.: A human plausible reasoning theory in the context of an active help system for unix1 users. PhD thesis, University of Sussex, Falmer, East Sussex, UK (1993). https://ethos.bl.uk/OrderDetails.do?uin=uk.bl.ethos.787162
63. Virvou, M.: Automatic reasoning and help about human errors in using an operating system. Interact. Comput. **11**(5), 545–573 (1999)
64. Virvou, M.: A domain-independent reasoning mechanism for an its authoring tool. In: EdMedia+ Innovate Learning, Association for the Advancement of Computing in Education (AACE), pp. 1762–1764 (2000)
65. Virvou, M.: Modelling the knowledge and reasoning of users in a knowledge-based authoring tool. Int. J. Bus. Perform. Manag. **5**(2–3), 399–412 (2003)
66. Virvou, M., Alepis, E.: Creating tutoring characters through a web-based authoring tool for educational software. In: SMC'03 Conference Proceedings. 2003 IEEE International Conference on Systems, Man and Cybernetics. Conference Theme-System Security and Assurance (Cat. No. 03CH37483), vol. 5, pp. 4884–4889. IEEE (2003)
67. Virvou, M., Alepis, E.: Mobile educational features in authoring tools for personalised tutoring. Comput. Educ. **44**(1), 53–68 (2005)
68. Virvou, M., Kabassi, K.: Reasoning about users' actions in a graphical user interface. Hum. Comput. Interact. **17**(4), 369–398 (2002)
69. Virvou, M., Kabassi, K.: Evaluating an intelligent graphical user interface by comparison with human experts. Knowl. Based Syst. **17**(1), 31–37 (2004)
70. Virvou, M., Manos, K.: A simulated student-player in support of the authoring process in a knowledge-based authoring tool for educational games. In: Proceedings 3rd IEEE International Conference on Advanced Technologies, pp. 338–339. IEEE (2003)
71. Virvou, M., Moundridou, M.: An authoring tool for algebra-related domains. In: HCI (2), Citeseer, pp. 647–651 (1999)
72. Virvou, M., Moundridou, M.: A web-based authoring tool for algebra-related intelligent tutoring systems. J. Educ. Technol. Soc. **3**(2), 61–70 (2000)
73. Virvou, M., Papadimitriou, S.: Use of digital 'guess who board game'in teaching english. In: IISA 2014, The 5th International Conference on Information, Intelligence, Systems and Applications, pp. 439–443. IEEE (2014)
74. Virvou, M., Papadimitriou, S.: Playing, learning, extending: educational "guess who" game that is renewable by a teacher. In: 2015 6th International Conference on Information, Intelligence, Systems and Applications (IISA), pp. 1–5. IEEE (2015)
75. Virvou, M., Stavrianou, A.: User modelling in a gui. In: Proceedings of HCI International (the 8th International Conference on Human-Computer Interaction) on Human-Computer Interaction: Ergonomics and User Interfaces, vol. I, pp. 261–265 (1999)
76. Virvou, M., Manos, K., Katsionis, G., Tourtoglou, K.: Vr-integate: A knowledge-based authoring tool for virtual reality educational games. In: Proceedings of ED-MEDIA 2002 World Conferences on Educational Multimedia, Hypermedia and Telecommunications (2002)
77. Virvou, M., Lampropoulos, A.S., Tsihrintzis, G.A.: Inma: a knowledge-based authoring tool for music education. In: International Conference on Knowledge-Based and Intelligent Information and Engineering Systems, pp. 376–383. Springer (2006)

78. Yang, C., Chiang, F.K., Cheng, Q., Ji, J.: Machine learning-based student modeling methodology for intelligent tutoring systems. J. Educ. Comput. Res. **59**(6), 1015–1035 (2021)
79. Yu, E.: Designing tools for 3d content authoring based on 3d sketching. Ph.D thesis, Université Côte d'Azur (2023)
80. Zadeh, L.A.: Fuzzy logic=computing with words. IEEE Trans. Fuzzy Syst. **4**(2), 103–111 (1996)
81. Zarei, N., Quek, F., Chu, S.L., Brown, S.A.: Towards designing enactment-scaffolded narrative authoring tools for elementary-school children. In: Proceedings of the 20th Annual ACM Interaction Design and Children Conference, pp. 387–395 (2021)

Chapter 8
Practicing Soft Skills for Programming Through an Intelligent Adventure Game

Abstract Hard skills refer to an individual's abilities related to their knowledge of a particular field. On the other hand, soft skills are a person's abilities related to their personality traits and capacity to establish relationships. Employers are now looking for both hard and soft skills in job candidates. Soft skills cannot be defined or proven through certifications and degrees; they are acquired through life experience. Educational institutions usually do not teach soft skills, so people need to work on developing them themselves. The game of the book helps students develop hard (learning HTML programming) and soft skills for developers, such as communication, teamwork, interpersonal skills, and customer relations. It follows a holistic approach and fills the gap between these two skill sets. The game's tasks include mutual help between non-playable characters (NPCs) and the player, individual tasks and collecting items, interdisciplinary cooperation, and developing customer-oriented skills. Furthermore, the adaptivity of the game scenario reinforces these skills by adding extra tasks when necessary.

Keywords Educational software · Personalized educational games · Artificial Intelligence in educational games · Fuzzy logic in education · User experience · AI-empowered software games · User modeling · Human-AI interaction · e-learning games · 21st-century skills · Soft skills for programming · Human-centred Artificial Intelligence

8.1 Introduction

Integrating Artificial Intelligence (AI) and personalization in interactive systems allows automatic reasoning about users resulting in more human-like interfaces [57, 58, 63, 65, 66, 74]. In the case of educational systems, AI has significantly improved learning experiences and outcomes. AI-based technologies have been thoroughly researched for their potential to improve education by offering customized learning experiences [51]. Developments in learning analytics demonstrate how Machine Learning (ML) models are used to personalize educational content to

S. Papadimitriou and M. Virvou, *Artificial Intelligence—Based Games as Novel Holistic Educational Environments to Teach 21st Century Skills*, Intelligent Systems Reference Library 93, https://doi.org/10.1007/978-3-031-77464-5_8

address the unique needs of individual students [78]. Personalized mobile learning, facilitated through object-oriented user interfaces, further illustrates the role of AI in creating adaptive learning environments [4].

Authoring tools significantly enhance educational content development by offering diverse functionalities that improve learning experiences across domains. Virvou et al. [75] describe VR-INTEGATE, a tool for creating Virtual Reality (VR) educational games that immerse students in engaging scenarios. A distinct example of combining education with entertainment is the INMA tool, a knowledge-based authoring tool for music education that aims to improve music perception and comprehension using intelligent educational methods [76], while personalization in mobile museum guides can also be used to combine entertainment with education [5]. Virvou and Moundridou [69] highlight a tool for algebra, enabling Intelligent Tutoring Systems (ITSs) that adapt to learners' needs, offering personalized feedback and paths. Moundridou and Virvou [31] discuss web-based ITS authoring, which expands educational access and personalization. Virvou and Alepis [64] emphasize mobile features in authoring tools for on-the-go learning, enhancing accessibility and continuous education. Virvou [60] explores modeling user knowledge and reasoning in authoring tools to align content with cognitive processes, improving educational system responsiveness. Virvou [59] introduces a domain-independent reasoning mechanism for ITS, increasing the versatility of authoring tools across subjects. These studies demonstrate the transformative impact of authoring tools in creating adaptive, personalized, and accessible educational experiences.

The recent advancements in education have primarily focused on cognitive aspects of learning, overlooking the importance of developing soft skills. There are some research works targeting soft skills, but these are not prevailing in comparison to research concerning students' cognitive abilities. For example, modeling soft skills involves understanding and enhancing interpersonal abilities such as collaboration and communication. The research conducted by Tourtoglou and Virvou [50] provides an example of how this can be achieved through student modeling. Their approach uses simulated annealing to create optimal groups of learners in a UML learning environment, balancing technical expertise with soft skills to improve group dynamics and learning outcomes. While not a definitive solution, this method illustrates how computational techniques can address the complex task of fostering soft skills in educational settings. These non-technical skills, including teamwork, communication, leadership, and problem-solving, are vital for how individuals operate, interact with others, manage their time, and tackle challenges [11, 30]. Soft skills enhance workplace dynamics and foster a positive working environment [23, 40, 46]. They are instrumental in enabling effective communication, promoting better understanding among team members, and creating a positive atmosphere.

Moreover, they are essential for leadership and management roles, guiding teams towards success [3, 9, 43]. In customer-facing positions, soft skills such as communication, empathy, and problem-solving are fundamental for delivering exceptional customer service and nurturing positive customer relationships [25, 33, 48, 56].

Emotional intelligence, which encompasses abilities like self-awareness and empathy, is essential for understanding and managing both one's own and others' emotions.

Hence, soft skills are essential throughout a person's career and must be continually developed and maintained [41, 47, 49]. Employers increasingly recognize the importance of these skills, understanding that they are crucial not only to individual success but also to the competitiveness of their organizations [28, 53]. Consequently, the criteria for assessing a candidate's professional intelligence have expanded beyond academic qualifications to include soft skills, which promote better interaction and cooperation and facilitate business development [16, 17].

Research has consistently emphasized that excelling as a programmer requires more than technical expertise, especially in collaborative software development environments [1, 2, 6]. Soft skills are equally important, as they enhance the likelihood of success and positively impact project outcomes. Conflicts may arise due to the close interaction between software engineers, clients, and project teams. Unfortunately, most academic and corporate training focuses solely on technical skills, often neglecting soft skills development. This oversight can hinder recruitment and job opportunities, resulting in many skilled programmers but few with developed soft skills.

In educational software, traditional user modeling has mainly focused on customizing tutoring based on students' cognitive and behavioral traits while overlooking soft skills [52, 58, 62]. This gap also exists in educational games, where user modeling has emphasized cognitive and affective aspects but often disregarded the development of soft skills [18, 35, 39, 67]. Some researchers suggest incorporating game-based learning methods or environments enhanced with gamification elements [13].

Even though there is extensive literature on using games to teach technical skills, only some games are available for practicing soft skills, and their potential is limited. Recently, educational games, along with other non-gaming approaches, have been used to improve soft skills [10, 26, 27, 29, 32]. In a study on educational games for software engineering, motivation, willingness to learn, competition, and problem-solving skills were the most frequently emphasized soft skills by digital games, with 80 occurrences (100%) [15].

In light of the above points, this chapter outlines the Fuzzy EPATHLO game of the EPATHLO Educational Game Suite described in the previous chapter and its engaging setting that helps students practice their soft skills. The game's captivating environment sustains the player's interest and falls within the adventure game category [13]. In the past, we created adaptive scenarios in which the narrative unfolds through exploration and puzzle-solving and customized the content for student-players using fuzzy logic [36, 38]. In the current section, we concentrate on aspects of the game that promote regular interaction with non-playable characters (NPCs), which provides an effective means of practicing soft skills such as communication, teamwork, and customer orientation. The game leverages typical adventure game features, including dialogues with NPCs. The NPCs encourage players to engage and solve problems through continuous dialogues and task assignments.

Simultaneously, this fosters empathy towards the NPCs and motivates players to seek resolutions to the issues posed by the NPCs.

The assessment of personalized educational software has predominantly centered on technical proficiency and cognitive skill enhancement while failing to address soft skill development adequately. Studies have primarily focused on evaluating the effectiveness of intelligent graphical user interfaces and multimedia educational tools using technical and cognitive metrics [66, 77]. Furthermore, evaluations of the gaming aspects within personalized educational software have delved into the impact of VR and adaptive features on improving engagement and cognitive outcomes but have overlooked the development of soft skills [37, 61, 68].

For the evaluation, we identified ten essential soft skills for software engineers. We employed the cognitive walkthrough method with human experts to evaluate the game's effectiveness in developing these skills. The identified skills are communication, teamwork, analytical, organizational and planning, interpersonal, learning, customer orientation, answer-finding, interdisciplinary adaptation, and coordination skills. The results indicate that the intelligent educational game (IEG) effectively facilitates the development of these soft skills to different extents through its diverse range of tasks.

8.2 Soft Skills Identification in EPATHLO Educational Game Suite

8.2.1 Preliminary Work

Various researchers identify the above problem and either study how some educational institutions deal with the issue or suggest methods to improve students' soft skills [7, 20, 81]. The methods are varied, but in both cases, they fall into the category of active learning [19, 45]. Active learning is an instructional approach involving students' active engagement with the subject matter through discussions, problem-solving, case studies, role-playing, or even new technologies and educational software.

The authors have created other adaptive educational games that offer various learning benefits [34, 35, 70–73]. However, some of these games belong to game genres different than adventure games and do not facilitate the development of social skills or only do so to a limited extent. For instance, they lack NPCs with whom players can interact, show empathy, cooperate, and assist each other. Nonetheless, as outlined in the earlier chapters, this study's main game leverages adventure games' unique characteristics to help students enhance their technical skills in HTML. The evaluation has demonstrated that the game is a valuable resource for students, enabling them to tackle ongoing challenges while achieving two objectives: acquiring knowledge and enjoying themselves. The game excels in its primary purpose and significantly reinforces students' soft skills. The inclusion of NPCs and tasks assigned to the protagonist serves to strengthen their soft skills.

A literature review was conducted to identify the most essential soft skills for upcoming software engineers, considering the wide range of such skills. Our study aimed to identify the soft skills crucial for the practice of software engineering. The software engineering industry frequently refers to 30 categories related to these skills. Among these, five categories of social skills were present in at least half of the items studied. These five social skills are communication, teamwork, analytical, organizational/planning, and interpersonal [2, 7, 14, 44].

Furthermore, we have discovered a set of soft skills essential for programmers and software developers that have not been previously included in research compilations found in the existing literature. One of these soft skills is the ability to adapt across disciplines [8, 54, 55, 79]. It is worth noting that effective interdisciplinary collaboration is a top priority for research funding agencies. Thus, it is important to acknowledge that interdisciplinary work necessitates translational efforts, specifically in adapting research findings to align with different scientific perspectives, to arrive at conclusions beneficial for all researchers and disciplines involved.

Based on our findings, professional and academic programmers must possess ten essential soft skills: communication, teamwork, analytical, organizational and planning, interpersonal, learning, customer orientation, problem-solving, interdisciplinary adaptability, and coordination.

8.2.2 The Cognitive Walkthrough Process

In our research, we employed the cognitive walkthrough method [21, 24, 42] to assess the skills acquired during each game task. The cognitive walkthrough is a technique primarily used to evaluate a system's learnability from the perspective of a new user [12, 22, 80]. Unlike user testing, it does not require the involvement of actual users, making it a relatively cost-effective approach. We conducted a workshop in which five soft-skills experts, including three educators from the Computer Science department of the University of Piraeus and two game designers, actively participated. They were provided with a comprehensive description of the game, its objectives, and instructions for gameplay.

In addition, we provided a list of all the game tasks, predefined soft skills, and task categories, instructing the experts to record each move that resulted in a specific outcome and note the NPC assigning the tasks. Whenever an expert encountered obstacles, we gave suggestions to help them advance. The game's adaptive hinting system supplied helpful cues for HTML quizzes when it detected that an expert was having difficulty. The task categories mentioned above include the following:

Category 1

Players meet NPCs who can help them complete a task. They can also make friends with them. In turn, they can receive help from them.

Category 2

The players are given tasks broken down into smaller actions and subtasks. To achieve their objectives, players need to obtain or discover various items from NPCs. These items will help them overcome challenges and progress further in the game.

Category 3

The player must carefully utilize each item gathered in the correct sequence to accomplish a specific task or subtask.

Category 4

Players are given diverse and interdisciplinary tasks, such as helping a child resolve a speed issue with their toy car by discovering and supplying the necessary oil.

After completing the game, the experts recorded the steps required for each task. We then presented the game on a large screen and played it with the experts. After completing each task, we discussed with the experts the category to which it belonged and the soft skills it helped to improve, leading to the creation of Table 8.1. This table provides a comprehensive summary of the study's methodology, detailing all the tasks, their respective categories, and the soft skills the experts felt they helped cultivate.

Table 8.2 presents the soft skills developed by each task category.

Each soft skill is described below, along with the task categories designed to help develop it.

1. Analytical skills

Gathering, analyzing, interpreting, and evaluating information to make informed decisions or solve complex problems requires analytical skills. Individuals with strong analytical skills can deconstruct complex issues into smaller components, recognize patterns, and draw significant conclusions. These abilities, often associated with critical thinking, are beneficial in various professional fields. Tasks falling into categories 2, 3, and 4 encourage the development of analytical skills by assessing the task at hand and identifying steps to resolve and accomplish them.

2. Answer-finding skills

The phrase "answer-finding skills" might not be an officially acknowledged term, but it pertains to abilities that empower a person to uncover information or solutions comprehensively and effectively. These abilities include the capacity for carrying out research, critical thinking, and practical problem-solving. The tasks in Category 4 enrich and strengthen these skills, rendering them advantageous and sought after in diverse environments.

3. Communication skills

As shown in Table 8.2, tasks that fall into at least one of categories 1, 2, and 4 develop communication skills. Effective communication involves conveying information, concepts, ideas, and feelings through spoken, nonverbal, and written methods. It

Table 8.1 The tasks, the categories they belong and the soft skills they promote

Task			
NPC	Objective	Categories	Soft skills
Businessman	Solve quiz	1	Teamwork, Learning, Answer finding, Collaboration with others
Chef	Solve quiz	1	Teamwork, Learning, Answer finding, Collaboration with others
Kid	Help the kid to play with his board game	1, 2, 3, 4	Communication, Teamwork, Analytical, Organizational/planning, Interpersonal, Learning, Customer orientation, Answer finding, Interdisciplinary adjustment, Coordination with others
Muscle man	Solve quiz	1	Teamwork, Learning, Answer finding, Collaboration with others
NPC with thick hair	Solve quiz	1	Teamwork, Learning, Answer finding, Collaboration with others
Older man	Find a music album so he can listen to music	1, 2	Communication, Teamwork, Analytical, Organizational/planning, Interpersonal, Customer orientation, Coordination with others
Thirsty man	Find a tasty and cool drink to cool him down	1, 2, 3	Communication, Teamwork, Analytical, Organizational/planning, Interpersonal, Customer orientation, Interdisciplinary adjustment, Coordination with others
Witch	Offer her a crystal ball so she can communicate with the spirits	1, 2, 3, 4	Communication, Teamwork, Analytical, Organizational/planning, Interpersonal, Learning, Customer orientation, Answer finding, Interdisciplinary adjustment, Coordination with others
	Give her money	1, 2, 3	Communication, Teamwork, Organizational/planning Interpersonal, Customer orientation, Collaboration with others
	Help her cool down	1, 2, 3, 4	Communication, Teamwork, Analytical, Organizational/planning, Interpersonal, Learning, Customer orientation, Answer finding, Interdisciplinary adjustment, Collaboration with others
	Solve quiz	1	Teamwork, Learning, Answer finding, Collaboration with others
Zombie	Solve quiz	1	Teamwork, Learning, Answer finding, Collaboration with others
Final examiner	Solve quiz	1	Teamwork, Learning, Answer finding, Collaboration with others

Table 8.2 Soft skills identification for each task category

Soft skills for programming	The players meet Non-Playable Characters (NPC) which may help to accomplish a task and make friends. In turn, they may be helped by them.	The players are given quests in the form of executing tasks, which involve sub-tasks and the use of objects that the players have to find or be given by an NPC and collect to use at a later stage	The players have to use the collected objects, one by one, in an appropriate way to complete a task or sub-task	The tasks assigned to players are highly interdisciplinary, such as the task of helping a child fix the problem with the speed of his toy-cars by finding and providing oil
1. Analytical skills		✓	✓	✓
2. Answer finding skills				✓
3. Communication skills	✓	✓		✓
4. Customer orientation skills				✓
5. Interdisciplinary adjustment skills	✓			✓
6. Interpersonal skills	✓	✓		✓
7. Learning skills		✓		✓
8. Organiza-tional/planning skills		✓	✓	✓
9. Teamwork skills	✓	✓		✓
10. Coordination with others skills	✓	✓	✓	✓

is crucial in many areas of life, including personal relationships and professional settings. The three categories require meaningful interaction between NPCs and the player, thereby enhancing the player's communication skills.

4. Customer orientation skills

Customer orientation skills are people's abilities and attitudes to meet customers' needs and expectations effectively. These skills are crucial in many industries and professions where customer interactions are a key part of the job. Tasks in Category 1 involve putting NPCs on the customer's side, who express their needs and desires to the hero. The player practices active listening, develops a deep sense of empathy toward the NPC, and looks for solutions to complete the mission, thereby improving problem-solving abilities, among other things.

5. Interdisciplinary adaptation skills

Interdisciplinary adaptation skills compose a practical set of abilities that empower people to navigate and adjust within an interdisciplinary environment or situation. Interdisciplinarity is not just about working together but about merging knowledge, methods, and insights from various disciplines or areas of study to solve real-world problems. People with solid interdisciplinary adaptation skills are effective and indispensable in diverse fields, as they can cooperate with experts from different backgrounds. Tasks in categories 1 and 4 are practical exercises to develop these crucial abilities.

6. Interpersonal skills

Interpersonal skills, including successful communication, attentive listening, and empathy, are not innate but can be developed with practice and effort. They are crucial in promoting practical collaboration, teamwork, and overall social skills in both personal and professional settings. Categories 1, 2, and 4 require using these skills to help the player improve them.

7. Learning skills

Mastering various learning skills is essential for individuals to acquire, process, retain, and apply knowledge effectively. These skills are crucial for success in education and remain valuable throughout one's life for continuous learning and personal development. Category 4 tasks enhance learning skills, and quiz-related activities can assist students in strengthening their knowledge of HTML and learning skills.

8. Organizational and planning skills

Effective organization and planning skills enable individuals to efficiently manage tasks, time, and resources to attain specific goals and objectives. Those with strong organizational and planning abilities can prioritize, plan, and coordinate activities to enhance productivity and meet deadlines. Tasks falling into categories 2, 3, and 4 promote these skills, requiring players to evaluate their goals and resources and arrange their subsequent steps accordingly.

9. Teamwork skills

Teamwork abilities are the skills and qualities necessary to accomplish common goals effectively. These skills are crucial in professional environments where people from different backgrounds collaborate on assignments and projects. In categories 1, 2, and 4, participants must show their capacity to cooperate with a team, help NPCs who play a crucial role in the game, and ensure their escape from the building where they are being held.

10. Coordination with other skills

The coordination of various skills involves the blending and integration of different abilities to achieve specific goals efficiently. This process incorporates problem-solving, decision-making, communication, creativity, time management, leadership, and other skills to enhance overall performance and productivity. Effective skill coordination is crucial in personal and professional spheres, enabling individuals to tackle complex tasks easily. Notably, the ability to coordinate skills improves with practice, and any task can contribute to honing this skill. Therefore, individuals who regularly engage in activities requiring the integration of multiple skills can enhance their coordination abilities, ultimately leading to improved performance and productivity.

8.3 Examples of Tasks

Task Example 1

The first example concerns a task where an older man requests a record from the hero (Fig. 8.1). If the hero provides the record to the older man, he will reward him with a stool. The hero needs the stool to reach the lemons on a lemon tree in the house's garden. This task falls into categories 1, 2, and 3. Players must employ analytical

Fig. 8.1 The older man asks the hero for a music record

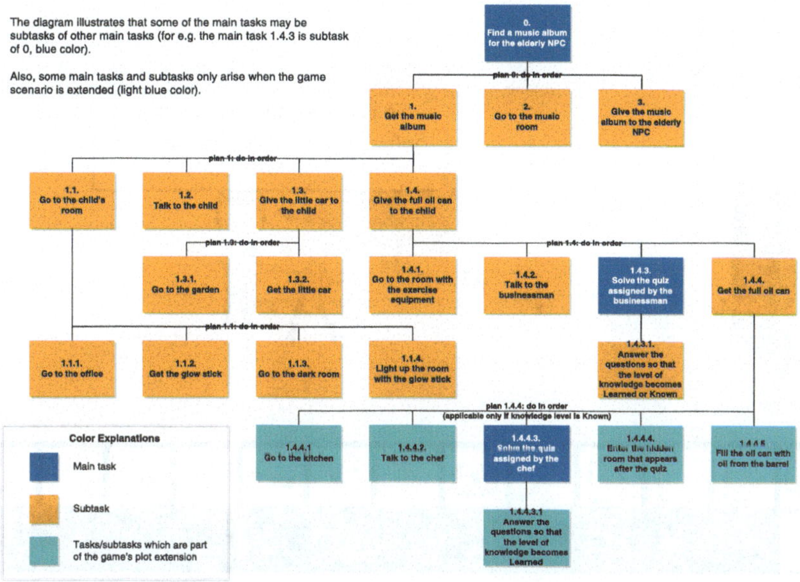

Fig. 8.2 The task analysis regarding the music album

and organizational skills, as this task comprises subtasks that necessitate following a specific sequence of steps.

As illustrated in Fig. 8.2, the ultimate objective at the top is to deliver the music album to the older man. By following the progression of nodes in the diagram, we can discern the sequence of actions required for the player to accomplish this task. Specifically, they need to locate the album, proceed to the music room where the older man is, and then present it to him. Locating the music album constitutes a subtask involving several actions, including visiting the child's room, conversing with the child, and giving the child the small car and the oil. Visiting the child's room encompasses a series of steps, which include going to the office, obtaining the glow stick, entering the dark room, and using the glow stick to illuminate it.

We can see how the player has to use their analytical and organizational skills to plan their actions and solve many small tasks to reach the goal. The task tests the player's ability to think critically and plan actions, making it a challenging but rewarding part of the game.

Task Example 2

A thirsty NPC requests a refreshing beverage from the hero (Fig. 8.3). The task falls under category 3, where the player must gather items and use them in the correct sequence. More specifically, in order to give the NPC a refreshing drink, the hero needs to go to the kitchen, figure out how to unlock the fridge to retrieve the soda, unlock a drawer to get a knife, and then head to the garden to pick a lemon from the tree. Using the knife, the hero slices the lemon and adds a piece to the soda. Hence,

Fig. 8.3 The thirsty man desires a refreshful and tasty drink

the players can practice their analytical and organizational skills by completing these steps and delivering the refreshing drink to the NPC.

This task also falls into category 1, allowing players to improve their communication, teamwork, and interpersonal skills. Indeed, the player must communicate and collaborate with the NPC to help each other and reap mutual benefits. After completing the task, the thirsty individual will grant the player a bucket.

Task Example 3

In task example 1, a task was divided into subtasks. One of the subtasks involved finding a car for the child's board game and obtaining oil to facilitate the car's movement (Figs. 8.4 and 8.5). The task with the oil is interdisciplinary, requiring the player-programmer to devise a means of moving the car along the track. Consequently, this task also falls under category 4, allowing the player to practice skills such as interdisciplinary adaptation and problem-solving while drawing upon concepts from the field of physics. Moreover, the child serves as a customer, articulating their needs and desires. The player must, therefore, refine their customer orientation skills to ensure satisfaction. This process involves communication skills (active listening, empathy) and teamwork. This example vividly illustrates the coordination of multiple skills.

The remaining tasks operate on similar principles. For instance, a witch will aid the protagonist once he fulfills three of her requests: locating a crystal ball, acquiring money, and providing air conditioning. To complete this objective, the player's soft skills are crucial. They must possess or develop effective communication skills with the wizard, collaborative teamwork, customer focus on understanding

Fig. 8.4 The child needs oil for the highway board game

Fig. 8.5 The player solves a sub-task and gets the oil to give it to the child

her requirements, and interpersonal abilities for engaging and attentive interactions. These three interwoven tasks are further divided into specific steps, enhancing the player's analytical and organizational skills.

8.4 Conclusions and Discussion

Individuals need to possess both hard skills, which pertain to their expertise in a specific subject, and soft skills, which primarily encompass interpersonal abilities and personal traits. Employers are increasingly seeking a combination of academic knowledge, field experience, and a diverse set of soft skills when evaluating potential candidates. Compared to hard skills, soft skills cannot be defined or proven through certifications and degrees, as they are developed through life experiences and are not always taught in educational institutions. Nonetheless, cultivating these skills is crucial for personal and professional growth.

The game of this work emphasizes the primary soft skills essential for developers, such as communication, teamwork, interpersonal skills, and approach to the customer and others. It follows a holistic approach to help students develop hard (learning HTML programming) and soft skills. In this way, it bridges the gap between these two sets of skills. It uses the main features of adventure games, such as dialogues with NPCs, which motivate the player to play and solve problems by constantly assigning tasks to them. At the same time, it creates the players' sympathy for the NPCs and motivates them to seek solutions to the problems they set.

We have divided the tasks into different categories. Some lead to mutual help between the NPCs and the player and help develop friendships. Others are divided into individual tasks, which require collecting items to complete other tasks later in the game. Some tasks support interdisciplinary cooperation, such as the player helping the child to play the highway board game. All these categories of tasks help, some more than others, develop the ten most critical soft skills developers need. For example, through dialogue, the game helps the player to develop communication and teamwork skills. The division of tasks into subtasks helps to develop analytical and organizational thinking. NPCs take on the role of customers, expressing needs and wants. Therefore, the player develops skills that revolve around the customer.

The above evaluation employed the cognitive walkthrough method, in which five experts participated, to identify the skills practiced in each task. The cognitive walkthrough was chosen because of the challenge of evaluating users' soft skills. Unlike hard skills, measuring and quantifying a user's proficiency in soft skills is complex. Therefore, it is difficult to conduct a study using a t-test to determine whether the game improves players' soft skills. Nevertheless, the cognitive walkthrough provides a reliable way to identify the soft skills essential to each task and thus highlights the skills players need to complete each task.

In conclusion, the fully developed innovative game incorporates adaptive learning and uses NPCs and task-oriented scenarios to promote hard and soft skills. The personalization and adaptability of the game scenario for as long as necessary for each student's cognitive progress contribute significantly to developing their skills. The structure of the educational adventure game keeps motivation and engagement high and allows for individualized learning.

References

1. Ahmed, F., Capretz, L.F., Campbell, P.: Evaluating the demand for soft skills in software development. It Professional **14**(1), 44–49 (2012)
2. Ahmed, F., Capretz, L.F., Bouktif, S., Campbell, P.: Soft skills and software development: A reflection from the software industry (2015). arXiv:1507.06873
3. Aldulaimi, S.H.: Leadership soft skills in higher education institutions. Soc. Sci. Learn. Educ. J. **3**(7), 01–08 (2018)
4. Alepis, E., Virvou, M.: Object-Oriented User Interfaces for Personalized Mobile Learning. Springer (2014)
5. Alepis, E., Kabassi, K., Virvou, M.: Personalized museum exploration by mobile devices. In: Auer, M.E., Tsiatsos, T. (eds.) Interactive Mobile Communication Technologies and Learning-Proceedings of the 11th IMCL Conference, 30 November–1 December 2017, Mediterranean Palace Hotel, Thessaloniki, Greece, Advances in Intelligent Systems and Computing, vol. 725, pp. 353–360. Springer (2017). https://doi.org/10.1007/978-3-319-75175-7_36
6. Bailey, J., Mitchell, R.B.: Industry perceptions of the competencies needed by computer programmers: technical, business, and soft skills. J. Comput. Inf. Syst. **47**(2), 28–33 (2006)
7. Caeiro-Rodríguez, M., Manso-Vázquez, M., Mikic-Fonte, F.A., Llamas-Nistal, M., Fernández-Iglesias, M.J., Tsalapatas, H., Heidmann, O., De Carvalho, C.V., Jesmin, T., Terasmaa, J., et al.: Teaching soft skills in engineering education: an european perspective. IEEE Access **9**, 29222–29242 (2021)
8. Caratozzolo, P., Alvarez-Delgado, A., Gonzalez-Pineda, Z., Hosseini, S.: Enhancing interdisciplinary skills in engineering with the cognitive tools of storytelling. In: Proceedings of the SEFI 47th Annual Conference, pp. 206–215 (2019)
9. Crosbie, R.: Learning the soft skills of leadership. Ind. Commer. Train. **37**(1), 45–51 (2005)
10. Dochie, E., Herman, C., Epure, C.: Using gamification for the development of soft skills. Skill generator assessment game case study. In: The International Scientific Conference eLearning and Software for Education, "Carol I" National Defence University, vol. 3, p. 610 (2017)
11. Emanuel, F., Ricchiardi, P., Sanseverino, D., Ghislieri, C.: Make soft skills stronger? An online enhancement platform for higher education. Int. J. Educ. Res. Open **2**, 100096 (2021)
12. Farzandipour, M., Nabovati, E., Tadayon, H., Jabali, M.S.: Usability evaluation of a nursing information system by applying cognitive walkthrough method. Int. J. Med. Inf. **152**, 104459 (2021)
13. Fernández-Vara, C.: Adventure. In: The Routledge Companion to Video Game Studies, pp. 232–240. Routledge (2014)
14. Fioravanti, M.L., de Oliveira Sestito, C.D., de Deus, W.S., Scatalon, L.P., Barbosa, E.F.: Role-playing games for fostering communication and negotiation skills. IEEE Trans. Educ. **65**(3), 384–393 (2021)
15. Garcia, I., Pacheco, C., Méndez, F., Calvo-Manzano, J.A.: The effects of game-based learning in the acquisition of "soft skillsâŁž on undergraduate software engineering courses: a systematic literature review. Comput. Appl. Eng. Educ. **28**(5), 1327–1354 (2020)
16. Helens-Hart, R., Engstrom, C.: Empathy as an essential skill for talent development consultants. J. Work. Learn. **33**(4), 245–258 (2021)
17. Hwang, S.: Effects of engineering students' soft skills and empathy on their attitudes toward curricula integration. Educ. Sci. **12**(7), 452 (2022)
18. Katsionis, G., Virvou, M.: A cognitive theory for affective user modelling in a virtual reality educational game. In: 2004 IEEE International Conference on Systems, Man and Cybernetics (IEEE Cat. No. 04CH37583), vol. 2, pp. 1209–1213. IEEE (2004)
19. Klegeris, A.: Mixed-mode instruction using active learning in small teams improves generic problem-solving skills of university students. J. Furth. High. Educ. **45**(7), 871–885 (2021)
20. Kostikova, I., Holubnycha, L., Girich, Z., Movmyga, N.: Soft skills development with university students at english lessons. Revista Romaneasca pentru Educatie Multidimensionala **13**(1), 398–416 (2021)

21. Lewis, C., Wharton, C.: Cognitive walkthroughs. In: Handbook of Human-Computer Interaction, pp. 717–732. Elsevier (1997)
22. Lyon, A.R., Coifman, J., Cook, H., McRee, E., Liu, F.F., Ludwig, K., Dorsey, S., Koerner, K., Munson, S.A., McCauley, E.: The cognitive walkthrough for implementation strategies (cwis): a pragmatic method for assessing implementation strategy usability. Implement. Sci. Commun. **2**, 1–16 (2021)
23. Lyu, W., Liu, J.: Soft skills, hard skills: What matters most? evidence from job postings. Appl. Energy **300**, 117307 (2021)
24. Mahatody, T., Sagar, M., Kolski, C.: State of the art on the cognitive walkthrough method, its variants and evolutions. Int. J. Hum. Comput. Interact. **26**(8), 741–785 (2010)
25. Mahdi, A.G., Al-Rabaiwi, S.H.: the role of soft skills in enhancing the impact of internal marketing on customer satisfaction an applied study of a sample of private bank managers in Baghdad. Int. J. Res Soc. Sci. Hum. **10**(2), 247–267 (2020)
26. Marques, A.B., Ferreira, B., Lopes, A., Silva, W.: Stimulating the development of soft skills in software engineering education through design thinking. In: SBES, pp. 690–699 (2020)
27. Masharova, T., Nimatulaev, M., Kharunzheva, E., Perevozchikova, M.: Formation of highly-demanded soft skills in the game environment of educational webquest. In: SLET-2020: International Scientific Conference on Innovative Approaches to the Application of Digital Technologies in Education, Stavropol, pp. 332–341 (2020)
28. Mattajang, R.: The importance of soft skills development in human resource management. Jurnal Ekonomi **12**(04), 2361–2368 (2023)
29. McGowan, N., López-Serrano, A., Burgos, D.: Serious games and soft skills in higher education: a case study of the design of compete! Electronics **12**(6), 1432 (2023)
30. Mitsea, E., Drigas, A., Mantas, P.: Soft skills & metacognition as inclusion amplifiers in the 21 st century. Int. J. Online Biomed. Eng. **17**(4) (2021)
31. Moundridou, M., Virvou, M.: Authoring intelligent tutoring systems over the world wide web. In: Proceedings First International IEEE Symposium Intelligent Systems, vol. 1, pp. 160–165. IEEE (2002)
32. Noguez, J., Espinosa, E.: Improving learning and soft skills using project oriented learning in software engineering courses. In: Proceedings of the 2nd International Workshop on Designing Computational Models of Collaborative Learning Interaction, Citeseer, pp. 83–88 (2004)
33. Oghenekaro, A.P., et al.: Deployment of soft skills for effective customer service in the 21st century library. J. Comput. Commun. **6**(03), 43 (2018)
34. Papadimitriou, S.: Adaptive scenarios and tutoring using fuzzy reasoning in an intelligent educational adventure game (2022)
35. Papadimitriou, S., Virvou, M.: An online adventure game for teaching math. In: 2016 7th International Conference on Information, Intelligence, Systems & Applications (IISA), pp. 1–5. IEEE (2016)
36. Papadimitriou, S., Virvou, M.: Adaptivity in scenarios in an educational adventure game. In: 2017 8th International Conference on Information, Intelligence, Systems & Applications (IISA), pp. 1–6. IEEE (2017)
37. Papadimitriou, S., Chrysafiadi, K., Virvou, M.: Evaluating the use of fuzzy logic in an educational game for offering adaptation. In: 2019 International Conference on Computer, Information and Telecommunication Systems (CITS), pp. 1–5. IEEE (2019)
38. Papadimitriou, S., Chrysafiadi, K., Virvou, M.: Fuzzeg: Fuzzy logic for adaptive scenarios in an educational adventure game. Multimed. Tools Appl. **78**(22), 32023–32053 (2019). https://doi.org/10.1007/s11042-019-07955-w
39. Papadimitriou, S., Kamitsios, M., Chrysafiadi, K., Virvou, M.: Learn-and-play personalised reasoning from point-and-click to virtual reality mobile educational games. Intell. Decis. Technol. **15**(2), 321–332 (2021)
40. Putra, A.S., Novitasari, D., Asbari, M., Purwanto, A., Iskandar, J., Hutagalung, D., Cahyono, Y.: Examine relationship of soft skills, hard skills, innovation and performance: the mediation effect of organizational learning. Int. J. Sci. Manag. Stud. (IJSMS) **3**(3), 27–43 (2020)

41. Qizi, K.N.U.: Soft skills development in higher education. Univ. J. Educ. Res. **8**(5), 1916–1925 (2020)
42. Rieman, J., Franzke, M., Redmiles, D.: Usability evaluation with the cognitive walkthrough. In: Conference Companion on Human Factors in Computing Systems, pp. 387–388 (1995)
43. Riggio, R.E., Tan, S.J.: Leader Interpersonal and Influence Skills: The Soft Skills of Leadership. Routledge (2013)
44. Robles, M.M.: Executive perceptions of the top 10 soft skills needed in today's workplace. Bus. Commun. Q. **75**(4), 453–465 (2012)
45. Rossi, I.V., de Lima, J.D., Sabatke, B., Nunes, M.A.F., Ramirez, G.E., Ramirez, M.I.: Active learning tools improve the learning outcomes, scientific attitude, and critical thinking in higher education: Experiences in an online course during the covid-19 pandemic. Biochem. Mol. Biol. Educ. **49**(6), 888–903 (2021)
46. Schulz, B.: The Importance of Soft Skills: Education Beyond Academic Knowledge (2008)
47. Tang, K.N.: The importance of soft skills acquisition by teachers in higher education institutions. Kasetsart J. Soc. Sci. **41**(1), 22–27 (2020)
48. Tanković, A.C., Kapeš, J., Kraljić, V.: Importance of soft skills and communication skills in tourism: viewpoint from tourists and future tourism employees. Tourism Southern Eastern Eur. **6**, 167–185 (2021)
49. Tem, S., Kuroda, A., Tang, K.N.: The importance of soft skills development to enhance entrepreneurial capacity. Int. Educ. Res. **3**(3), p1–p1 (2020)
50. Tourtoglou, K., Virvou, M. (2010) Simulated annealing in finding optimum groups of learners of uml. In: Intelligent Interactive Multimedia Systems and Services, pp. 147–156. Springer
51. Tsihrintzis, G.A., Virvou, M., Phillips-Wren, G.: Surveys in artificial intelligence-based technologies. Intell. Decis. Technol. **13**(4), 393–394 (2019)
52. Tsiriga, V., Virvou, M.: Modelling the student to individualise tutoring in a web-based icall. Int. J. Contin. Eng. Educ. Life Long Learn. **13**(3–4), 350–365 (2003)
53. Varava, I.P., Bohinska, A.P., Vakaliuk, T.A., Mintii, I.: Soft skills in software engineering technicians education. J. Phys. Conf. Ser. **1946**, 012012 (2021). (IOP Publishing)
54. Vasconcelos, S., Melo, A., Melo, C., Liberato, D., Lopes, M.C.: Soft skills in action: developing tourism students skills through interdisciplinarity. In: Advances in Tourism, Technology and Systems: Selected Papers from ICOTTS 2021, vol. 2, pp. 203–213. Springer (2022)
55. Vasconcelos, S., Melo, C., Melo, A., Liberato, D.: Interdisciplinarity in action: developing students' soft skills through project-based learning and field work. In: INTED2022 Proceedings, IATED, pp. 4852–4859 (2022)
56. Vielma, R.D.E., Romo, A.S., Pérez, J.F.L.: Soft skills in customer service: sales area in automobile dealers. Contaduría y administración **67**(2), 10 (2022)
57. Virvou, M.: A human plausible reasoning theory in the context of an active help system for unix1 users. Ph.D thesis, University of Sussex, Falmer, East Sussex, UK (1993). https://ethos.bl.uk/OrderDetails.do?uin=uk.bl.ethos.787162
58. Virvou, M.: Automatic reasoning and help about human errors in using an operating system. Interact. Comput. **11**(5), 545–573 (1999)
59. Virvou, M.: A domain-independent reasoning mechanism for an its authoring tool. In: EdMedia+ Innovate Learning, Association for the Advancement of Computing in Education (AACE), pp. 1762–1764 (2000)
60. Virvou, M.: Modelling the knowledge and reasoning of users in a knowledge-based authoring tool. Int. J. Bus. Perform. Manag. **5**(2–3), 399–412 (2003)
61. Virvou, M.: On the evaluation of the combined role of virtual reality games and animated agents in edutainment. Intell. Comput. Graph. **2011**, 79–95 (2012)
62. Virvou, M.: A new era towards more engaging and human-like computer-based learning by combining personalisation and artificial intelligence techniques. In: Proceedings of the 23rd Annual ACM Conference on Innovation and Technology in Computer Science Education, pp. 2–3 (2018)
63. Virvou, M.: Artificial intelligence and user experience in reciprocity: Contributions and state of the art. Intell. Decis. Technol. **17**(Preprint), 1–53 (2023)

64. Virvou, M., Alepis, E.: Mobile educational features in authoring tools for personalised tutoring. Comput. Educ. **44**(1), 53–68 (2005)
65. Virvou, M., Kabassi, K.: Reasoning about users' actions in a graphical user interface. Hum. Comput. Interac. **17**(4), 369–398 (2002)
66. Virvou, M., Kabassi, K.: Evaluating an intelligent graphical user interface by comparison with human experts. Knowl. Based Syst. **17**(1), 31–37 (2004)
67. Virvou, M., Katsionis, G.: Web services for an intelligent tutoring system that operates as a virtual reality game. In: SMC'03 Conference Proceedings. 2003 IEEE International Conference on Systems, Man and Cybernetics. Conference Theme-System Security and Assurance (Cat. No. 03CH37483), vol. 1, pp. 378–383. IEEE (2003)
68. Virvou, M., Manos, K.: A simulated student-player in support of the authoring process in a knowledge-based authoring tool for educational games. In: Proceedings 3rd IEEE International Conference on Advanced Technologies, pp. 338–339. IEEE (2003)
69. Virvou, M., Moundridou, M.: An authoring tool for algebra-related domains. In: HCI (2), Citeseer, pp. 647–651 (1999)
70. Virvou, M., Papadimitriou, S.: Learning addition by playing an adaptive game with coins. In: IISA 2013, pp. 1–4. IEEE (2013)
71. Virvou, M., Papadimitriou, S.: Simple arithmetic lessons through an adaptive snake game. In: 2013 International Conference on Computer, Information and Telecommunication Systems (CITS), pp. 1–4. IEEE (2013)
72. Virvou, M., Papadimitriou, S.: Use of digital 'guess who board game'in teaching english. In: IISA 2014, The 5th International Conference on Information, Intelligence, Systems and Applications, pp. 439–443. IEEE (2014)
73. Virvou, M., Papadimitriou, S.: Playing, learning, extending: educational "guess who" game that is renewable by a teacher. In: 2015 6th International Conference on Information, Intelligence, Systems and Applications (IISA), pp. 1–5. IEEE (2015)
74. Virvou, M., Stavrianou, A.: User modelling in a gui. In: Proceedings of HCI International (the 8th International Conference on Human-Computer Interaction) on Human-Computer Interaction: Ergonomics and User Interfaces, vol. I, pp. 261–265 (1999)
75. Virvou, M., Manos, K., Katsionis, G., Tourtoglou, K.: Vr-integate: a knowledge-based authoring tool for virtual reality educational games. In: Proceedings of ED-MEDIA 2002 World Conferences on Educational Multimedia, Hypermedia and Telecommunications (2002)
76. Virvou, M., Lampropoulos, A.S., Tsihrintzis, G.A.: Inma: a knowledge-based authoring tool for music education. In: International Conference on Knowledge-Based and Intelligent Information and Engineering Systems, pp. 376–383. Springer (2006)
77. Virvou, M., Alepis, E., Mpalasis, K.: Evaluation of a multimedia educational tool for geography in elementary schools. In: Proceedings of International Conference on Information Communication Technologies in Education, ICICTE, pp. 364–374 (2013)
78. Virvou, M., Alepis, E., Tsihrintzis, G.A., Jain, L.C.: Machine Learning Paradigms: Advances in Learning Analytics. Springer (2020)
79. Vogler, J.S., Thompson, P., Davis, D.W., Mayfield, B.E., Finley, P.M., Yasseri, D.: The hard work of soft skills: augmenting the project-based learning experience with interdisciplinary teamwork. Instr. Sci. **46**, 457–488 (2018)
80. Weninger, M., Grünbacher, P., Gander, E., Schörgenhumer, A.: Evaluating an interactive memory analysis tool: findings from a cognitive walkthrough and a user study. Proc. ACM Hum. Comput. Interact. **4**(EICS):1–37 (2020)
81. Younis, A.A., Sunderraman, R., Metzler, M., Bourgeois, A.G.: Developing parallel programming and soft skills: a project based learning approach. J. Parallel Distrib. Comput. **158**, 151–163 (2021)

Chapter 9
General Data Protection Regulation and Adaptive Educational Games

Abstract Intelligent gaming as an educational tool is becoming increasingly popular thanks to its ability to provide a fun and engaging way for students to learn. These games often incorporate Artificial Intelligence (AR), which enables data collection and processing that is subject to the guidelines of the GDPR, the new EU data privacy framework. As a result, experts emphasize the importance of obtaining consent for personal data processing while balancing the benefits of gaming and education with regulatory compliance. In this chapter, we describe the features of the Privacy and Ethics Module of the EPATHLO Educational Game Suite, which ensure that the suite's educational games are GDPR-compliant. Guided by these features, we combine legal and computer science expertise to offer practical solutions in the form of guidelines for gaming stakeholders, particularly those in the educational sector. These guidelines aim to help stakeholders navigate the complexities of GDPR compliance concerning intelligent gaming while guaranteeing that students continue to benefit from these games' educational advantages.

Keywords Educational software · Personalized educational games · Artificial Intelligence in educational games · Fuzzy logic in education · User experience · AI-empowered software games · User modeling · Human-AI interaction · e-learning games · 21st-century skills · Human-centred Artificial Intelligence · GDPR · AI ethics

9.1 Introduction to GDPR

The rapid advancements in Artificial Intelligence (AI), driven by the accumulation of big data and internet technologies, have significantly impacted various fields, leading to more sophisticated Machine Learning (ML) and Deep Learning (DL) models, data mining, and data science with the respective methodologies that enhance decision-making and software engineering processes [95]. These advancements allow for the development of AI-empowered decision support systems that leverage vast datasets to improve accuracy and efficiency across industries, thereby augmenting human intelligence in complex scenarios [2, 96, 104]. Furthermore, the integration of AI with

S. Papadimitriou and M. Virvou, *Artificial Intelligence—Based Games as Novel Holistic Educational Environments to Teach 21st Century Skills*, Intelligent Systems Reference Library 93, https://doi.org/10.1007/978-3-031-77464-5_9

internet technologies has facilitated real-time data analysis and robust applications, enabling innovative solutions in intelligent decision technologies [93] and user experience [92, 104]. Advances in AI are reshaping software engineering by automating tasks and enhancing the capabilities of cyber-physical systems, as highlighted by discussions at international conferences and in research publications [95]. However, these rapid advancements bring ethical concerns, such as potential biases in AI models, transparency issues in decision-making processes, and privacy concerns associated with data collection and usage [113]. Addressing these ethical challenges ensures that AI technologies contribute positively to society while safeguarding individual rights [96, 121].

For the above reasons, the EU devised the GDPR, which stands for General Data Protection Legislation. It is a European Union (EU) law that came into effect on 25 May 2018. GDPR governs how we can use, process, and store personal data (information about an identifiable, living person).

In the context of AI systems, meeting software requirements related to ethics and GDPR involves ensuring transparency, explainability, consent, trust, and privacy. Virvou et al. [122] emphasize the importance of trust dynamics models like VIRTSI, which enhance AI collaboration by fostering user trust and improving explainability. This is very important, especially in sensitive applications such as healthcare or healthcare education [69, 70]. Furthermore, the work of Mougiakou and Virvou [61, 80] highlights the critical need for AI systems to comply with GDPR by implementing robust privacy and data protection measures, ensuring user consent, and maintaining transparency in data handling processes. These requirements collectively ensure that AI systems are ethical, secure, and aligned with regulatory standards.

9.2 GDPR Compliance in Software Development: A Review

1. Embedding Security Controls in Software Design

Ayala-Rivera et al. [5] emphasize the importance of integrating security controls into software design to achieve GDPR compliance. Their research advocates for embedding privacy and security measures throughout the software development lifecycle rather than retrofitting them after development. This proactive approach directly incorporates GDPR principles such as data minimization and user consent into the software architecture, enhancing overall security and compliance.

2. Open-Source Software Challenges

Franke et al. [29] explore the unique challenges that open-source software (OSS) projects face, such as diverse contributor bases and decentralized development models. These factors complicate consistent data protection measures. Their study reveals that many OSS projects struggle with GDPR compliance due to lacking resources and expertise. They suggest incorporating automated tools and templates for privacy notices and Data Protection Impact Assessments (DPIAs) to help OSS communities align with GDPR requirements more effectively.

Further investigation by Franke, Farzanehpour, et al. [30] uses a mixed-methods approach to examine compliance practices within OSS. The study found that while GDPR awareness is high among developers, its practical application is inconsistent due to varying interpretations and challenges in adapting GDPR requirements to the open-source model. This study emphasizes the need for clearer guidelines and support mechanisms to aid OSS projects in achieving compliance.

3. Integration into Software Development Processes

Gómez-Martínez et al. [36] discuss integrating GDPR principles into the Unified Software Development Process (USDP). Their work suggests embedding GDPR considerations into established software development methodologies to streamline compliance. Incorporating GDPR requirements into design phases helps developers anticipate potential data protection issues, addressing them early in the development process. This integration helps create privacy-aware software systems, reducing the risk of non-compliance.

4. Case Studies and Architectural Approaches

Horák et al. [42] present a case study on GDPR compliance in cybersecurity software, focusing on DPIAs in an information-sharing platform. Their findings highlight the importance of DPIAs in identifying potential risks and implementing appropriate safeguards, ensuring software systems comply with GDPR, and enhancing user trust through robust data protection measures.

Kittmann et al. [49] propose a distributed software architecture designed to comply with GDPR, particularly for automation services. This architecture incorporates privacy-by-design principles, ensuring data protection is integral to the system's functionality. This approach is relevant for automation services, which process large volumes of personal data. By embedding privacy measures into the architecture, developers can create systems that automatically enforce GDPR requirements, reducing the burden on users and administrators.

5. Overall Challenges and Recommendations

Integrating GDPR compliance into software development is a multifaceted challenge requiring technical, organizational, and procedural changes. The reviewed studies highlight the importance of incorporating privacy and security measures into the design and development phases, ensuring software systems are built with GDPR requirements. OSS projects face unique challenges due to their collaborative nature, necessitating tailored solutions to achieve compliance. Emphasizing privacy by design and proactively embedding GDPR principles into software systems can significantly enhance compliance efforts, leading to more secure and privacy-respecting applications. These findings underscore the critical role of evolving software processes to address regulatory requirements effectively and sustainably.

9.3 GDPR Intelligent Educational Games

Intelligent Tutoring Systems (ITSs) utilize AI techniques to deliver personalized instruction, inspiring students to be more interested and engaged, ultimately leading to improved learning outcomes [3, 13–15, 20, 21, 23, 24, 46, 52, 65, 77, 97, 106, 114, 115, 125]. Learning analytics significantly enhances education by using ML and AI to analyze educational data, providing insights that personalize and improve learning experiences [94, 120]. Analytics is useful in student modeling which provides a representation of many cognitive aspects of students or preferences and levels of knowledge [16, 44, 45, 51, 88, 91, 107, 108, 118, 119]. These tools support educators in tailoring their teaching strategies to meet individual student needs, thereby improving learning outcomes. For instance, learning analytics tools help teachers make informed decisions by offering insights into student performance and enabling adaptive learning environments. ITSs use these insights to deliver personalized instruction, adjusting content and pacing to match each student's progress and learning style.

Additionally, there has been a growing focus on digital games as educational tools, with the educational approach centered around entertainment applications leading to improved educational outcomes [17, 19, 40, 41, 47, 54, 71, 84, 99, 109–112, 116, 117]. Interactions enriched with challenges and rewards [43, 59, 72, 74, 76] enhance student involvement by stimulating their interest and motivation and contribute to the efficiency of the tutor by reducing tutoring time and workload [18, 73, 105]. Research supports the notion that the impact of digital games as an educational tool is comparable to that of private tutoring. Adaptive educational games primarily focus on adjusting the learning content according to the student's characteristics and knowledge level, as well as modifying the game context to attract the student's attention and trigger engagement [40, 84, 100]. Adaptive educational games also employ intelligent techniques to provide a personalized experience, similar to ITSs. Initially, the game (or ITS) automatically collects learner data and processes it to make inferences about the learning and game content it will generate, discreetly providing a personalized, immersive experience that aligns with the student's characteristics without explicitly indicating the background processes.

This is particularly the case with students' data accumulated for learning analytics in the context of AI in education, ITSs, and educational games. Ethical concerns are related to data privacy and the General Data Protection Regulation (GDPR), highlighting the need to ensure student data is handled responsibly and transparently to protect individual privacy and rights. The GDPR, developed in response to the growing prevalence of online personal data sharing, is designed to empower individuals to exert greater control over their data [1, 5, 85, 89, 123, 131]. Enforced across the EU as of May 25th, 2018, the GDPR impacts both EU citizens and visitors, as it regulates the products and services offered by entities based within and outside the EU. Integrating ITSs with adaptive educational games presents a challenge for examining data protection, particularly within the GDPR framework.

The review of the EPATHLO game suite shows that the latter has contributed significantly to the integration of ethical and data protection considerations in ITSs. A key focus has been on ensuring transparency, particularly for children and adolescents, as highlighted by Mougiakou, Papadimitriou, and Virvou [62]. Building on this foundation, Mougiakou, Papadimitriou, and Virvou [63] developed comprehensive design guidelines for automated decision-making processes that prioritize personal data protection, ensuring compliance with GDPR while maintaining the efficacy of educational tools. Furthermore, Papadimitriou, Mougiakou, and Virvou [75] explored the role of smart educational games in navigating GDPR's requirements for consent, devising strategies for effectively obtaining and managing user consent. Moreover, Mougiakou, Papadimitriou, and Virvou [64] examined the implications of GDPR for synchronous and asynchronous learning methods, ensuring that data protection principles are upheld across diverse educational delivery modes. Collectively, these efforts have advanced the development of ITSs that prioritize user privacy and trust, paving the way for more ethical and secure educational technologies.

In the following sections, we highlight the advancements of the EPATHLO project, focusing on transparency, GDPR compliance, and data protection in educational technologies. More specifically, we merge computer science and legal science to achieve a dual purpose: identifying the disparities between ITS, adaptive educational games, and the GDPR concerning automated decision-making and profiling while providing practical recommendations in the form of guidance for developers. The Fuzzy EPATHLO and EPATHLO Student Profile Creator serve as testing environments, allowing the evaluation of GDPR compliance and applying theory to real-world scenarios. Eirini Mougiakou, a legal sciences PhD candidate, played a crucial role in implementing this chapter and fostering interdisciplinary collaboration between computer science and legal science [75].

9.4 GDPR and Automated Profiling/Decision-Making in the EPATHLO Educational Game Suite

The GDPR introduces significant provisions regarding the External Personal Data Definition[1] [7, 32, 35, 81, 98], which refers to information that can either identify an individual on its own or in combination with other data, as described in GDPR articles 4.1 and 9. The regulation also addresses Data Protection rights related to automated decision-making and user modeling [11, 25, 38, 68, 82, 127]. This includes the right to object to profiling[2] [78, 90, 103, 128], as well as core principles concerning processing, consent, accountability, and transparency [4, 10, 86, 129]. Profiling

[1] External Personal Data Definition stands for information which, either jointly with other information, or stand-alone, can identify an individual as combinedly depicted within GDPR articles 4.1 and 9.

[2] Data protection rights while re-ascertaining DPD originating ones, such as access, consent, portability, and personal data profiling, map the following rights: (a) to be informed, (b) of access, (c) to

under Article 4(4) of the GDPR involves processing personal data to assess specific attributes, analyze or predict aspects of preferences, performance, behavior, interests, and trustworthiness, and influence decisions concerning individuals [22, 28].

The EPATHLO Student Profile Creator adapts quiz questions based on students' knowledge levels. Fuzzy EPATHLO adjusts the extension of the game scenario based on the student's knowledge level to keep them engaged. These adaptations occur automatically using data collected about the student. Per GDPR, it is vital to communicate these functions to the user. The Privacy and Ethics Module provides users with information about data collection and processing, ensuring their right to be informed about creating their profile. Users are informed that an intelligent technique assesses their knowledge level based on their quiz performance. Failing to disclose data collection and processing information in an adaptive educational game would violate the student's right to be informed about creating their profile. Specifically, a disclaimer or notice should inform students that an AI process will evaluate and categorize their knowledge level after each quiz.

The Privacy and Ethics Module of the EPATHLO Educational Game Suite, following Articles 12–14 of the GDPR [33, 34, 37, 53], provides students with a clear and understandable account of the methodologies employed for making AI-driven conclusions. This ensures that the average user can comprehend the processes involved. Moreover, the system provides more detailed information for AI specialists to adhere to the provisions of Recital 71 of the GDPR, emphasizing the necessity for fair and transparent processing. The initial description is supplemented by a technical and comprehensive analysis, allowing experts to assess the accuracy and validity of the processes thoroughly.

When using Fuzzy EPATHLO, users know that the suite utilizes fuzzy logic, employing simple verbal rules to evaluate their understanding. However, a more detailed explanation of the process, including the entire flow of student data through the fuzzy inference system (FIS), is also provided, primarily targeting AI experts. This allows them to raise concerns about the method's accuracy and capability to make dependable and accurate assessments of the student's understanding.

As per articles 13–15 of the GDPR, students must comprehend the reasons behind the collection and processing of their data by the EPATHLO Educational Game Suite [26, 35, 39, 48, 53, 55, 124]. The Privacy and Ethics Module elucidates that these processes are primarily aimed at enriching the student's learning experience and enjoyment. Customizing learning experiences and gaming activities are designed to support these objectives effectively. Moreover, the description emphasizes the benefits of using student data to deliver personalized content.

The Privacy and Ethics Module requires the provision of the above information. This approach enhances transparency, a fundamental GDPR principle that aims to empower students' rights, ensure accountability, and promote legitimate and fair data processing [87]. The transparency makes students feel controlled and respected, as they are fully informed about the system's processes. However, to enable automated

rectification, (d) to erasure, (e) to restrict processing, (f) to data portability (g) to object and (h) in relation to automated decision making and profiling (GDPR articles 13–22).

profiling and decision-making, the Privacy and Ethics Module requires explicit consent from the students after providing them with the descriptions above. Consent may be sought for both data collection and processing procedures simultaneously or separately for each procedure. Decision-making is only possible if the system has generated the appropriate student model. Students who permit the system to process their data retain the right to withdraw their consent. Students playing Fuzzy EPATHLO have access to all information about themselves, including the complete history of the quizzes taken and the progression of their knowledge level as determined by the FIS. Therefore, another aspect that creators of educational games should consider is that adaptive games must obtain explicit consent from students to collect and process their data automatically.

The Privacy and Ethics Module in the suite enables students to access a complete history of their quiz participation and the knowledge level calculated after each quiz. This feature allows students to exercise their right to access their data, as mandated by regulations [8, 25]. Students can view their data on their screen and export all student data in popular, machine-readable formats such as CSV (comma-separated values) files [60]. This functionality is essential for students who want to transfer their data to another data processor without hindrance from the EPATHLO Educational Game Suite [6].

The student should have the right to request the modification of their data if they believe it is inaccurate. This right results from the student's ability to access their data, allowing them to review the information maintained by the system and determine its accuracy. Additionally, the student should be able to provide additional data to the data controller to improve the system's understanding and create a more accurate student model. In this context, students should be informed about data processing purposes, including assessing their knowledge level and providing appropriate exercises. The Privacy and Ethics Module fulfills these requirements by enabling students to correct or supplement their data, thereby fostering a sense of validation and respect.

In the system, students can control the collection and processing of their data. However, suppose a student initially consents to data collection and processing but later wishes to withdraw that consent. In that case, they should be able to request the deletion of all their data from the system, even if it has already been collected [12, 31, 50, 56, 57, 79, 101, 130]. The Privacy and Ethics Module provides explicitly the capability for students to delete their data promptly.

Additionally, it sends a corresponding notification to the administrator so they can proceed with the necessary procedures of deleting any backup data regarding the student in a reasonable time. The Privacy and Ethics Module notifies the student that their data has been deleted and informs them that the procedure for deleting the backup is pending. The Privacy and Ethics Module lets the system administrator notify the student when the backup is deleted.

9.5 Software Engineers Guidelines on Automated Decision-Making and Profiling

In order to guarantee that automated decision-making systems are compliant, the creator of an ITS or a personalized educational game needs to consider the following recommendations, grouped by the specific provisions of the GDPR that are addressed:

Consent

Before seeking consent, the system must provide comprehensive and sufficiently detailed information on profiling. This step is vital to ensure that the data subject fully comprehends the data collection and processing procedures. Once consent for profiling is given, the individual should have the right to revoke their consent at any time. Separate and specific consent should be sought for different processing purposes, and new processing purposes also necessitate separate and specific consent. The procedure for revoking consent should be as straightforward as giving consent.

Right to be informed

Notices should be disseminated comprehensively at various process stages. The system should provide timely and progressively detailed notifications to the student individually or collectively. This approach, including alerts containing concise information, links to detailed versions, and visual and interactive tools, is designed to prevent students from being overwhelmed with information and improve their understanding.

Right of access

Notices should be communicated multi-layered throughout the proceedings. The system should promptly deliver specific and incremental notifications to the student, individually or collectively, in a single communication. The system can utilize alerts with summarized information, links to detailed versions, and visual and interactive techniques to avoid overwhelming students with information and improve their understanding.

Ensuring valuable information involves the system revealing the data utilized in automated decision-making. This includes the data source, the categories used to create profiles, statistics, and other analytical information. It should also outline the connection between profiling and automated decision-making.

Appropriate safeguards also include: (a) Pairing data minimization with the minimum required by regulations entails the system notifying the student about data retention periods. Once these periods expire, the system must either delete the student's data, obtain further consent to retain it or continue collecting and processing it. (b) To enhance privacy, the system must use anonymization or pseudonymization as privacy-enhancing factors [9, 27, 39, 58, 66, 67, 83, 102, 126] and avoid revealing students' real names. (c) Use appropriate tools to handle data subject to automated decision-making in a well-documented manner. (d) Securing certifications for processing as an improved second and third layer of defense involves carrying out routine sampling inspections of the systems of data controllers to guarantee that they are

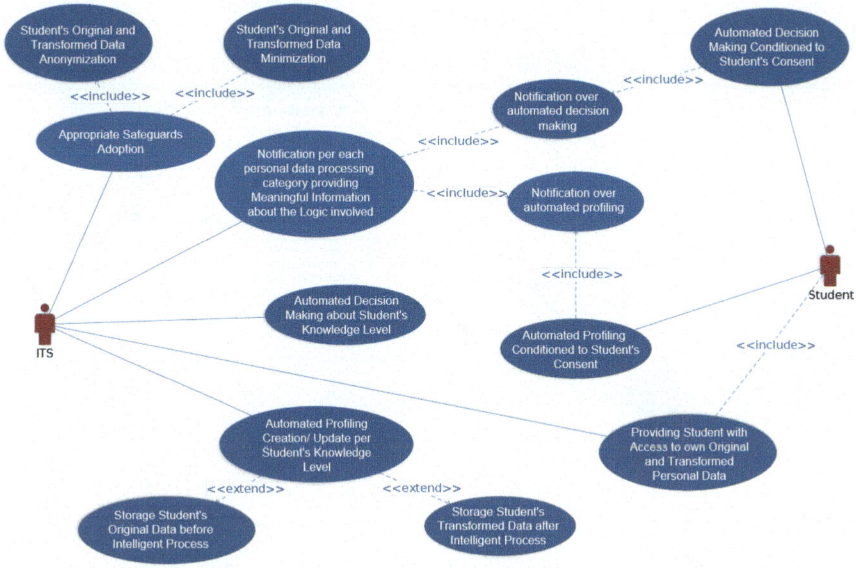

Fig. 9.1 The guidelines use-case diagram

free from discrimination. Additionally, specific ML controls must be implemented to ensure the correct configurations.

Figure 9.1 illustrates a UML case diagram representing the guidelines above.

9.6 Adaptive Educational Games' Components Affected by GDPR

The GDPR has implications for the Fuzzy EPATHLO and EPATHLO Student Profile Creator components. The diagram in Fig. 9.2 illustrates the integrated system involving these components. This visual also depicts the overall architecture of an adaptive educational game, showcasing the flow of student data among the system's components and the junctures at which the student's consent is required for data collection and processing. The system is structured into four levels (tiers): AI level, application level, database level, and user level. The user tier is linked to the application tier, while the database and AI layers, ranked first and third, are not directly visible to the user. The user engages with these layers through the game interface, initially during student data collection and subsequently during the implementation of intelligent decision-making techniques.

The primary audience for educational games is students who play for learning and entertainment. These games display educational content directly on the student's

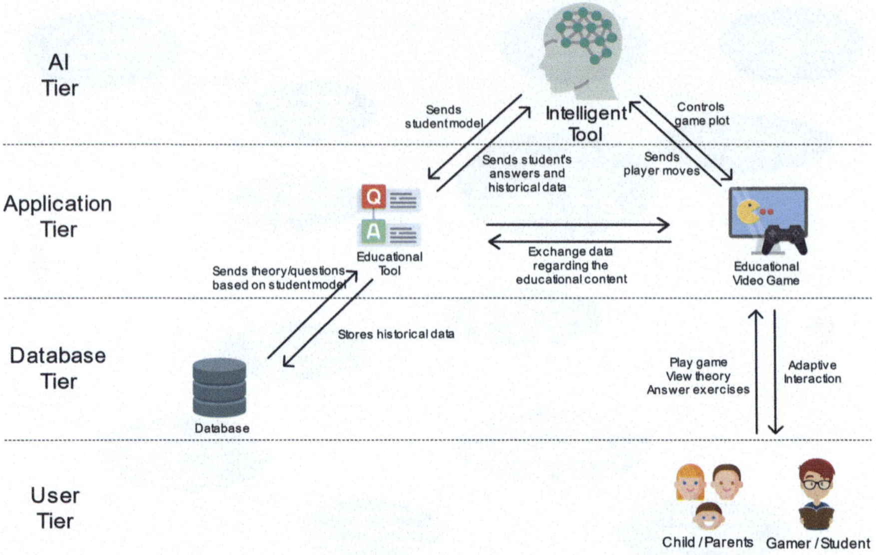

Fig. 9.2 A top-level overview of intelligent educational games architecture

screen and include activities to improve their knowledge and enhance their comprehension of the subject matter. The games consist of the interactive component, promoting student engagement and generating enthusiasm to complete the tasks, making learning a fun experience. Moreover, the games can function independently from the educational tool, allowing these components to operate autonomously. This setup enables the integration of educational content at different stages of the game, facilitating the learning of various subjects such as mathematics, history, science, and more.

Students can access teaching resources and activities throughout the game to progress through the story. The educational module pulls theory and questions from the database and incorporates them into the game, displaying them visually on the screen. After students complete an exercise, their responses are sent to the educational tool. The tool verifies the accuracy of the input and then stores it in the database, creating a historical record of the student's interactions.

The database contains student information, including first name, family name, email, age, additional data collected during the game, and theory and exercises. It allows for storing a comprehensive historical record of the tasks the student has completed, their responses, and their corresponding scores. This feature enables individuals such as the tutor or game administrator to monitor the student's performance and knowledge level evolution. Through the use of AI, the game delivers a personalized, game-based learning environment tailored to each student's unique characteristics. The AI tool identifies students' needs and adjusts the game scenario and educational context accordingly. This educational tool facilitates various operations for each task, connecting the database with the game by retrieving questions, sending them to the game environment, collecting responses, and updating the database with the results.

The AI tool enables seamless integration between the game, the educational tool, and the database, which gathers and analyses data to generate valuable insights. The data comes from different sources, such as the educational tool (e.g., quizzes and responses) or a combination of factors, including performance on the game's quizzes. The AI tool performs computations related to a specific concept, such as assessing the distribution of a student's knowledge level based on their performance and adjusting the difficulty of exercises accordingly. The educational component then uses the student's model to select educational materials and suggest exercises from the AI component, followed by presenting questions within the game's storyline. The student engages with customized exercises based on their model created by the AI tool without realizing that each response triggers a dual process in modeling and question assignment.

The automated decision-making process significantly influences the educational tool and the game. Depending on the game's genre, the AI tool uses the student's progress to adapt the game elements. For instance, in a war game, players may receive additional weapons as rewards, while in an exploration game, their performance in exercises may earn them keys and maps. Other features may involve integrating virtual agents to address incorrect responses, pinpointing areas in the educational content that need further review, or providing supportive advice. If a student shows areas of weakness, the game may offer rewards as incentives to encourage continued engagement, leading to enhanced learning opportunities.

9.7 Adaptive Educational Games Guidelines on GDPR Consent

In our current project, we use the Privacy and Ethics Module GDPR-related features to provide direction on developing an adaptive educational game that complies with the new regulation. The guidance focuses on consent, data collection, processing, transparency, and revocation. We evaluate how GDPR affects the system's different parts and then establish general guidelines for adaptive educational games based on our findings. These guidelines ensure that data flow through the system's components and functions adheres to the consent requirements outlined in the GDPR.

The comprehensive structure's diagrams demonstrate consent distribution across various categories. The data flow for students emphasizes the GDPR-affected system elements, denoted by asterisks. The representation of guardian pairs with asterisked padlocks pertains to situations involving students under 16. In these cases, the student's consent requires corresponding guardian consent. Within this framework, the application level includes age-based differentiation among students.

The guidelines encompass a wide range of topics, particularly emphasizing educational games that incorporate personalized learning. While these games' primary purpose is educational, they are also designed to be enjoyable for the players. Integrating an agent into the game's storyline ensures that the player's enjoyment and engagement remain uninterrupted. The agent must be strategically placed within the game's context before any personal data collection or processing takes place, and it cannot be overridden. They obtain the player's explicit consent, whether affirmative or negative, as a prerequisite for the game's progress. If separate consents are needed for initial data collection and subsequent data processing, the agent must inform the player that these are distinct operations.

1. For Initial Data Collection

The game must obtain the student's consent before collecting and processing their data and making automated decisions. This consent can be requested separately or as a single consent for all operations. The first option, which involves requesting separate consent for each operation, is shown in Fig. 9.3 and pertains to the data collection operation and system setup. The system elements affected by the GDPR element are the following:

- Video Game: It represents the main element that provides the user interface for students to interact and submit their data. Throughout the gaming session, the system captures and processes the students' actions (moves) to model the student and carry out the automated decision-making process.
- Educational Tool: Integrated into the game, it evaluates the student's answers to the given exercises and sends the relevant information to the database.
- Database: It stores the gathered information and maintains the records of the exercises completed by the student. The player must give consent, and the system should not permit the student to evade responding to the consent request.

2. For Subsequent Data Processing

The data processing that follows the data collection process must also have the student's consent. Consequently, the AI tool's processing and drawing of conclusions from the student's data are impacted by the student's consent status. Figure 9.4 illustrates the system's components impacted by the GDPR element:

- Database: It is in charge of relaying student data to other system components. It receives student data and acknowledges decisions from the educational tool and information about the student model generated by the AI tool. Subsequently, the database houses unprocessed data, encompassing the student's actions and responses to exercises, and preserves the results of automated processing, specifically the automated decisions made by the AI.
- AI Tool: The system populates the database using the educational tool to make automated decisions (inferences), which includes the model automatically generated from the student's input on the game's moves and exercise responses. Additionally, it encompasses the automated decisions that govern the game scenario's progression, which aligns with the learner's model.

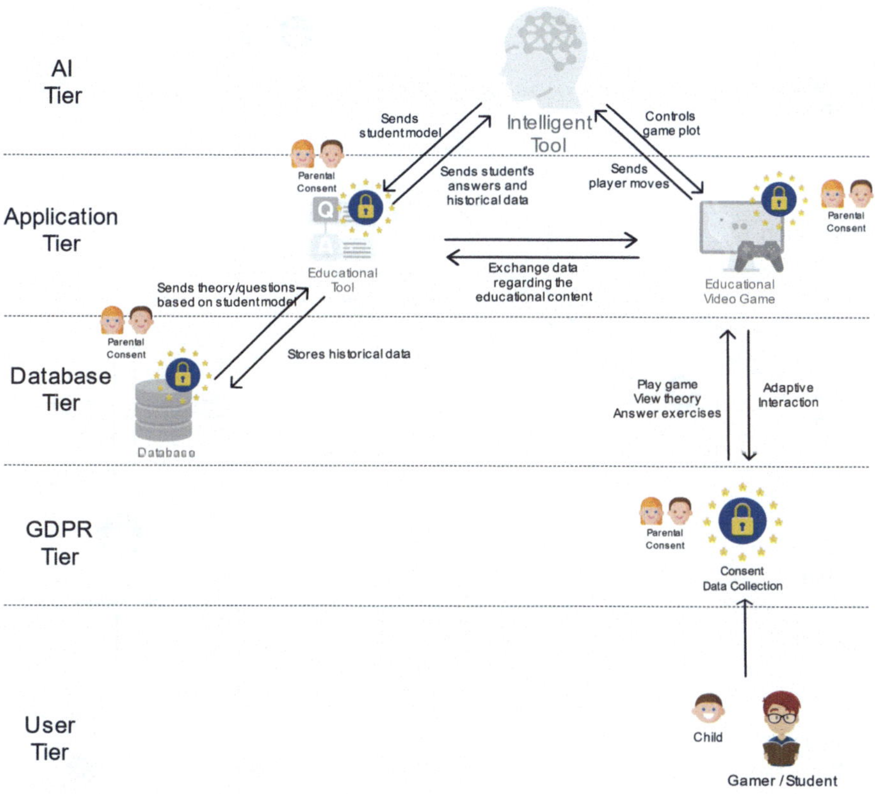

Fig. 9.3 The components affected by the initial data collection

It is important for the system to clearly communicate to the student that data collection and processing are separate processes despite being interconnected. Processing is reliant on data collection. The system must obtain the player's consent and ensure the student cannot bypass this requirement.

3. For Initial Data Collection and Subsequent Data Processing (one-off consent for all functions)

Consent should be requested separately for each functionality or as a one-time request covering all functionalities. Figure 9.5 depicts the elements impacted when the system seeks a single consent for all functionalities. In this scenario, the requirement for a single consent is expected to affect all system components. The system must inform the player about the initial data collection and subsequent processing to provide a personalized gaming and learning experience. Furthermore, the system must obtain the player's consent to collect and process data. Once the student's approval is received, the system directs the student's data to all components, activating their respective functions based on the student's role.

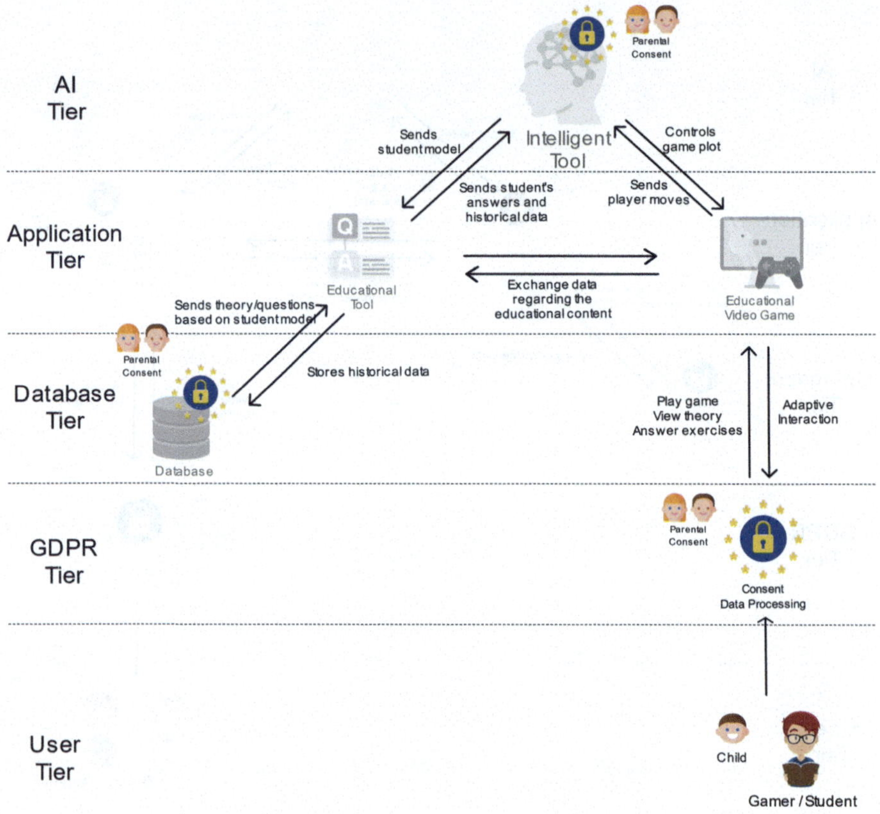

Fig. 9.4 The components affected by the subsequent data processing

4. Revoking Consent (one-off revocation for all functions)

Once the student consents to collect and process their data, the game must offer the option to withdraw their consent at any time. The revocation may apply separately to each function or cumulatively to both functions. Figure 9.6 depicts the impacted elements in case of withdrawal, which apply to all functions:

- Video game: The student interacts directly with this component. Therefore, it should enable the student to withdraw their consent at any time.
- Educational tool: consent revocation impacts the component, which stops saving the student's responses in the database.
- Database: New student data must not be added to the database, regardless of its source (e.g., student exercise responses or conclusions based on the student's model). Existing data is held in the student's log file unless the student has requested to have their data erased (i.e., the right of erasure).
- AI tool: When consent is withdrawn, the AI tool is deactivated because it no longer receives student data or makes inferences and recommendations.

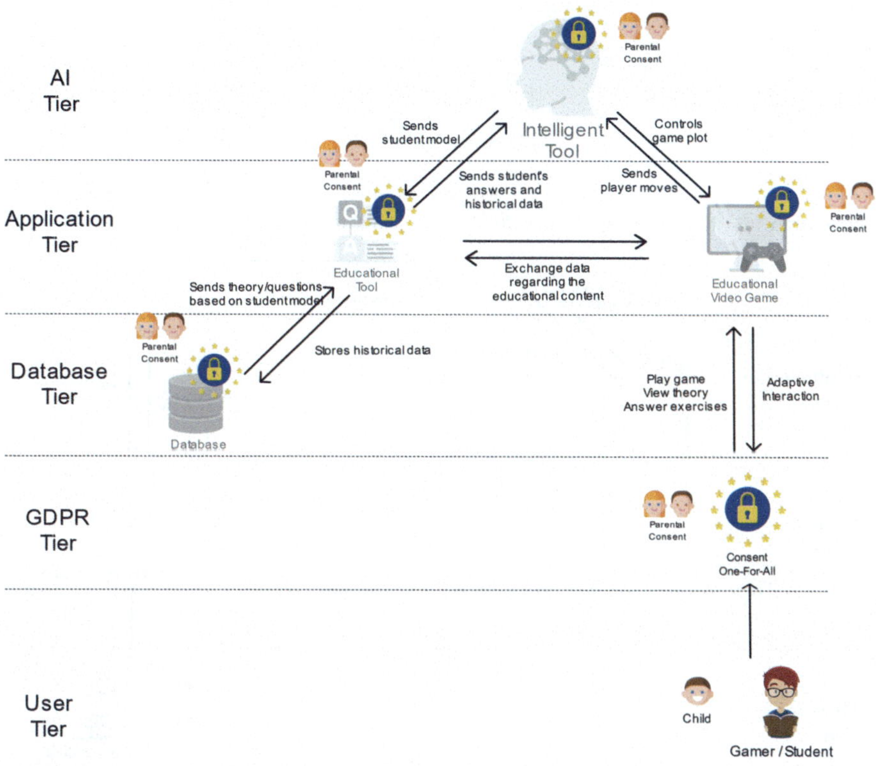

Fig. 9.5 Initial Data Collection and Subsequent Data Processing (one-off all functions)

9.8 Conclusions and Discussion

This chapter outlines the functionalities of the Privacy and Ethics Module in the EPATHLO Educational Game Suite and details how it ensures compliance with GDPR. It offers a comprehensive overview of data collection and processing procedures, the intended use of personal data, and the associated advantages for the student. Students can exercise their rights as specified in the regulation, including accessing their data, requesting amendments or deletions, and doubting the accuracy of assessments regarding their knowledge level.

The above features are translated into general guidelines, aimed to be a reference point for those involved in implementing intelligent educational games (IEGs) that are entirely GDPR compliant. When an adaptive educational game provides complete transparency about data collection and decision-making functions and clearly describes the benefits of a personalized experience at the educational and gameplay

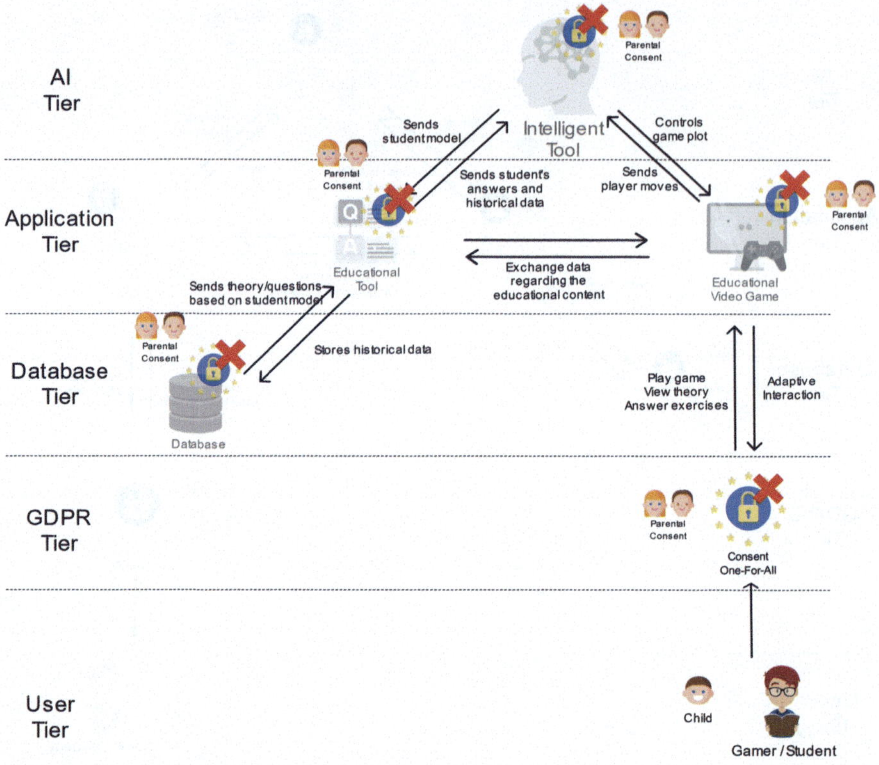

Fig. 9.6 The components affected by the revoking consent action

level, students can more easily consent to the system's use of their data. Moreover, if they know at any time what data is being collected about them and, more importantly, that they can ask for their deletion, they will be more likely to opt for a personalized experience. The student needs to know that they are in complete control of their desire to have their data collected and processed and that they can easily and quickly withdraw this wish and request the complete deletion of their data as if they had never used the game.

This chapter underscores the necessity of user consent in collecting and processing their data. It provides an architectural description of the components of an adaptive educational game that are influenced by student consent and illustrates how student data flows between these components.

References

1. Albrecht, J.P.: How the gdpr will change the world. Eur. Data Prot. L. Rev. **2**, 287 (2016)
2. Alogogianni, E., Virvou, M.: Data mining for targeted inspections against undeclared work by applying the crisp-dm methodology. In: 2021 12th International Conference on Information, Intelligence, Systems & Applications (IISA), pp. 1–8. IEEE (2021)
3. Alrakhawi, H.A., Jamiat, N., Abu-Naser, S.S.: Intelligent tutoring systems in education: a systematic review of usage, tools, effects and evaluation. J. Theor. Appl. Inf. Technol. **101**(4), 1205–1226 (2023)
4. Ausloos, J., et al.: Gdpr transparency as a research method. SSRN Electron. J. 1–23 (2019)
5. Ayala-Rivera, V., Pasquale, L.: The grace period has ended: an approach to operationalize gdpr requirements. In: 2018 IEEE 26th International Requirements Engineering Conference (RE), pp. 136–146. IEEE (2018)
6. Bentzen, H.B., Olav, H.K., Ursin, G.: Maximizing the gdpr potential for data transfers: first in europe. Lancet Reg. Health Eur. **27**, 100600 (2023)
7. Borking, J.: Der identity protector. Datenschutz und Datensicherheit **20**(11), 654–658 (1996)
8. Bottis, M., Panagopoulou-Koutnatzi, F., Michailaki, A., Nikita, M.: The right to access information under the gdpr. Int. J. Technol. Policy Law **3**(2), 131–142 (2019)
9. Brasher, E.A.: Addressing the failure of anonymization: guidance from the european union's general data protection regulation. Colum. Bus. L. Rev. 209 (2018)
10. Breen, S., Ouazzane, K., Patel, P.: Gdpr: Is your consent valid? Bus. Inf. Rev. **37**(1), 19–24 (2020)
11. Brkan, M.: Do algorithms rule the world? algorithmic decision-making and data protection in the framework of the gdpr and beyond. Int. J. Law Inf. Technol. **27**(2), 91–121 (2019)
12. Bunn, A.: The curious case of the right to be forgotten. Comput. Law Secur. Rev. **31**(3), 336–350 (2015)
13. Chrysafiadi, K., Virvou, M.: Evaluating the integration of fuzzy logic into the student model of a web-based learning environment. Expert Syst. Appl. **39**(18), 13127–13134 (2012)
14. Chrysafiadi, K., Virvou, M.: A knowledge representation approach using fuzzy cognitive maps for better navigation support in an adaptive learning system. Springerplus **2**(1), 1–13 (2013)
15. Chrysafiadi, K., Virvou, M.: Fuzzy logic for adaptive instruction in an e-learning environment for computer programming. IEEE Trans. Fuzzy Syst. **23**(1), 164–177 (2014)
16. Chrysafiadi, K., Virvou, M.: Create dynamically adaptive test on the fly using fuzzy logic. In: 2018 9th International Conference on Information, Intelligence, Systems and Applications (IISA), pp. 1–8. IEEE (2018)
17. Chrysafiadi, K., Papadimitriou, S., Virvou, M.: Fuzzy states for dynamic adaptation of the plot of an educational game in relation to the learner's progress. In: 2020 11th International Conference on Information, Intelligence, Systems and Applications (IISA), pp. 1–7. IEEE (2020)
18. Chrysafiadi, K., Papadimitriou, S., Virvou, M.: Cognitive-based adaptive scenarios in educational games using fuzzy reasoning. Knowl. Based Syst. **250**, 109111 (2022)
19. Chrysafiadi, K., Kamitsios, M., Virvou, M.: Fuzzy-based dynamic difficulty adjustment of an educational 3d-game. Multimed. Tools Appl. 1–25 (2023)
20. Chrysafiadi, K., Virvou, M., Tsihrintzis, G., Hatzilygeroudis, I.: An adaptive learning environment for programming based on fuzzy logic and machine learning. Int. J. Artif. Intell, Tools (2023)
21. Chrysafiadi, K., Virvou, M., Tsihrintzis, G.A., Hatzilygeroudis, I.: Evaluating the user's experience, adaptivity and learning outcomes of a fuzzy-based intelligent tutoring system for computer programming for academic students in greece. Educ. Inf. Technol. **28**(6), 6453–6483 (2023)
22. Clarke, N., Vale, G., Reeves, E.P., Kirwan, M., Smith, D., Farrell, M., Hurl, G., McElvaney, N.G.: Gdpr: an impediment to research? Irish J. Med. Sci. (1971-) **188**, 1129–1135 (2019)

23. Corbett, A.T., Anderson, J.R.: (2021) Lisp intelligent tutoring system: research in skill acquisition. In: Computer Assisted Instruction and Intelligent Tutoring Systems, pp 73–109. Routledge
24. Corbett, A.T., Koedinger, K.R., Anderson, J.R.: Intelligent tutoring systems. In: Handbook of Human-Computer Interaction, pp. 849–874. Elsevier (1997)
25. Custers, B., Heijne, A.S.: The right of access in automated decision-making: the scope of article 15 (1)(h) gdpr in theory and practice. Comput. Law Secur. Rev. **46**, 105727 (2022)
26. Dumas, J.: General data protection regulation (gdpr): prioritizing resources. Seattle UL Rev. **42**, 1115 (2018)
27. Esayas, S.: The role of anonymisation and pseudonymisation under the eu data privacy rules: beyond the 'all or nothing'approach. Eur. J. Law Technol. **6**(2) (2015)
28. Finck, M., Pallas, F.: They who must not be identified-distinguishing personal from non-personal data under the gdpr. Int. Data Privacy Law **10**(1), 11–36 (2020)
29. Franke, L., Liang, H., Brantly, A., Davis, J.C., Brown, C.: A first look at the general data protection regulation (GDPR) in open-source software. In: Proceedings of the 2024 IEEE/ACM 46th International Conference on Software Engineering: Companion Proceedings, ICSE Companion 2024, Lisbon, Portugal, 14–20 Apr 2024, pp. 268–269. ACM (2024). https://doi.org/10.1145/3639478.3643077
30. Franke, L., Liang, H., Farzanehpour, S., Brantly, A., Davis, J.C., Brown, C.: An exploratory mixed-methods study on general data protection regulation (GDPR) compliance in open-source software (2024b). CoRR, arXiv:2406.14724, https://doi.org/10.48550/ARXIV.2406.14724
31. Frosio, G.F.: The right to be forgotten: much ado about nothing. Colo. Tech. L. J. **15**, 307 (2016)
32. Gal, M.S., Aviv, O.: The competitive effects of the gdpr. J. Compet. Law Econ. **16**(3), 349–391 (2020)
33. Gerl, A., Meier, B.: The layered privacy language art. 12–14 gdpr extension-privacy enhancing user interfaces. Datenschutz und Datensicherheit-DuD **43**, 747–752 (2019)
34. Gerl, A., Pohl, D.: Critical analysis of lpl according to articles 12-14 of the gdpr. In: Proceedings of the 13th International Conference on Availability, Reliability and Security, pp. 1–9 (2018)
35. Gil González, E., De Hert, P.: Understanding the legal provisions that allow processing and profiling of personal data-an analysis of gdpr provisions and principles. Era Forum, vol. 19, pp. 597–621. Springer (2019)
36. Gómez-Martínez, E., Marroyo, M., Acuña, S.T.: Towards the integration of the gdpr in the unified software development process (s). In: SEKE, pp. 199–204 (2021)
37. Gupta, I., Naithani, P.: Transparent communication under article 12 of the gdpr: Advocating a standardised approach for universal understandability. J. Data Prot. Privacy **5**(2), 149–161 (2022)
38. Hänold, S.: Profiling and automated decision-making: legal implications and shortcomings. In: Robotics, AI and the Future of Law, pp. 123–153 (2018)
39. Hintze, M., El Emam, K.: Comparing the benefits of pseudonymisation and anonymisation under the gdpr. J. Data Prot. Privacy **2**(2), 145–158 (2018)
40. Hooshyar, D., Malva, L., Yang, Y., Pedaste, M., Wang, M., Lim, H.: An adaptive educational computer game: effects on students' knowledge and learning attitude in computational thinking. Comput. Hum. Behav. **114**, 106575 (2021)
41. Hooshyar, D., Pedaste, M., Yang, Y., Malva, L., Hwang, G.J., Wang, M., Lim, H., Delev, D.: From gaming to computational thinking: an adaptive educational computer game-based learning approach. J. Educ. Comput. Res. **59**(3), 383–409 (2021)
42. Horák, M., Stupka, V., Husák, M.: Gdpr compliance in cybersecurity software: a case study of dpia in information sharing platform. In: Proceedings of the 14th International Conference on Availability, Reliability and Security, pp. 1–8 (2019)
43. Howard-Jones, P.A., Jay, T.: Reward, learning and games. Curr. Opin. Behav. Sci. **10**, 65–72 (2016)

44. Jones, J., Millington, M., Virvou, M.: An assumption-based truth maintenance system in active aid for unix users. In: Intelligent Help Systems for UNIX, pp. 229–252 (2000)
45. Kabassi, K., Virvou, M.: Combination of a cognitive theory with the multi-attribute utility theory. In: International Conference on Knowledge-Based and Intelligent Information and Engineering Systems, pp. 944–950. Springer (2003)
46. Kabassi, K., Virvou, M.: Using web services for personalised web-based learning. J. Educ. Technol. Soc. **6**(3), 61–71 (2003)
47. Katsionis, G., Virvou, M.: Personalised e-learning through an educational virtual reality game using web services. Multimed. Tools Appl. **39**, 47–71 (2008)
48. Kingston, J.: Using artificial intelligence to support compliance with the general data protection regulation. Artif. Intell. Law **25**(4), 429–443 (2017)
49. Kittmann, T., Lambrecht, J., Horn, C.: A privacy-aware distributed software architecture for automation services in compliance with GDPR. In: 23rd IEEE International Conference on Emerging Technologies and Factory Automation, ETFA 2018, Torino, Italy, 4–7 Sept 2018, pp. 1067–1070. IEEE (2018) https://doi.org/10.1109/ETFA.2018.8502545
50. Kocharyan, H., Vardanyan, L., Hamul'ák, O., Kerikmäe, T.: Critical views on the right to be forgotten after the entry into force of the gdpr: Is it able to effectively ensure our privacy? Int. Comp. Law Rev. **21**(2), 96–115 (2021)
51. Krouska, A., Troussas, C., Virvou, M.: Computerized adaptive assessment using accumulative learning activities based on revised bloom's taxonomy. In: Knowledge-Based Software Engineering: 2018: Proceedings of the 12th Joint Conference on Knowledge-Based Software Engineering (JCKBSE 2018) Corfu, Greece, vol. 12, pp. 252–258. Springer (2019)
52. Lin, C.C., Huang, A.Y., Lu, O.H.: Artificial intelligence in intelligent tutoring systems toward sustainable education: a systematic review. Smart Learn. Environ. **10**(1), 41 (2023)
53. Liu, S., Zhao, B., Guo, R., Meng, G., Zhang, F., Zhang, M.: Have you been properly notified? Automatic compliance analysis of privacy policy text with gdpr article 13. In: Proceedings of the Web Conference, vol. 2021, pp. 2154–2164 (2021)
54. Liu, Z., Moon, J., Kim, B., Dai, C.P.: Integrating adaptivity in educational games: A combined bibliometric analysis and meta-analysis review. Educ. Tech. Res. Dev. **68**, 1931–1959 (2020)
55. Malgieri, G.: The concept of fairness in the gdpr: a linguistic and contextual interpretation. In: Proceedings of the 2020 Conference on Fairness, Accountability, and Transparency, pp. 154–166 (2020)
56. Mangini, V., Tal, I., Moldovan, A.N.: An empirical study on the impact of gdpr and right to be forgotten-organisations and users perspective. In: Proceedings of the 15th International Conference on Availability, Reliability and Security, ACM, ARES 2020 (2020)
57. Mantelero, A.: The eu proposal for a general data protection regulation and the roots of the 'right to be forgotten'. Comput. Law Secur. Rev. **29**(3), 229–235 (2013)
58. Marques, J.F., Bernardino, J.: Analysis of data anonymization techniques. In: KEOD, pp. 235–241 (2020)
59. McKernan, B., Martey, R.M., Stromer-Galley, J., Kenski, K., Clegg, B.A., Folkestad, J.E., Rhodes, M.G., Shaw, A., Saulnier, E.T., Strzalkowski, T.: We don't need no stinkin'badges: the impact of reward features and feeling rewarded in educational games. Comput. Hum. Behav. **45**, 299–306 (2015)
60. Mitlöhner, J., Neumaier, S., Umbrich, J., Polleres, A.: Characteristics of open data csv files. In: 2016 2nd International Conference on Open and Big Data (OBD), pp. 72–79. IEEE (2016)
61. Mougiakou, E., Virvou, M.: Based on gdpr privacy in uml: Case of e-learning program. In: 2017 8th International Conference on Information, Intelligence, Systems & Applications (IISA), pp. 1–8. IEEE (2017)
62. Mougiakou, E., Papadimitriou, S., Virvou, M.: Intelligent tutoring systems and transparency: the case of children and adolescents. In: 2018 9th International Conference on Information, Intelligence, Systems and Applications (IISA), pp. 1–8. IEEE (2018)
63. Mougiakou, E., Papadimitriou, S., Virvou, M.: Automated decision making and personal data protection in intelligent tutoring systems: design guidelines. In: Knowledge-Based Software Engineering: 2018: Proceedings of the 12th Joint Conference on Knowledge-Based Software Engineering (JCKBSE 2018) Corfu, Greece, vol. 12, pp. 231–241. Springer (2019)

64. Mougiakou, E., Papadimitriou, S., Virvou, M.: Synchronous and asynchronous learning methods under the light of general data protection regulation. In: 2020 11th International Conference on Information, Intelligence, Systems and Applications (IISA), pp. 1–7. IEEE (2020)
65. Moundridou, M., Virvou, M.: Authoring and delivering adaptive web-based textbooks using wear. In: Proceedings IEEE International Conference on Advanced Learning Technologies, pp. 185–188. IEEE (2001)
66. Mourby, M., Mackey, E., Elliot, M., Gowans, H., Wallace, S.E., Bell, J., Smith, H., Aidinlis, S., Kaye, J.: Are 'pseudonymised'data always personal data? Implications of the gdpr for administrative data research in the UK. Comput. Law Secur. Rev. **34**(2), 222–233 (2018)
67. Münch, D., Grosselfinger, A.K., Krempel, E., Hebel, M., Arens, M.: Data anonymization for data protection on publicly recorded data. In: Computer Vision Systems: 12th International Conference, ICVS 2019, Thessaloniki, Greece, 23–25 Sept 2019, Proceedings, vol. 12, pp. 245–258. Springer (2019)
68. Nišević, M., Sears, A.M., Fosch-Villaronga, E., Custers, B.: Understanding the legal bases for automated decision-making under the gdpr. In: Research Handbook on EU Data Protection Law, pp. 435–454. Edward Elgar Publishing (2022)
69. Panagoulias, D.P., Virvou, M., Tsihrintzis, G.A.: Tailored explainability in medical artificial intelligence-empowered applications: personalisation via the technology acceptance model. In: 2023 IEEE 35th International Conference on Tools with Artificial Intelligence (ICTAI), pp. 486–490. IEEE (2023)
70. Panagoulias, D.P., Virvou, M., Tsihrintzis, G.A.: A novel framework for artificial intelligence explainability via the technology acceptance model and rapid estimate of adult literacy in medicine using machine learning. Expert Syst. Appl. **248**, 123375 (2024)
71. Papadimitriou, S., Virvou, M.: An online adventure game for teaching math. In: 2016 7th International Conference on Information, Intelligence, Systems & Applications (IISA), pp. 1–5. IEEE (2016)
72. Papadimitriou, S., Virvou, M.: Adaptivity in scenarios in an educational adventure game. In: 2017 8th International Conference on Information, Intelligence, Systems & Applications (IISA), pp. 1–6. IEEE (2017)
73. Papadimitriou, S., Chrysafiadi, K., Virvou, M.: Evaluating the use of fuzzy logic in an educational game for offering adaptation. In: 2019 International Conference on Computer, Information and Telecommunication Systems (CITS), pp. 1–5. IEEE (2019)
74. Papadimitriou, S., Chrysafiadi, K., Virvou, M.: Fuzzeg: Fuzzy logic for adaptive scenarios in an educational adventure game. Multimed. Tools Appl. **78**(22), 32023–32053 (2019). https://doi.org/10.1007/s11042-019-07955-w
75. Papadimitriou, S., Mougiakou, E., Virvou, M.: Smart educational games and consent under the scope of general data protection regulation. In: 2019 10th International Conference on Information, Intelligence, Systems and Applications (IISA), pp. 1–8. IEEE (2019)
76. Papadimitriou, S., Kamitsios, M., Chrysafiadi, K., Virvou, M.: Learn-and-play personalised reasoning from point-and-click to virtual reality mobile educational games. Intell. Decis. Technol. **15**(2), 321–332 (2021)
77. Papadimitriou, S., Chrysafiadi, K., Virvou, M.: Adaptive quizzes using fuzzy genetic algorithm. In: 2023 14th International Conference on Information, Intelligence, Systems & Applications (IISA), pp. 1–8. IEEE (2023)
78. Perera, H., Hussain, W., Mougouei, D., Shams, R.A., Nurwidyantoro, A., Whittle, J.: Towards integrating human values into software: mapping principles and rights of gdpr to values. In: 2019 IEEE 27th International Requirements Engineering Conference (RE), pp. 404–409. IEEE (2019)
79. Politou, E., Alepis, E., Virvou, M., Patsakis, C.: The "Right to Be Forgotten" in the GDPR: Implementation Challenges and Potential Solutions, pp. 41–68. Springer International Publishing (2021)
80. Politou, E., Alepis, E., Virvou, M., Patsakis, C.: Privacy and Data Protection Challenges in the Distributed Era. Springer International Publishing (2022)

81. Robaldo, L., Bartolini, C., Palmirani, M., Rossi, A., Martoni, M., Lenzini, G.: Formalizing gdpr provisions in reified i/o logic: the dapreco knowledge base. J. Logic Lang. Inf. **29**, 401–449 (2020)
82. Scarcella, L.: Tax compliance and privacy rights in profiling and automated decision making. Internet Policy Rev. **8**(4) (2019)
83. Senavirathne, N., Torra, V.: On the role of data anonymization in machine learning privacy. In: 2020 IEEE 19th International Conference on Trust, Security and Privacy in Computing and Communications (TrustCom), pp. 664–675. IEEE (2020)
84. Shute, V., Rahimi, S., Smith, G., Ke, F., Almond, R., Dai, C.P., Kuba, R., Liu, Z., Yang, X., Sun, C.: Maximizing learning without sacrificing the fun: stealth assessment, adaptivity and learning supports in educational games. J. Comput. Assist. Learn. **37**(1), 127–141 (2021)
85. Sirur, S., Nurse, J.R., Webb, H.: Are we there yet? Understanding the challenges faced in complying with the general data protection regulation (gdpr). In: Proceedings of the 2nd International Workshop on Multimedia Privacy and Security, pp. 88–95 (2018)
86. Smith, M., Torres-Agüero, A., Grossman, R., Sen, P., Chen, Y., Borcea, C.: A study of gdpr compliance under the transparency and consent framework. In: Proceedings of the ACM on Web Conference, vol. 2024, pp. 1227–1236 (2024)
87. Spagnuelo, D., Ferreira, A., Lenzini, G.: Accomplishing transparency within the general data protection regulation. In: ICISSP, pp. 114–125 (2019)
88. Stathopoulou, I., Alepis, E., Tsihrintzis, G., Virvou, M.: On assisting a visual-facial affect recognition system with keyboard-stroke pattern. In: Research and Development in Intelligent Systems XXVI: Incorporating Applications and Innovations in Intelligent Systems XVII, p. 451 (2009)
89. Tankard, C.: What the gdpr means for businesses. Netw. Secur. **2016**(6), 5–8 (2016)
90. Tosoni, L.: The right to object to automated individual decisions: resolving the ambiguity of article 22 (1) of the general data protection regulation. Int. Data Privacy Law **11**(2), 145–162 (2021)
91. Triantafyllou, A.M., Tsihrintzis, G.A., Virvou, M., Alepis, E.: A bimodal system for emotion recognition via computer of known or unknown persons in normal or fatigue situations. In: Advances in Core Computer Science-Based Technologies: Papers in Honor of Professor Nikolaos Alexandris, pp. 9–35 (2021)
92. Tsihrintzis, G.A., Virvou, M.: Multimedia Services in Intelligent Environments: Integrated Systems, vol 3. Springer Science & Business Media (2010)
93. Tsihrintzis, G.A., Virvou, M., Phillips-Wren, G.: Surveys in artificial intelligence-based technologies. Intell. Decis. Technol. **13**(4), 393–394 (2019)
94. Tsihrintzis, G.A., Virvou, M., Sakkopoulos, E., Jain, L.C.: Applications of learning and analytics in intelligent systems, pp. 1–6. Applications of Learning and Analytics in Intelligent Systems, Machine Learning Paradigms (2019)
95. Tsihrintzis, G.A., Virvou, M., Saruwatari, T.: Special Collection of Extended Selected Papers on "Novel Research Results Presented at the 14th International Joint Conference on Knowledge-Based Software Engineering (jckbse2022), 22–24 Aug 2022, Larnaca, Cyprus" (2022). https://easyconferences.eu/jckbse2022/
96. Tsihrintzis, G.A., Virvou, M., Doukas, H., Jain, L.C.: Introduction to advances in artificial intelligence-empowered decision support systems. In: Advances in Artificial Intelligence-Empowered Decision Support Systems: Papers in Honour of Professor John Psarras, pp. 1–11. Springer (2024)
97. Tsiriga, V., Virvou, M.: Initializing the student model using stereotypes and machine learning. In: Proceedings of the IEEE International Conference on Systems, Man and Cybernetics, Citeseer (2002)
98. Van Ooijen, I., Vrabec, H.U.: Does the gdpr enhance consumers' control over personal data? An analysis from a behavioural perspective. J. Consum. Policy **42**, 91–107 (2019)
99. Vanbecelaere, S., Van den Berghe, K., Cornillie, F., Sasanguie, D., Reynvoet, B., Depaepe, F.: The effectiveness of adaptive versus non-adaptive learning with digital educational games. J. Comput. Assist. Learn. **36**(4), 502–513 (2020)

100. Vanbecelaere, S., Cornillie, F., Sasanguie, D., Reynvoet, B., Depaepe, F.: The effectiveness of an adaptive digital educational game for the training of early numerical abilities in terms of cognitive, noncognitive and efficiency outcomes. Br. J. Edu. Technol. **52**(1), 112–124 (2021)
101. Vavra, A.N.: The right to be forgotten: an archival perspective. Am. Arch. **81**(1), 100–111 (2018)
102. Verhenneman, G., Claes, K., Derèze, J., Herijgers, P., Mathieu, C., Rademakers, F., Reyda, R., Vanautgaerden, M.: How gdpr enhances transparency and fosters pseudonymisation in academic medical research. Eur. J. Health Law **27**(1), 35–57 (2020)
103. Veronese, A.: The right of explanation and the right to object automatic decisions: comparing the european union general data protection regulation (eu/gdpr) with the brazilian general data protection federal act (bgdpfa) (2019)
104. Virvou, M.: Artificial intelligence and user experience in reciprocity: Contributions and state of the art. Intelli. Decis. Technol. **17**(Preprint), 1–53 (2023)
105. Virvou, M., Katsionis, G.: On the usability and likeability of virtual reality games for education: the case of vr-engage. Comput. Educ. **50**(1), 154–178 (2008)
106. Virvou, M., Manos, K.: Individualizing a cognitive model of students' memory in intelligent tutoring systems. In: Knowledge-Based Intelligent Information and Engineering Systems: 7th INternational Conference, KES 2003, Oxford, UK, September 2003. Proceedings, Part I, vol. 7, pp. 893–897. Springer (2003)
107. Virvou, M., Manos, K.: A simulated student-player in support of the authoring process in a knowledge-based authoring tool for educational games. In: Proceedings 3rd IEEE International Conference on Advanced Technologies, pp. 338–339. IEEE (2003)
108. Virvou, M., Maras, D.: Error diagnosis in an english tutor. In: Proceedings of the HCI International'99 (The 8th International Conference on Human-Computer Interaction) on Human-Computer Interaction: Communication, Cooperation, and Application Design, vol. 2, pp. 657–661 (1999)
109. Virvou, M., Papadimitriou, S.: Learning addition by playing an adaptive game with coins. In: IISA 2013, pp. 1–4. IEEE (2013a)
110. Virvou, M., Papadimitriou, S.: Simple arithmetic lessons through an adaptive snake game. In: 2013 International Conference on Computer, Information and Telecommunication Systems (CITS), pp. 1–4. IEEE (2013)
111. Virvou, M., Papadimitriou, S.: Use of digital 'guess who board game' in teaching english. In: IISA 2014, The 5th International Conference on Information, Intelligence, Systems and Applications, pp. 439–443. IEEE (2014)
112. Virvou, M., Papadimitriou, S.: Playing, learning, extending: educational "guess who" game that is renewable by a teacher. In: 2015 6th International Conference on Information, Intelligence, Systems and Applications (IISA), pp. 1–5. IEEE (2015)
113. Virvou, M., Tsihrintzis, G.A.: Is chatgpt beneficial to education? A holistic evaluation framework based on intelligent tutoring systems. In: 2023 14th International Conference on Information, Intelligence, Systems & Applications (IISA), pp. 1–8. IEEE (2023)
114. Virvou, M., Maras, D., Tsiriga, V.: Student modelling in an intelligent tutoring system for the passive voice of english language. J. Educ. Technol. Soc. **3**(4), 139–150 (2000)
115. Virvou, M., Tsiriga, V., Maras, D.: Evaluation of an its for the passive voice of the english language using the ciao! framework. In: EdMedia+ Innovate Learning, Association for the Advancement of Computing in Education (AACE), pp. 1757–1758 (2000)
116. Virvou, M., Manos, C., Katsionis, G., Tourtoglou, K.: Vr-engage: a virtual reality educational game that incorporates intelligence. In: Proceedings of IEEE International Conference on Advanced Learning Technologies, pp 16–19 (2002)
117. Virvou, M., Katsionis, G., Manos, K.: Combining software games with education: evaluation of its educational effectiveness. J. Educ. Technol. Soc. **8**(2), 54–65 (2005)
118. Virvou, M., Tsihrintzis, G., Alepis, E., Stathopoulou, I.O.: Designing a multi-modal affective knowledge-based user interface: combining empirical studies. In: Knowledge-Based Software Engineering, pp. 250–259. IOS Press (2008)

119. Virvou, M., Alepis, E., Sidiropoulos, S.C.: A learning analytics tool for supporting teacher decision. In: 2015 6th International Conference on Information, Intelligence, Systems and Applications (IISA), pp. 1–3. IEEE (2015)
120. Virvou, M., Alepis, E., Tsihrintzis, G.A., Jain, L.C.: Machine Learning Paradigms: Advances in Learning Analytics. Springer (2020)
121. Virvou, M., Tsihrintzis, G.A., Bourbakis, N.G., Jain, L.C.: Handbook on Artificial Intelligence-Empowered Applied Software Engineering: Smart Software Applications in Cyber-Physical Systems, vols. 2 and 3. Springer (2022)
122. Virvou, M., Tsihrintzis, G.A., Tsichrintzi, E.A.: Virtsi: A novel trust dynamics model enhancing artificial intelligence collaboration with human users-insights from a chatgpt evaluation study. Inf. Sci. **675**, 120759 (2024)
123. Voigt, P., Von dem Bussche, A.: The eu general data protection regulation (gdpr). In: A Practical Guide, 1st ed., vol. 10, no. 3152676, pp. 10–5555. Springer International Publishing, Cham (2017)
124. Wachter, S.: The gdpr and the internet of things: a three-step transparency model. Law Innov. Technol. **10**(2), 266–294 (2018)
125. Wang, H., Tlili, A., Huang, R., Cai, Z., Li, M., Cheng, Z., Yang, D., Li, M., Zhu, X., Fei, C.: Examining the applications of intelligent tutoring systems in real educational contexts: A systematic literature review from the social experiment perspective. Educ. Inf. Technol. **28**(7), 9113–9148 (2023)
126. Weitzenboeck, E.M., Lison, P., Cyndecka, M., Langford, M.: The gdpr and unstructured data: is anonymization possible? Int. Data Privacy Law **12**(3), 184–206 (2022)
127. Wiedemann, K.: Profiling and (automated) decision-making under the gdpr: a two-step approach. Comput. Law Secur. Rev. **45**, 105662 (2022)
128. Wolters, P.: The control by and rights of the data subject under the gdpr (2018)
129. Wulf, A.J., Seizov, O.: "Please understand we cannot provide further informationâŁž: evaluating content and transparency of gdpr-mandated ai disclosures. AI & Soc. **39**(1), 235–256 (2024)
130. Zaman, R., Hassani, M.: Process mining meets gdpr compliance: the right to be forgotten as a use case. In: 2019 International Conference on Process Mining Doctoral Consortium, ICPM-DC (2019)
131. Zarsky, T.Z.: Incompatible: the gdpr in the age of big data. Seton Hall L. Rev. **47**, 995 (2016)

Chapter 10
Contributions and Discussions on Advancing 21st Century Skills Through EPATHLO: Prosocial Educational Games with Artificial Intelligence

Abstract This chapter highlights the contributions of EPATHLO, an innovative AI-based educational adventure game, in fostering 21st-century skills. The game utilizes advanced AI techniques to assess students' knowledge dynamically, adapting educational content and scenarios to achieve dual objectives: improving knowledge acquisition and maintaining engagement. By involving students in interdisciplinary tasks through non-playable characters (NPCs), EPATHLO creates a prosocial gaming environment that cultivates critical soft skills essential for modern careers, including communication, problem-solving, teamwork, time management, and adaptability. The chapter concludes by discussing the challenges and future directions for intelligent educational games, highlighting the potential for integrating emerging technologies to enhance learning outcomes.

Keywords Educational software · Personalized educational games · Artificial Intelligence in educational games · Fuzzy logic in education · User experience · AI-empowered software games · User modeling · Human-AI interaction · e-learning games · 21st-century skills · Human-centred Artificial Intelligence

10.1 Overview of Contributions and Discussion

Artificial intelligence (AI) has rapidly evolved, driving significant advancements in AI-empowered software, profoundly impacting various domains. AI technologies have been integrated into intelligent decision-making systems, enhancing their ability to analyze complex data and make informed decisions efficiently [153, 158, 159]. The latest research showcases novel AI applications in software engineering, emphasizing their role in improving software development processes and outcomes [157]. Machine Learning (ML) and Deep Learning (DL) techniques have been fundamental in advancing AI, leading to the creation of more sophisticated and autonomous systems [156]. AI-empowered software applications have found significant use in cyber-physical systems, demonstrating their capability to adapt and respond to dynamic environments [188]. The integration of AI in user experience design has also been

S. Papadimitriou and M. Virvou, *Artificial Intelligence—Based Games as Novel Holistic Educational Environments to Teach 21st Century Skills*, Intelligent Systems Reference Library 93, https://doi.org/10.1007/978-3-031-77464-5_10

highlighted, focusing on creating more intuitive and personalized interactions [168]. As these technologies continue to evolve, they promise to enhance decision support systems further, underscoring the potential for AI to transform industries and improve efficiencies [187, 190]. The integration of AI technologies, such as fuzzy logic, ML, and clustering techniques [131] in educational games enhances personalization and adaptivity, creating tailored experiences that take into account individual learning styles and paces [29]. AI-driven adaptive learning systems provide real-time feedback and support, facilitating more effective learning processes and helping students achieve better outcomes [160, 175].

As such, the primary aim of the research described in this book was to create an innovative AI-empowered educational game, EPATHLO (or Fuzzy EPATHLO), with a dual focus: enhancing both hard and soft skills needed for 21st-century careers in students while continuously engaging and motivating them to participate in additional exercises. EPATHLO is designed to adapt to each student's unique characteristics, successfully meeting the research objectives. The system customizes exercises based on student's performance and understanding of the educational material, adjusting the game content to maintain players' interest and motivation as they advance. This ensures a sustained sense of engagement and curiosity.

To achieve these dual objectives, it has been essential to address each aspect separately. The first step involved developing a student model to personalize educational content effectively. Extensive research highlights the success of various student modeling techniques, as evidenced by notable results [13, 21, 22, 45, 64, 70, 104, 112, 136, 145, 152]. AI-driven student modeling and learning analytics have significantly enhanced how educational software personalizes learning by employing various original approaches. One such approach is the use of cognitive theories to develop Intelligent Tutoring Systems (ITSs) that individualize students' memory models [173], thereby customizing instructional strategies to enhance learning outcomes [30]. Another approach involves rule-based systems for error diagnosis, which are particularly effective in language learning applications. These systems identify and correct student misconceptions, providing tailored feedback to improve students' proficiency [175]. ML techniques, such as fuzzy logic, enable the creation of dynamically adaptive tests that adjust to a student's performance in real time, offering a personalized assessment experience [23]. Neural networks also play a crucial role in student modeling by integrating affective computing. These systems interpret emotional responses through multimodal inputs, such as facial expressions and keystroke patterns, to adapt to the learning environment dynamically [144, 184]. Furthermore, emotion recognition systems utilizing neural networks can adapt to students' emotional states, even under fatigue or stress, ensuring educational content is delivered optimally [151]. These advanced AI techniques enable more accurate and responsive student modeling, resulting in more effective and engaging educational experiences [60, 62].

AI-driven student modeling is based on learning analytics. Learning analytics involves using data to enhance educational outcomes by analyzing learner behaviors and contexts. Virvou et al. [186] highlight that advances in ML have greatly improved learning analytics, enabling personalized and effective teaching strategies. Tsihrintzis et al. [155] show many research works on how intelligent systems use learning analytics to adapt to individual learner needs, enhancing educational outcomes through data-driven decision-making. Virvou, Alepis, and Sidiropoulos [185] developed a learning analytics tool that aids teachers by providing insights and feedback, allowing them to tailor their instruction to meet student needs better.

In EPATHLO, the implementation of student modeling relies on successfully applying fuzzy logic techniques.

Our research has focused on developing innovative adaptive educational technologies that enhance personalized learning experiences by applying fuzzy logic and interactive gaming environments.

1. FSP Creator

We introduced the FSP Creator, a novel web service API designed to create fuzzy student progress (FSP) profiles. FSP Creator has been integrated into the suite we present in the book and renamed EPATHLO Student Profile Creator. This tool utilizes fuzzy logic to interpret complex data about student performance, enabling educators to gain deeper insights into individual learning trajectories and adapt their teaching strategies accordingly. This approach facilitates more personalized educational interventions and supports a nuanced understanding of student progress [112].

2. FuzzEG

We developed FuzzEG, a system incorporating fuzzy logic to design adaptive scenarios in educational adventure games. This system dynamically adjusts the game environment and challenges based on real-time assessments of a learner's skills and progress. By personalizing the learning experience, FuzzEG increases student engagement and effectiveness in achieving educational outcomes [114]. FuzzEG was incorporated into the suite and renamed Fuzzy EPATHLO (or EPATHLO).

3. Learn-and-Play Personalized Reasoning

We explored the integration of personalized reasoning in mobile educational games, evolving from traditional point-and-click interfaces to immersive Virtual Reality (VR) environments. This advancement provides learners with interactive and adaptive learning experiences that cater to diverse learning styles and preferences. Incorporating VR enhances learner engagement and offers a unique platform for experiential learning [116].

Through these projects, our research has significantly contributed to the field of adaptive learning technologies, demonstrating the potential of fuzzy logic and interactive gaming environments to create more personalized and effective educational experiences.

More specifically, fuzzy logic is a multi-valued logic system in which the truth values of variables can range from 0 to 1. It emulates human thought processes by permitting ambiguous or imprecise statements. For example, it can express "the student knows the lesson perfectly" or utilize linguistic expressions such as "the student knows the lesson satisfactorily" and "the student knows the lesson but not to a satisfactory degree." When student data is clear and accurate, teachers may require assistance concluding a student's level of knowledge. Fuzzy logic proficiently models and represents the uncertainties and challenges that teachers face in assessing a student's knowledge level.

EPATHLO employs a fuzzy inference system (FIS) to evaluate students' performance in exercises and infer their level of knowledge. Determining knowledge level serves as the groundwork for a rule-based system that guides the selection of future exercises. These exercises are customized to assist students in improving their learning and reaching a point where they have consolidated all essential knowledge. After examining the student's recent quiz scores, the rule-based system applies pertinent criteria to ascertain the difficulty level of upcoming exercises. Consequently, as students advance, the quizzes incorporate more challenging questions. In contrast, inadequate performance leads to selecting simpler questions or a blend of easy and medium-difficulty questions, enabling students to expand their knowledge gradually. Additionally, each question corresponds to a specific aspect of the curriculum. Hence, if a student consistently answers specific questions incorrectly, the system recommends focusing on the relevant parts of the theory to address gaps and fortify weaknesses.

Another research objective was to create a game that would entertain students while enhancing their engagement. To achieve this, we conducted a thorough literature review to identify the most suitable game genres for educational purposes. This comprehensive review confirmed that adventure games are among the most effective options for educational use, reassuring educators and game developers of the game's strong foundation. The interactive and narrative nature of adventure games and the flexibility for students to progress at their own pace facilitate the discreet integration of educational content and lead to valuable learning outcomes. As a result, the current project involves creating an adventure game from the ground up, where students assume the role of a hero trying to escape from a locked building. Throughout the game, students encounter characters who assign tasks to the hero, requiring them to find specific objects to advance the storyline or complete educational quizzes based on their performance.

Every game, including educational games, is time-limited due to the finite number of exercises and educational challenges. It is crucial to determine whether a student has genuinely learned the material after completing the game.

For instance, if the game evaluates the player tolerantly and does not require an exceptional success rate, the learner may complete the exercises with satisfactory but imperfect performance, potentially leaving learning gaps unfilled upon completing the game. In contrast, if the game quizzes consistently demand high success, students who struggle to achieve perfect scores might encounter a static game story. This could lead to frustration and disappointment, potentially causing the student to give up on the game, resulting in inadequate knowledge retention. These potential challenges must be considered when designing the game. However, careful design and adaptive features can mitigate these issues. For example, the game can provide hints or additional resources to help students overcome challenges, thereby reducing frustration. It can also adjust the difficulty level based on student performance, ensuring the game remains challenging but not overwhelming. These strategies help maintain student engagement and ensure continued effective learning.

The above challenges necessitated the integration of adaptive features into the game itself, in addition to the adaptivity associated with the educational component. Existing literature highlights the positive impact of adaptive digital games on learning. This underscores the importance of adaptivity in our research. Many educational games require refinement, particularly in adjusting elements like difficulty levels, for instance, in an obstacle avoidance game. Furthermore, numerous games only adapt the educational aspects rather than elements intrinsic to the game context. EPATHLO, however, stands out with its adaptivity tailored to its unique features. For example, adventure games are distinguished by their narrative and the player's mission to complete diverse tasks, such as collecting objects and interacting with characters. Adding customization to one of the main features of an adventure game, such as the scenario, increases player engagement and fun and enhances learning outcomes.

This research introduces an innovative approach that employs narration characteristics and customizes them to meet students' individual needs. The fuzzy logic theory is used to evaluate the student's needs and communicate the student's current knowledge level to the game, determining whether improvements are needed. The game includes a rule-based system that assesses the student's knowledge level and decides on expanding the game scenario. The expansion may include new rooms, characters, puzzles, tasks, and exercises, integrating additional challenges into the game without distracting the student from the plot and overall enjoyment. The rule-based system ensures that the game adapts to the student's performance by offering more challenges to those who excel and supporting those who encounter difficulties. This approach rewards players for good quiz performance and assigns extra exercises when necessary, ensuring they effectively learn the curriculum by completing the game.

The research involved managing two distinct objectives as independent processes. As a result, we developed two sub-systems, each dedicated to fulfilling a specific goal. The EPATHLO Student Profile Creator focuses on the educational aspect [116], using fuzzy logic to assess the student's knowledge level and a rule-based system to select exercise difficulty. EPATHLO is an engaging adventure game designed for students' enjoyment. These two systems are intended to work collaboratively, allowing EPATHLO to receive personalized exercises and information about the student's educational profile from the EPATHLO Student Profile Creator. Additionally, EPATHLO forwards the student's responses to the EPATHLO Student Profile Creator, which then assesses them and updates the student's profile accordingly.

The use of web services technology facilitates communication between different systems and is often employed for front-end/back-end isolated systems or third-party interactions. In our work, web services are employed to grant external applications access to students' educational profiles. This approach enables developers to streamline their focus on designing and building their applications while benefiting from the capabilities and advantages offered by the EPATHLO Student Profile Creator, which is a novel web service API that creates platform-independent and technological-independent student models using the technology of web services [112]. The API's innovative features include its ability to embed the student profile into any educational application–whether web-based, mobile, or game–regardless of the system's technological characteristics or programming language. Additionally, the API ensures that all applications can recognize and interact with the student profile by calling the same web services. The time saved can then be dedicated to adding new features to the external application, thereby enhancing the learning experience for students. This emphasis on time-saving will make educational game developers feel efficient and productive in their work.

The use of web services technology is important in addressing the issue related to student modeling. Web services technology ensures continuity in educational games by centralizing student models, allowing all games to access up-to-date information about a student's progress, preferences, and learning history. This centralization means that their performance and progress data are updated in real time whenever a student interacts with a game. The system can then retrieve this information whenever the student returns to the game or switches to a different one, ensuring they can resume their learning experience without interrupting the gaming entertainment and learning procedure.

The games use this data to dynamically adapt their content to match the student's current knowledge level. If a student demonstrates mastery, the game presents more advanced challenges; if not, it provides remedial exercises. This ensures that the learning experience remains personalized and relevant. Moreover, web services enable the educational content to be coherent and progressive by maintaining a consistent educational experience across multiple platforms, whether on a desktop or mobile device. This adaptability helps educators align educational content with each student's evolving needs, enhancing overall learning outcomes.

By removing the educational elements from Fuzzy EPATHLO, the platform can be seamlessly reconfigured into a purely entertaining game intended to captivate and amuse players without the educational aspect. External applications that link with the EPATHLO Student Profile Creator can encompass various software types and platforms, such as educational platforms, web or mobile games, and more.

Fuzzy EPATHLO can be combined with Surviving Businessman. This game differentiates itself from EPATHLO Student Profile Creator in terms of gameplay mechanics and visual style. Unlike Fuzzy EPATHLO, Surviving Businessman is an exploration game with 3D graphics. Both games, however, use the EPATHLO Student Profile Creator web services to access the student's model and provide relevant exercises.

Students can select their preferred game while still engaging with the educational content from EPATHLO Student Profile Creator. An evaluation of this method shows that incorporating educational material into various personalized games presents several educational and entertainment advantages. It increases students' enjoyment and fosters greater involvement in the learning experience.

In Fuzzy EPATHLO, students can seamlessly transition from one quiz to another within Surviving Businessman and vice versa. This means that the educational content remains accessible across all external applications, providing students with a comfortable and familiar learning environment. Moreover, the system provides an authoring tool that enables instructors to incorporate new courses, theories, and exercises without programming skills. This flexibility allows the EPATHLO Student Profile Creator to be utilized across various domains such as programming, physics, mathematics, and history. Each game utilizes the web services of EPATHLO Student Profile Creator to access the desired course. Students can select their preferred domain within the game environment if the feature is available. This feature enables them to apply their knowledge across multiple courses by playing the same game.

Furthermore, the literature covers a range of authoring tools and emphasizes their significance [55, 98, 99, 171, 177, 182, 193]. These tools eliminate the need for programmer intervention to modify educational content, saving substantial time and effort. Many authoring tools enable educators to monitor students' progress by accessing their complete exercise history. However, the Fuzzy Author, which is described in this book, goes beyond these capabilities. It allows educators to intervene and control the AI question selection mechanism. Educators can define membership functions (MFs) for each fuzzy set representing the student's knowledge level. The tool supports triangular, trapezoidal, and Gaussian functions, and educators can select the type of function and adjust the relevant parameters. The tutor selects the type of function and adjusts the corresponding parameters. Moreover, they can also modify the fuzzy sets and their associated membership functions for extra fuzzy variables.

Additional fuzzy variables include:

1. The error rate on a question.
2. The frequency of help requested.
3. The extra time it took the student to answer a question.
4. The reduction rate in the student's knowledge level.

Tutors can select specific fuzzy variables to impact the ultimate determination of the student's level of understanding.

Our current work also explores the intersection of the General Data Protection Regulation (GDPR) and adaptive educational games. The GDPR, which became enforceable on May 25, 2018 [198], requires data controllers to obtain explicit consent from data subjects before collecting and processing their data. Given that adaptive educational games rely on advanced techniques to collect and process student data for personalized learning and gaming experiences, our research, which is also described in [93–96, 115], provides guidelines for game developers to ensure compliance with the regulation while effectively tailoring content to student needs.

Educational applications should ensure transparency regarding the handling of student data. This includes clearly explaining the purposes and benefits of data processing and allowing students to exercise their data rights, such as accessing, correcting, deleting, or transferring their data to another controller. A clear description of the application's functions makes it easier for students to consent to data usage and enjoy a personalized experience.

Further effort is needed to identify and address potential conflicts between educational games and GDPR, as well as to develop relevant guidance. By identifying these conflicts, we can establish a comprehensive set of guidelines for creating innovative adaptive educational games that fully comply with GDPR.

10.2 Contribution to Development of Intelligent Educational Software Games

Computer games constitute pieces of software with rich content, including visuals, animations, sound, and video, making them incredibly appealing [12, 18, 20, 43, 50, 92, 121]. The appealing characters in these games can effectively be used to enhance student learning, motivation, and engagement across various subjects [54, 68, 134].

The development of educational games has evolved significantly, focusing on creating engaging and educational experiences through various design methodologies and tools. Mehm, Göbel, and Steinmetz [87] introduced an authoring tool for creating educational adventure games, highlighting the importance of structured game models and authoring processes to enhance educational content delivery. Amory [4] emphasized the integration of educational theories into the design of adventure games, advocating for an approach that balances educational objectives with engaging gameplay mechanics. Zahari et al. [196] reviewed different modeling languages used in designing educational adventure games, underscoring the need for adaptable and robust design frameworks to cater to diverse educational goals. Chiang et al. [19] developed a safety educational adventure game, illustrating the potential of games to teach critical skills such as safety awareness in an interactive environment. Neto et al. [101] addressed inclusivity in educational game design by considering the needs of visually impaired learners, ensuring that games are accessible to all students. These developments demonstrate a trend towards more sophisticated and inclusive educational games that leverage technology to provide meaningful learning experiences.

Intelligent educational games (IEGs) transform modern education by combining gaming elements with AI to create personalized and engaging learning experiences. These games integrate ITSs and adaptive learning technologies to enhance educational outcomes across various subjects.

Early research laid the foundation for IEGs by developing authoring tools that enable educators to create content tailored to specific learning objectives, allowing for greater flexibility and customization [197]. Further studies explored modeling students' emotions within games, which enhanced the adaptivity of educational games to improve learning outcomes [32].

Research in the mid-2000s evaluated the educational impact of intelligent games and developed tools that facilitate adaptive learning environments responsive to individual student needs [59, 81]. The potential of IEGs to enhance learning through data-driven insights was also highlighted, integrating educational data mining with ITS to optimize learning strategies and game design [72].

Recent advancements have focused on specific educational contexts. A review of a decade of research demonstrated the effectiveness of IEGs in teaching subjects like spelling and mathematics, emphasizing the need for continuous research and development [142]. Comparisons between educational and non-educational games have shown that IEGs promote cognitive development more effectively [52].

Platforms like iMoodle use gamification and learning analytics to predict at-risk students, illustrating the integration of intelligent games with existing educational technologies [34]. The design of educational games for drill-based training has also been explored, highlighting the benefits of combining repetitive practice with intelligent feedback [191].

Research has also demonstrated that IEGs can improve social skills, showing their broader impact beyond cognitive learning [143]. This expands the potential of IEGs to enhance various learning aspects, including emotional and social development.

In summary, IEGs offer a powerful approach by integrating AI, adaptive learning, and engaging gameplay. As research advances, these games will likely play an increasingly significant role in the educational landscape, providing diverse learners with personalized and effective learning experiences.

Furthermore, it is essential for educational applications, including educational games, to provide a personalized and adaptive experience based on each student's unique characteristics [61, 88, 146, 195].

The main game of the book, Fuzzy EPATHLO, is an educational adventure game that stands out with its innovative features. This game, designed to introduce students to HTML coding, takes them through various rooms, where they interact with objects and engage with NPCs. What makes Fuzzy EPATHLO unique is its adaptive features, which provide each student with a personalized educational experience based on their knowledge level. This is possible through a student modeling component that utilizes fuzzy logic, a concept that challenges traditional quantitative description.

EPATHLO incorporates adaptivity into both the learning and the game. Regarding the learning part, it adjusts the difficulty of the questions to assist students in addressing any weaknesses and enhancing their knowledge. As for the game, it expands it with extra puzzles, rooms, and quizzes to motivate students, encouraging their participation in additional quizzes to extend the time required to complete the game. Additionally, it offers assistance through agents whenever necessary.

The research has significantly advanced the field of educational games by introducing an innovative mechanism that utilizes fuzzy logic to create personalized student models and adapt the content of an educational game in real time. This means that the educational game can tailor its quizzes, scenarios, and storyline to meet the specific educational needs of each student, delivering both educational value and the enjoyment of an unpredictable game. The evaluation of the study indicates that an educational game incorporating elements of adventure games, fuzzy logic-based intelligence, and adaptive scenarios can positively impact the learning process. Students find the game's adaptive nature and dynamic expansion appealing, leading to increased engagement.

Students can engage with and learn from various media platforms, a concept known as multi-platform learning. This study focuses on implementing cross-platform learning in educational games by integrating a learn-and-play mechanism called EPATHLO Student Profile Creator. The system is highly adaptable and is designed to facilitate the transfer of educational content across different types of games. The research demonstrates the adaptability of students' educational profiles and related exercises in two distinct genres of educational games: Fuzzy EPATHLO (a 2D point-and-click adventure game) and Surviving Businessman (a 3D exploration game). EPATHLO Student Profile Creator utilizes web services technology to seamlessly integrate educational material and update student profiles from any external application. It also ensures that students' educational progress and knowledge are retained and accessible across all games they choose to play.

The evaluation of the cross-platform learn-and-play system across different platforms revealed significant benefits. This method allows students to choose their favorite game genre while accessing the same educational content, enhancing their learning experience, increasing enjoyment, and encouraging greater engagement in the learning process. A noteworthy discovery from this research is that the educational content can be seamlessly integrated into any game genre and platform, leveraging the unique characteristics of each game to capture learners' interest in the learning process. Additionally, web services technology enables interoperability among applications and the sharing of educational components, significantly reducing the time and effort required for developers to create them.

The versatility of the student's profile and exercises extends beyond educational games. A significant advancement in cross-platform learning is the integration of web services technology. This technology allows any external educational application to utilize its functionalities through RESTFul web services and data exchange in JSON format. The third-party application is distinct from the EPATHLO Student Profile Creator's core logic and is focused solely on web service calls and processing their responses. The development technology at the core is independent of the educational tool's technology, making it possible for the resulting application to target any platform, including desktop, web, or mobile. Educators can import educational resources using the authoring tool Fuzzy Author, which accompanies the EPATHLO Student Profile Creator. These resources, along with the student's profile and quiz performance, are accessible at a single point of access and available to any authorized external application consuming the web services. As a result, the EPATHLO Student Profile Creator maintains a unique profile for each student and seamlessly distributes it to all applications, eliminating the need for programmers to re-implement fuzzy logic techniques for student modeling and adaptive question selection.

10.3 Contribution to Interoperability and Authoring Tools of Intelligent Educational Games

Interoperability is very important in educational games over the Web so that students are not deprived of efficient access irrespective of place and the device they use. Web services provide a flexible and scalable framework for integrating diverse applications and systems, enabling seamless communication, data exchange, and interoperability across different platforms and devices, whether mobile [2] or other, which enhances efficiency and collaboration in both business and educational environments [75, 117, 125]. In our work, we have employed Web Services to create quizzes accessible to many educational gaming environments [63], and for keeping unique user models for each student-player which are always efficiently updated by the respective users' actions irrespective of the location or the device, that they use [113].

Numerous researchers and educational software developers are creating authoring tools for Intelligent Tutoring Systems that allow tutors to input and modify educational content easily [35, 49, 66, 107, 108]. Authoring tools also exist for complex automatic reasoning about users' actions and help with error diagnosis [24, 162, 163, 165, 172, 178]. This significantly reduces the costs associated with developing such processes. Many of these authoring tools also provide educators with valuable student statistics, enabling tutors to monitor their student's progress and adjust the educational material accordingly. In this respect, authoring tools are needed for reducing the development cost of serious games or educational games [31, 48, 77, 170, 174].

Moreover, authoring tools have significantly advanced, providing educators and developers with powerful platforms to create personalized and effective educational experiences. Virvou et al. [182] introduced VR-INTEGATE, a knowledge-based authoring tool that leverages VR to enhance educational games, offering a more immersive and interactive learning environment. The INMA tool is an example of integrating education with entertainment, functioning as a knowledge-based authoring tool for music education that enhances music perception and understanding through intelligent educational techniques [183]. Earlier, Virvou and Moundridou [176] developed an authoring tool tailored for algebra-related domains, highlighting the potential for subject-specific customization in educational software. Moundridou and Virvou made further advancements [97], who explored authoring ITSs on the World Wide Web (WWW), thus extending the accessibility and reach of educational technologies. As discussed by Virvou and Alepis [171], the introduction of mobile educational features in authoring tools emphasized the importance of portability and accessibility in personalized tutoring systems. Virvou's work [165] on modeling user knowledge and reasoning within a knowledge-based authoring tool underscored the importance of understanding learner behavior to enhance the educational process. Additionally, Virvou [164] proposed a domain-independent reasoning mechanism for Intelligent Tutoring System authoring tools, emphasizing the flexibility and adaptability of these tools across different subject areas. These developments in authoring tools have contributed to more dynamic, user-friendly, and effective educational environments catering to diverse learning needs and preferences.

The literature explores authoring tools that enhance interactive educational content using VR and Augmented Reality (AR).

VR Authoring Tools

Virvou et al. introduced VR-INTEGATE, which integrates Intelligent Tutoring Systems with VR games, offering advanced diagnostics of student interactions [182]. Horst et al. highlighted modular VR content design for flexible learning scenarios [57], while Müser et al. emphasized VR's potential to enhance learning through simulations, stressing usability and accessibility [100]. De Oliveira et al. found that ease of use and social influence are crucial for VR tool adoption [105], and Quintana et al. showcased VR's potential in healthcare education through tailored rehabilitation exercises [123].

AR Authoring Tools

AR tools have transformed interactive content creation, emphasizing accessibility and user-centric design. Blattgerste focused on AR for procedural training, highlighting intuitive interfaces [9]. Silva et al. developed AR design principles from teacher feedback [141], while Ez-zaouia et al. advocated for democratizing AR in education [40]. Bräker et al. designed no-code AR tools for users without programming knowledge [11], and Kim et al. presented RealityBrush, an AR system that enhances realism by using kinetic properties of objects [67].

User Interaction and Playfulness

Schäfer examined hand gesture tools for intuitive VR interactions [132], and Liapis et al. incorporated playful elements in human-AI tools to boost creativity [74].

Collaborative and 3D Content Authoring

Hossain and Zaman's NCCollab supports collaborative game development [58]. Yu's 3D sketching tools enhance creativity in designing models [194], while Lam et al. explored tangible interaction in kitchen design for immersive experiences [71].

These tools integrate AR and VR in education, providing innovative solutions for creating interactive and personalized learning environments. Emphasis on usability and accessibility ensures they meet diverse educational needs, fostering engagement and enhancing outcomes.

3D Content Authoring and Tangible Interaction

Yu [194] discusses 3D sketching tools that simplify complex model creation, boosting creativity and precision. Lam et al. [71] explore tangible interaction techniques for kitchen design, integrating physical and digital interactions for immersive experiences.

Evaluating and Improving Authoring Tools

Evaluating authoring tools is crucial for meeting user needs and accessibility standards. Salmen and Schroeder [128] offer a framework for assessing usability and effectiveness, while Baldiris et al. [6] emphasize accessibility standards under ATAG and WCAG guidelines.

Interactive Narratives and Cultural Heritage

Massidda et al. [84] highlight the role of authoring tools in creating interactive digital narratives (IDN) to preserve cultural heritage. Gómez-Albarrán et al. [46] discuss web-based tools for interactive fiction, emphasizing flexibility and accessibility. Zarei et al. [199] focus on narrative tools for young learners, enhancing creativity and critical thinking.

Adaptive Learning and Mulsemedia

Olney et al. [106] highlight the importance of smart authoring tools for adaptive learning, tailoring content to learner needs. De Mattos et al. [85] advocate for mulsemedia tools integrating multiple sensory modalities for enriched learning experiences. Coelho et al. [31] evaluate tools for creating 360-degree multisensory videos, emphasizing intuitive interfaces.

Evaluation and Design Guidelines

Shiels and Tanner [138] study learning design tools among South African teachers, identifying factors influencing effectiveness. Chamusca et al. [17] evaluate VR authoring tools using NVIDIA Omniverse, highlighting user-centered design and iterative testing.

Accessibility and Interaction Design

Roig et al. [126] provide recommendations for improving the accessibility of word processors and interaction design. Ez-zaouia et al. [39] focus on accessible educational AR tools for user-friendly integration into educational practices.

Comparative Analysis and Resources

Shibolet and Lombardo [137] offer resources for comparing IDN authoring tools, aiding developers and educators in tool selection based on specific needs.

Data Visualization and Cognitive Interaction

Rubab et al. [127] examine data visualization tools, emphasizing design that aligns with cognitive processes to create impactful visualizations for decision-making.

AI and Deep Authoring for Multimedia

Takács and Vincze [147] introduce AI-driven tools for creating complex multimedia experiences, simplifying content creation with features like automated generation and optimization.

XR Applications and Educational Workshops

Masneri et al. [83] discuss workshop-authoring tools for XR applications, highlighting XR technologies' potential to enhance educational environments through immersive learning.

Procedural Content Creation for Gaming

Halperin [51] examines procedural tools in game development, showcasing their ability to automate content generation and allow developers to focus on creativity and innovation.

The EPATHLO authoring tool's new feature is tailored to the needs of tutors. In addition to monitoring student progress and creating courses, the tutor can actively engage with the AI system and fine-tune the fuzzy mechanism's operation. For instance, the tutor can define the membership functions that gauge the student's knowledge level using standard function types like triangular, trapezoidal, and Gaussian. Moreover, the tutor can introduce additional fuzzy variables to assess the student's knowledge level beyond the quiz success rate. These variables encompass the frequency of a question's appearance, the percentage of incorrect answers, and the time the student takes to respond to a question beyond the expected timeframe. These variables leverage fuzzy logic to determine the degree of knowledge level adjustment, leading to more realistic results.

The feature has significant potential. Allowing the tutor to customize the intelligent mechanisms allows the tool to gain even more precise insights into the student's educational performance or the material being presented. This paves the way for a more personalized learning experience for the student, ultimately enhancing their learning abilities and fostering a sense of optimism about the tool's potential impact among educators.

10.4 Contribution to 21st Century Skill Development Through Intelligent Educational Games

The rapid advancement of technology has led to a growing demand for software engineering, which requires software engineers to possess a broad range of soft skills. Unlike the technical skills acquired through academic education, soft skills are more closely related to an individual's character and personality. Although academic institutions equip students with the necessary knowledge for their fields, they often do not adequately teach soft skills. However, these skills are just as crucial for software engineers to succeed. When evaluating potential employees, companies look for candidates who demonstrate collaboration, teamwork, analytical and organizational thinking, and effective communication with colleagues and clients [38, 76, 89, 120, 122, 133, 148, 149].

21st-century skills are important for thriving in today's complex world, encompassing competencies like critical thinking, collaboration, creativity, communication, digital literacy, adaptability, and problem-solving. Recent research emphasizes various strategies for developing these skills, highlighting the role of technology, innovative pedagogies, and experiential learning.

Educational robotics and gamification have shown promise in fostering 21st-century skills. Robotics engages young learners in problem-solving and critical thinking through hands-on activities, effectively building these essential competencies [7]. Similarly, gamification, such as DOTA 2, enhances critical thinking, collaboration, and communication, demonstrating how game-based learning environments can nurture these skills [129].

The COVID-19 pandemic accelerated the shift to digital learning, presenting both opportunities and challenges in enhancing students' digital skills. E-learning platforms have effectively improved digital competencies in higher education [15]. Integrating digital skills into language teaching has also proven beneficial, enhancing the readiness of both educators and students for digital challenges [124].

Immersive technologies like VR or AR significantly support spatial abilities and 21st-century skills in K-12 education. These technologies offer engaging learning experiences that boost creativity, collaboration, and problem-solving abilities [135]. Additionally, interactive e-books have facilitated critical thinking and digital literacy among high school students, offering an innovative medium for skill acquisition [130].

Agile methodologies, such as Scrum, have been successfully integrated into engineering education to develop collaboration, communication, and problem-solving skills [118]. Game development projects, like escape room templates, provide experiential learning opportunities that require students to apply and assess their 21st-century skills [161].

Design and project-based learning initiatives support the development of 21st-century skills while aligning with sustainable development goals. Incorporating design projects into educational curricula fosters creativity and innovation, particularly in early schooling [1, 8]. Digital tools like Minecraft Education Edition enhance creativity and collaboration in higher education settings [73].

The digital economy has a growing demand for professional skills that align with 21st-century competencies. Integrating these skills into educational programs is crucial for preparing students for the evolving workforce [33]. Design sprints can enhance digital skills and promote innovation and adaptability in professional settings, underscoring the importance of these skills for career readiness [91].

Despite various strategies for developing 21st-century skills, challenges remain, particularly in ensuring equitable access to technology and resources. The ongoing discussion about digital competence and emergent educational issues highlights the need for a "new normal" in education that prioritizes these skills [139].

Soft skills are increasingly vital across industries, enhancing individual and organizational performance. Innovative assessment techniques, such as fuzzy logic, provide nuanced insights into skill levels, supporting tailored educational interventions [103]. Modern technology can enhance traditional training methods, improving soft skills development, especially in healthcare [192]. Soft skills training reduces work-related stress and is crucial for effective leadership and management [90, 140]. Addressing gaps between academic curricula and industry requirements is essential for preparing students for the workforce, particularly regarding soft skills in software engineering and testing [10, 53]. AI offers personalized training programs, enhancing the precision of skill assessments and providing adaptive learning pathways tailored to individual needs [44, 150]. Engaging learning methods like role-play and gamification support soft skills development through interactive and enjoyable approaches [16, 78]. The adaptability of soft skills training across various sectors and cultures underscores their broad relevance [14, 79]. In virtual work environments, developing soft skills is crucial for productivity and collaboration, highlighting the need for specific metrics to assess these skills in remote settings [65]. Empowering underrepresented groups through soft skills training promotes inclusivity and diversity in tech industries, emphasizing the role of these skills in empowering women and other marginalized groups in digital projects [36].

While the use of games to teach technical skills has been extensively studied, the potential of educational games to enhance soft skills is a promising but often overlooked area. These games have shown promise in improving soft skills when used alongside non-gaming methods. Several studies have explored this potential, particularly focusing on how games can enhance motivation, willingness to learn, competitiveness, and problem-solving abilities [37, 80, 82, 86, 102]. One study specifically examined educational games for software engineering and found that these games consistently emphasized the development of soft skills, with all 80 instances (100%) highlighting their importance.

The Educational Game Suite of Fuzzy EPATHLO incorporates features designed to help players practice essential soft skills for software engineers. Using NPCs fosters communication, teamwork, analytical thinking, problem solving, empathy development, and customer orientation skills. Also, while most educational game evaluations typically focus on technical learning outcomes and recreational experiences, the evaluation of Fuzzy EPATHLO emphasized soft skills development. Five experts engaged in a workshop played the game, and in a cognitive walkthrough, they documented the steps required to complete each task and the soft skills exercised. The evaluation indicated that Fuzzy EPATHLO is a valuable resource for software engineers seeking to enhance their programming technical and soft skills.

As such, Fuzzy EPATHLO is an advanced and comprehensive game designed to enhance hard and soft skills by integrating adaptive learning, NPCs, and task-oriented scenarios. Its adaptive and personalized approach is tailored to each student's cognitive development, making it highly effective for soft skill development and instilling confidence in its ability to meet individual needs. The game's profound impact on cultivating essential soft skills, which employers demand, allows it to play a role in workforce preparation.

10.5 Contribution to GDPR and Intelligent Educational Games

AI-based games transform educational software and settings by offering immersive and personalized learning environments designed to teach 21st-century skills. These environments engage learners through interactive and adaptive content, fostering critical thinking, problem-solving, and collaboration–skills essential for modern learners.

Despite these advancements, challenges remain regarding the reliability and explainability of AI systems in education. It is important to ensure transparency and trust in AI-driven educational tools for their successful adoption. Recent research highlights the importance of trust dynamics in AI-human interactions, emphasizing the need for AI systems that foster collaboration and understanding between users and machines [190]. To ensure GDPR compliance, a coordinated effort between Legal and IT teams is essential. This partnership focuses on modifying or enhancing existing applications to ensure adherence to new and existing regulations. Importantly, this collaboration establishes software engineering standards to support the creation of educational applications that safeguard student data following GDPR principles and guidelines.

In particular, more comprehensive research needs to outline the specific considerations for designers when creating educational applications that lawfully handle student data, especially in ITS and adaptive educational games. The popularity of adaptive educational applications stems from their personalized learning experiences, which can significantly enhance students' cognitive development. Given such applications' ongoing development and numerous benefits, designers must adhere to a standardized framework to streamline the creation of GDPR-compliant adaptive educational applications.

The General Data Protection Regulation emphasizes the concepts of automated profiling and automated decision-making [38, 76, 89, 120, 133]. Both processes involve the collection and processing of students' data. Implementing the General Data Protection Regulation by the European Union in 2018 has significantly influenced software development, particularly regarding data protection and privacy. This review examines the integration of GDPR compliance into software systems, focusing on security controls, challenges in open-source software (OSS), and evolving software processes.

Ayala-Rivera et al. [5] highlight the necessity of embedding security controls into software design to achieve GDPR compliance. Their research suggests that integrating privacy and security measures throughout the software development lifecycle, from the outset, is more effective than retrofitting them after development. This proactive approach directly incorporates GDPR principles, such as data minimization and user consent, into the software architecture, enhancing overall security and compliance.

In the context of OSS, Franke et al. [41] explore these projects' unique challenges, such as diverse contributor bases and decentralized development models. These factors complicate consistent data protection measures. Their study reveals that many OSS projects struggle with GDPR compliance due to lacking resources and expertise. They suggest incorporating automated tools and templates for privacy notices and Data Protection Impact Assessments (DPIAs) to help OSS communities align with GDPR requirements more effectively.

Further investigation by Franke et al. [42] uses a mixed-methods approach to examine compliance practices within OSS. The study found that while GDPR awareness is high among developers, its practical application is inconsistent due to varying interpretations and challenges in adapting GDPR requirements to the open-source model. This study emphasizes the need for clearer guidelines and support mechanisms to aid OSS projects in achieving compliance.

Gómez-Martínez et al. [47] discuss integrating GDPR principles into the Unified Software Development Process (USDP). Their work suggests embedding GDPR considerations into established software development methodologies to streamline compliance. Incorporating GDPR requirements into design phases helps developers anticipate potential data protection issues, addressing them early in development. This integration helps create privacy-aware software systems, reducing the risk of non-compliance.

Horák et al. [56] present a case study on GDPR compliance in cybersecurity software, focusing on DPIAs in an information-sharing platform. Their findings highlight the importance of DPIAs in identifying potential risks and implementing appropriate safeguards, ensuring software systems comply with GDPR, and enhancing user trust through robust data protection measures.

Kittmann et al. [69] propose a distributed software architecture designed to comply with GDPR, particularly for automation services. This architecture incorporates privacy-by-design principles, ensuring data protection is integral to the system's functionality. This approach is relevant for automation services, which process large volumes of personal data. By embedding privacy measures into the architecture, developers can create systems that automatically enforce GDPR requirements, reducing the burden on users and administrators.

Integrating GDPR compliance into software development is a multifaceted challenge requiring technical, organizational, and procedural changes. The reviewed studies highlight the importance of incorporating privacy and security measures into the design and development phases, ensuring software systems are built with GDPR requirements. OSS projects face unique challenges due to their collaborative nature, necessitating tailored solutions to achieve compliance. Emphasizing privacy by design and proactively embedding GDPR principles into software systems can significantly enhance compliance efforts, leading to more secure and privacy-respecting applications. These findings underscore the critical role of evolving software processes to address regulatory requirements effectively and sustainably.

Examining Fuzzy EPATHLO and EPATHLO Student Profile Creator as paradigms highlights the EPATHLO Suite's contributions to compliance with the General Data Protection Regulation. This research focuses on the two essential features of adaptive educational games or Intelligent Tutoring Systems: student modeling and content adaptivity. By analyzing these components, the research outlines the architecture of an IEG, illustrating the flow of student data between system components and identifying the operations requiring explicit student consent under the GDPR.

Mougiakou and Virvou [93] emphasize the importance of integrating privacy considerations into the design of e-learning programs using Unified Modeling Language (UML). This approach ensures that data protection is addressed early in software development. The EPATHLO suite exemplifies these principles by implementing robust student data management practices aligned with GDPR requirements.

Furthermore, the research examines students' rights when engaging in educational games, including accessing, deleting, rectifying, or exporting their data in a machine-readable format to facilitate potential transfer to another data controller [115]. This aspect is significant, as Mougiakou, Papadimitriou, and Virvou [95] stress the need for automated decision-making processes in ITS to effectively incorporate personal data protection guidelines. These guidelines ensure that intelligent systems operate transparently and provide students with control over their personal information.

The significant contribution of this work lies in its provision of guidelines for designing ITS or adaptive educational games that adhere to GDPR. Specifically, it focuses on student modeling, automatic decision-making, and students' rights concerning their data. Mougiakou, Papadimitriou, and Virvou [96] extend this understanding by examining synchronous and asynchronous learning methods and their compliance with GDPR, highlighting the necessity for explicit consent and transparent data handling practices in all educational contexts.

The research also emphasizes the need for complete transparency regarding the intelligent techniques used in the game and the mechanism for assessing the student's performance and educational or gaming progress. Papadimitriou, Mougiakou, and Virvou [115] highlight that smart educational games must inform users about the data collected, how it is used, and the logic behind automated decisions to build trust and empower students.

As a result, this work lays the groundwork for further research to establish a unified access point and a standardized framework containing guidelines. This framework will ensure that all designers use a consistent reference point when developing applications to ensure GDPR compliance. The transparency and user control offered by adaptive educational applications lead to a more personalized experience with numerous benefits for both learning and gaming. This transparency empowers students, making them fully informed about the data stored by the system and its usage, resulting in a better understanding and utilization of their rights under GDPR.

By providing these guidelines, the EPATHLO suite ensures compliance with legal standards and enhances the educational experience by fostering an environment where students feel secure and informed. This approach aligns with the broader trend toward integrating data privacy into educational technology, setting a precedent for future developments in the field.

10.6 Conclusions

The book "Artificial Intelligence-based Games as Novel Holistic Educational Environments to Teach 21st Century Skills" explores the potential of AI-driven educational games to teach both technical and soft skills essential for software engineers in today's fast-paced technological world. While educational games have traditionally focused on imparting technical skills, the book underscores the growing importance of soft skills, such as collaboration, teamwork, analytical thinking, and communication, which employers increasingly value but often neglect in traditional academic curricula.

A central theme of the research is the role of fuzzy logic in maintaining student engagement and motivation through adaptive scenarios, particularly in the EPATHLO suite. Fuzzy logic enables the system to make nuanced adjustments to both educational content and game narratives based on individual learner progress. As discussed by Chrysafiadi, Papadimitriou, and Virvou [28], this approach allows for continuous monitoring of students' cognitive aspects in technical learning and enables adaptive teaching methods. By employing fuzzy reasoning, as demonstrated in FuzzEG [114], the EPATHLO suite offers a dynamic platform that adjusts the plot and tasks of the game to cultivate both technical skills and soft skills effectively.

The EPATHLO suite also allows student and player models to be transferred between different games, allowing students to choose the type of game they want to play while benefiting from personalized learning. This feature enhances engagement by allowing learners to select games that match their interests and learning preferences.

The Educational Game Suite of Fuzzy EPATHLO exemplifies this approach by integrating features that allow players to practice essential soft skills through interactions with NPCs. These interactions help develop communication, teamwork, analytical thinking, problem-solving, empathy, and customer orientation skills. The adaptive nature of these games is highlighted in studies by Papadimitriou and Virvou [111] and Chrysafiadi et al. [27], which describe how dynamic plot adaptations and fuzzy states adjust the game environment based on learner progress.

Evaluations of educational games compared to web-based applications by Chrysafiadi, Papadimitriou, and Virvou [26] found that games provide more engaging and effective learning environments due to their interactive and adaptive nature. The shift from point-and-click to immersive VR experiences, as explored by Papadimitriou et al. [116], further enhances engagement and experiential learning opportunities.

Additionally, the book addresses data privacy concerns, as discussed by Papadimitriou, Mougiakou, and Virvou [115], emphasizing the importance of complying with the General Data Protection Regulation in the development of smart educational games. Ensuring transparency and consent in data handling is critical for creating trust and maintaining ethical standards in educational technology.

In conclusion, Fuzzy EPATHLO is a sophisticated and comprehensive game designed to foster both hard and soft skills through adaptive learning and task-oriented scenarios. By leveraging fuzzy logic, it provides personalized experiences that cater to each student's cognitive development, making it effective for enhancing soft skills and instilling confidence. By focusing on skills employers demand, Fuzzy EPATHLO plays a significant role in workforce preparation, bridging the gap between academic education and professional requirements. This research underscores the potential of combining AI with educational games to create holistic learning environments that prepare learners for the challenges of the modern world. Such innovations in educational technology represent a significant shift towards more engaging and impactful learning environments that align with the needs of 21st-century learners [167, 169].

10.7 Future Work and Open Problems

Future research should aim to broaden the scope of AI-based educational games across various educational levels and disciplines. Developing more advanced AI models that can accurately capture the complexities of human learning and provide even more precise personalization is essential. Exploring the integration of emerging AI techniques, such as Natural Language Processing (NLP) [119] and Large Language Models (LLMs), could further enhance the interactivity and engagement of educational games [154].

Long-term evaluation studies are necessary to assess the sustained impact of AI-based games on educational outcomes and skill development [166]. Such studies could provide valuable insights into how these technologies affect students' learning trajectories and their ability to acquire and apply 21st-century skills over time.

Ethical considerations, such as data privacy, algorithmic fairness, and the potential displacement of human educators, must also be explored. Developing frameworks and guidelines to ensure that AI systems are used responsibly in educational settings will be critical to their successful adoption with control on human experts [180, 181].

Finally, fostering interdisciplinary collaboration among educators, AI researchers, and policymakers is crucial for the continued evolution of AI-based educational games. This collaboration can ensure that technological innovations are pedagogically sound, culturally relevant, and aligned with the broader goals of education, ultimately creating more inclusive and effective learning environments for all students.

Integration of Educational Games with LLMs

One promising area for future research is the integration of LLMs like ChatGPT into educational games to enhance interactivity and personalized learning. LLMs have demonstrated significant potential in facilitating human-like conversations, understanding nuanced contexts, and providing adaptive responses tailored to individual learners' needs [109, 189]. By embedding LLMs into educational games, we can create dynamic learning environments where students engage in meaningful dialogues, receive contextualized feedback, and explore complex scenarios through natural language interactions.

This integration can also support the development of critical thinking and communication skills as learners interact with AI-driven characters capable of simulating real-world conversations. Future work should optimize the interaction between LLMs and educational content, ensuring that the models are fine-tuned to provide accurate and pedagogically relevant information. Researchers should explore methods to incorporate LLMs in games to dynamically adjust content and difficulty levels based on real-time assessment of student understanding, thereby fostering a more individualized and engaging learning experience [114].

Development Processes and Evaluation Frameworks

As educational games evolve to incorporate advanced AI technologies, establishing efficient development processes and robust evaluation frameworks is crucial to ensuring their usability, utility, and educational effectiveness. The software life cycle of an educational game that incorporates learning analytics and student modeling can be enhanced through several methodologies. Panagoulias et al. [109] highlight microservices, which can enable modular and scalable game development, allowing developers to implement updates and new features with ease. By employing the CRISP-DM methodology, as discussed by Alogogianni and Virvou [3], developers can systematically mine and analyze student data to refine game content and provide personalized feedback. Additionally, utilizing the Rational Unified Process (RUP) in iterative cycles, as described by Savvopoulos et al. [131], ensures that educational games can adapt over time by incorporating user feedback and learning insights to improve student engagement and learning outcomes continuously. These approaches facilitate rapid prototyping, testing, and deployment of game features, allowing developers to integrate user feedback continuously and refine AI functionalities to meet educational objectives.

To evaluate the effectiveness of AI-based educational games, future research should develop comprehensive frameworks that assess not only learning outcomes but also user engagement, satisfaction, and trust in AI systems, including explainability and transparency [110] of AI-driven interactions within games, where there has not been sufficient research yet. Such frameworks should also incorporate methods to measure the alignment of game activities with curricular goals, ensuring that the educational content remains relevant and impactful.

Moreover, evaluating the learning outcomes of AI-enhanced games using fuzzy logic and adaptive learning techniques, as demonstrated by Chrysafiadi and Virvou [25], can provide insights into how these systems can be optimized to support diverse learning styles and preferences. Educators and developers can use a holistic evaluation framework to ensure that educational games effectively facilitate skill development and knowledge acquisition [179].

Ethical Considerations and Trust Building

As AI-based educational games increasingly leverage LLMs and other advanced technologies, addressing ethical considerations such as data privacy, algorithmic fairness, and user trust is essential. Future work should focus on developing guidelines and best practices for the ethical deployment of AI in educational contexts, ensuring that AI systems are transparent, fair, and aligned with educational values.

Building trust between learners and AI systems is crucial for their acceptance and effectiveness. Research should explore developing trust dynamics models that enhance collaboration between AI and human users, ensuring that AI-driven educational games are perceived as reliable and supportive learning tools [190]. By fostering trust and transparency, AI-based educational games can become valuable assets in modern educational environments.

References

1. Abidin, A.F.B., Ahmad, A.: The element of integrated design project towards fulfilling the 21st century skills set and sustainable development goal (sdg). In: 2023 IEEE 12th International Conference on Engineering Education (ICEED), pp. 28–31. IEEE (2023)
2. Alepis, E., Virvou, M.: Object-Oriented User Interfaces for Personalized Mobile Learning. Springer (2014)
3. Alogogianni, E., Virvou, M.: Addressing the issue of undeclared work-part i: applying associative classification per the crisp-dm methodology. Intell. Decis. Technol. **15**(4), 721–747 (2021)
4. Amory, A.: Building an educational adventure game: theory, design, and lessons. J. Interact. Learn. Res. **12**(2), 249–263 (2001)
5. Ayala-Rivera, V., Portillo-Dominguez, A.O., Pasquale, L.: Gdpr compliance via software evolution: Weaving security controls in software design. J. Syst. Softw. 112144 (2024)
6. Baldiris, S., Vargas, D., Garzón, J., Ávila-Garzón, C., Burgos, D.: Evaluation of authoring tools under atag and wcag recommendations. Univ. Access Inf. Soc. **22**(3), 919–930 (2023)
7. Bano, S., Atif, K., Mehdi, S.A.: Systematic review: potential effectiveness of educational robotics for 21st century skills development in young learners. Educ. Inf. Technol. 1–19 (2023)
8. Bhandari, U., Mathew, D.J.: Development of design education curriculum for grade 5 indian cbse schools to facilitate 21st century skills. In: 2023 IEEE 12th International Conference on Engineering Education (ICEED), pp. 165–170. IEEE (2023)
9. Blattgerste, J.: The design space of augmented reality authoring tools and its exploration for the procedural training context. Ph.D thesis, Bielefeld University (2023)
10. Borges, G.G., de Souza, R.C.G.: Skills development for software engineers: systematic literature review. Inf. Softw. Technol. 107395 (2024)
11. Bräker, J., Hertel, J., Semmann, M.: Empowering users to create augmented reality-based solutions–deriving design principles for no-code ar authoring tools (2023)
12. Braun, B., Stopfer, J.M., Müller, K.W., Beutel, M.E., Egloff, B.: Personality and video gaming: Comparing regular gamers, non-gamers, and gaming addicts and differentiating between game genres. Comput. Hum. Behav. **55**, 406–412 (2016)
13. Brusilovsky, P.: Developing adaptive educational hypermedia systems: from design models to authoring tools. In: Authoring Tools for Advanced Technology Learning Environments: Toward Cost-Effective Adaptive, Interactive and Intelligent Educational Software, pp. 377–409. Springer (2003)
14. Calhau, R.F., Almeida, J.P.A., Kokkula, S., Guizzardi, G.: Modeling competences in enterprise architecture: from knowledge, skills, and attitudes to organizational capabilities. Softw. Syst. Model. 1–40 (2024)
15. Carabregu-Vokshi, M., Ogruk-Maz, G., Yildirim, S., Dedaj, B., Zeqiri, A.: 21st century digital skills of higher education students during covid-19-is it possible to enhance digital skills of higher education students through e-learning? Educ. Inf. Technol. **29**(1), 103–137 (2024)
16. Cerezo, E., González-González, C.S., Bonillo, C.: Empowering soft skills in children with adhd through the co-creation of tangible tabletop games. Univ. Access Inf. Soc. **23**(1), 3–21 (2024)
17. Chamusca, I.L., Santos, F.V.D.J., Ferreira, C.V., Murari, T.B., Junior, A.L.A., Winkler, I.: Evaluation of design guidelines for the development of intuitive virtual reality authoring tools: a case study with nvidia omniverse. In: 2022 IEEE International Symposium on Mixed and Augmented Reality Adjunct (ISMAR-Adjunct), pp. 357–362. IEEE (2022)
18. Chesham, A., Wyss, P., Müri, R.M., Mosimann, U.P., Nef, T., et al.: What older people like to play: genre preferences and acceptance of casual games. JMIR Serious Games **5**(2), e7025 (2017)
19. Chiang, F.K., Chun-Hao, C., Hu, D., Zhang, G., Liu, Y.: Design and development of a safety educational adventure game. Int. J. Emerg. Technol. Learn. (Online) **14**(3), 201 (2019)

20. Choi, E., Shin, S.H., Ryu, J.K., Jung, K.I., Kim, S.Y., Park, M.H.: Commercial video games and cognitive functions: video game genres and modulating factors of cognitive enhancement. Behav. Brain Funct. **16**, 1–14 (2020)
21. Chrysafiadi, K., Virvou, M.: Evaluating the integration of fuzzy logic into the student model of a web-based learning environment. Expert Syst. Appl. **39**(18), 13127–13134 (2012)
22. Chrysafiadi, K., Virvou, M.: Student modeling approaches: a literature review for the last decade. Expert Syst. Appl. **40**(11), 4715–4729 (2013)
23. Chrysafiadi, K., Virvou, M.: Create dynamically adaptive test on the fly using fuzzy logic. In: 2018 9th International Conference on Information, Intelligence, Systems and Applications (IISA), pp. 1–8. IEEE (2018)
24. Chrysafiadi, K., Virvou, M.: Enhancing automatic reasoning of human errors in an operating system using fuzzy logic. In: 2019 10th International Conference on Information, Intelligence, Systems and Applications (IISA), pp. 1–4. IEEE (2019)
25. Chrysafiadi, K., Virvou, M.: Evaluating the user experience of a fuzzy-based intelligent tutoring system. In: 2021 12th International Conference On Information, Intelligence, Systems & Applications (IISA), pp 1–7. IEEE (2021)
26. Chrysafiadi, K., Papadimitriou, S., Virvou, M.: Which is better for learning: a web-based educational application or an educational game? In: 2019 International Symposium on Performance Evaluation of Computer and Telecommunication Systems (SPECTS), pp. 1–6. IEEE (2019)
27. Chrysafiadi, K., Papadimitriou, S., Virvou, M.: Fuzzy states for dynamic adaptation of the plot of an educational game in relation to the learner's progress. In: 2020 11th International Conference on Information, Intelligence, Systems and Applications (IISA, pp. 1–7. IEEE (2020)
28. Chrysafiadi, K., Papadimitriou, S., Virvou, M.: Cognitive-based adaptive scenarios in educational games using fuzzy reasoning. Knowl. Based Syst. **250**, 109111 (2022)
29. Chrysafiadi, K., Virvou, M., Tsihrintzis, G., Hatzilygeroudis, I.: An adaptive learning environment for programming based on fuzzy logic and machine learning. Int. J. Artif. Intell, Tools (2023)
30. Chrysafiadi, K., Virvou, M., Tsihrintzis, G.A.: A fuzzy-based evaluation of e-learning acceptance and effectiveness by computer science students in greece in the period of covid-19. Electronics **12**(2), 428 (2023)
31. Coelho, H., Monteiro, P., Gonçalves, G., Melo, M., Bessa, M.: Authoring tools for virtual reality experiences: a systematic review. Multimed. Tools Appl. **81**(19), 28037–28060 (2022)
32. Conati, C., Zhou, X.: Modeling students' emotions from cognitive appraisal in educational games. In: International Conference on Intelligent Tutoring Systems, pp 944–954. Springer (2002)
33. Demchenko, Y., Degeler, V., Opresu, A., Brewer, S.: Professional and 21st century skills for data driven digital economy. In: 2023 IEEE Global Engineering Education Conference (EDUCON), pp. 1–8. IEEE (2023)
34. Denden, M., Tlili, A., Essalmi, F., Jemni, M., Chang, M., Kinshuk, Huang, R. (2019) imoodle: an intelligent gamified moodle to predict "at-risk" students using learning analytics approaches. In: Data Analytics Approaches in Educational Games and Gamification Systems, pp. 113–126
35. Dermeval, D., Paiva, R., Bittencourt, I.I., Vassileva, J., Borges, D.: Authoring tools for designing intelligent tutoring systems: a systematic review of the literature. Int. J. Artif. Intell. Educ. **28**, 336–384 (2018)
36. Desidério, S.B., Lelis, M.R.L., Rodrigues, M.E., Marques, A.B.: How ready for hci? a qualitative analysis of the practice of soft skills related to hci by women involved in the digital girls program partners projects. J. Interact. Syst. **15**(1), 504–516 (2024)
37. Dochie, E., Herman, C., Epure, C.: Using gamification for the development of soft skills. Skill generator assessment game case study. In: The International Scientific Conference eLearning and Software for Education, "Carol I" National Defence University, vol. 3, p. 610 (2017)

38. Emanuel, F., Ricchiardi, P., Sanseverino, D., Ghislieri, C.: Make soft skills stronger? an online enhancement platform for higher education. Int. J. Educ. Res. Open **2**, 100096 (2021)
39. Ez-Zaouia, M., Marfisi-Schottman, I., Oueslati, M., Mercier, C., Karoui, A., George, S.: A design space of educational authoring tools for augmented reality. In: International Conference on Games and Learning Alliance, pp. 258–268. Springer (2022)
40. Ez-Zaouia, M., Marfisi-Schottman, I., Mercier, C.: Authoring tools: the road to democratizing augmented reality for education. In: Proceedings of the 15th International Conference on Computer Supported Education, vol. 1, pp. 115–127 (2023)
41. Franke, L., Liang, H., Brantly, A., Davis, J.C., Brown, C.: A first look at the general data protection regulation (GDPR) in open-source software. In: Proceedings of the 2024 IEEE/ACM 46th International Conference on Software Engineering: Companion Proceedings, ICSE Companion 2024, Lisbon, Portugal, 14–20 April 2024, pp. 268–269. ACM (2024). https://doi.org/10.1145/3639478.3643077
42. Franke, L., Liang, H., Farzanehpour, S., Brantly, A., Davis, J.C., Brown, C.: An exploratory mixed-methods study on general data protection regulation (GDPR) compliance in open-source software (2024). CoRR, arXiv:abs/2406.14724. https://doi.org/10.48550/ARXIV.2406.14724
43. Gabbiadini, A., Greitemeyer, T.: Uncovering the association between strategy video games and self-regulation: a correlational study. Personal. Individ. Differ. **104**, 129–136 (2017)
44. Gafni, R., Aviv, I., Kantsepolsky, B., Sherman, S., Rika, H., Itzkovich, Y., Barger, A.: Objectivity by design: the impact of ai-driven approach on employees' soft skills evaluation. Inf. Softw. Technol. **170**, 107430 (2024)
45. Gao, H., Zeng, Y., Ma, B., Pan, Y.: Improving knowledge learning through modelling students' practice-based cognitive processes. Cogn. Comput. 1–18 (2023)
46. Gómez-Albarrán, M., Sarasa-Cabezuelo, A., Sierra-Rodríguez, J.L., Temprado-Battad, B.: Authoring and playing interactive fiction with conventional web technologies. Multimed. Tools Appl. **81**(11), 14705–14747 (2022)
47. Gómez-Martínez, E., Marroyo, M., Acuña, S.T.: Towards the integration of the gdpr in the unified software development process (s). In: SEKE, pp. 199–204 (2021)
48. Gordillo, A., Barra, E., Gallego, D., Quemada, J.: An online e-learning authoring tool to create interactive multi-device learning objects using e-infrastructure resources. In: 2013 IEEE Frontiers in Education Conference (FIE), pp. 1914–1920. IEEE (2013)
49. Gordillo, A., Barra, E., Quemada, J.: An easy to use open source authoring tool to create effective and reusable learning objects. Comput. Appl. Eng. Educ. **25**(2), 188–199 (2017)
50. Greenberg, B.S., Sherry, J., Lachlan, K., Lucas, K., Holmstrom, A.: Orientations to video games among gender and age groups. Simul. Gaming **41**(2), 238–259 (2010)
51. Halperin, X.: Marvel's spider-man: Miles morales procedural tools for playstation 5 content authoring. In: SIGGRAPH Asia 2021 Technical Communications, pp. 1–4 (2021)
52. Hammedi, S., Essalmi, F., Jemni, M., Qaffas, A.A.: An investigation of ai in games: educational intelligent games vs non-educational games. In: 2020 International Multi-Conference on: "Organization of Knowledge and Advanced Technologies" (OCTA), pp. 1–4. IEEE (2020)
53. Hanna, S.: Analysis of the gap between software testing courses at universities and the needed skills by industry. Int. J. Learn. Technol. **19**(2), 203–242 (2024)
54. Hawlitschek, A., Joeckel, S.: Increasing the effectiveness of digital educational games: the effects of a learning instruction on students' learning, motivation and cognitive load. Comput. Hum. Behav. **72**, 79–86 (2017)
55. Herbert, B., Billinghurst, M., Weerasinghe, A., Ens, B., Wigley, G.: A generalized, rapid authoring tool for intelligent tutoring systems. In: Proceedings of the 30th Australian Conference on Computer-Human Interaction, pp. 368–373 (2018)
56. Horák, M., Stupka, V., Husák, M.: Gdpr compliance in cybersecurity software: a case study of dpia in information sharing platform. In: Proceedings of the 14th International Conference on Availability, Reliability and Security, pp. 1–8 (2019)

57. Horst, R., Gerstmeier, S., Naraghi-Taghi-Off, R., Wagner, J., Rau, L., Dörner, R.: Virtual reality content creation based on self-contained components in the e-learning domain: Re-using pattern-based vr content in different authoring toolkits. Multimed. Tools Appl. **83**(15), 46557–46594 (2024)
58. Hossain, M.Y., Zaman, L.: Nccollab: collaborative behavior tree authoring in game development. Multimed. Tools Appl. **82**(3), 4671–4708 (2023)
59. Isomoto, Y.: Educational evaluation of intelligent and creative ability with computer games: a case study for e-learning. In: International Conference on Knowledge-Based and Intelligent Information and Engineering Systems, pp. 985–993. Springer (2006)
60. Jones, J., Millington, M., Virvou, M.: An assumption-based truth maintenance system in active aid for unix users. Intelligent Help Systems for UNIX, pp. 229–252 (2000)
61. Jumaa, Y.M., Moussa, S.M., Khalifa, M.E.: The main aspects of adaptive educational games for normal and disabled/disordered learners: a comprehensive study. In: 2017 Eighth International Conference on Intelligent Computing and Information Systems (ICICIS), pp. 348–355. IEEE (2017)
62. Kabassi, K., Virvou, M.: Combination of a cognitive theory with the multi-attribute utility theory. In: International Conference on Knowledge-Based and Intelligent Information and Engineering Systems, pp. 944–950. Springer (2003)
63. Kamitsios, M., Papadimitriou, S., Chrysafiadi, K., Virvou, M.: Incorporating quizzes into a 3d educational game calling web services. In: EdMedia+ Innovate Learning, Association for the Advancement of Computing in Education (AACE), pp. 1170–1179 (2018)
64. Kay, J.: Stereotypes, student models and scrutability. In: Intelligent Tutoring Systems: 5th International Conference, ITS 2000 Montréal, Canada, June 19–23, 2000 Proceedings, vol. 5, pp. 19–30. Springer (2000)
65. Kee, K., Jankowski, H., Knepper, R., Snapp-Childs, W., Brady, R.: Fostering a thriving virtual organizational culture: the development of metrics for ten soft skills in the case of xsede. SN Comput. Sci. **5**(5), 615 (2024)
66. Khademi, M., Haghshenas, M., Kabir, H.: A review on authoring tools. In: Proceedings of the 5th International Conference on Distance Learning and Education, IPCSIT, vol. 12, pp. 40–44 (2011)
67. Kim, H., Hong, S., Kim, J., Jang, T., Woo, W., Heo, S., Lee, B.: Correction to: realitybrush: an ar authoring system that captures and utilizes kinetic properties of everyday objects. Multimed. Tools Appl. **81**(22), 32703–32703 (2022)
68. Kim, S., Chang, M., Deater-Deckard, K., Evans, M.A., Norton, A., Samur, Y.: Educational games and students' game engagement in elementary school classrooms. J. Comput. Educ. **4**, 395–418 (2017)
69. Kittmann, T., Lambrecht, J., Horn, C.: A privacy-aware distributed software architecture for automation services in compliance with GDPR. In: 23rd IEEE International Conference on Emerging Technologies and Factory Automation, ETFA 2018, Torino, Italy, 4–7 Sept 2018, pp. 1067–1070. IEEE (2018). https://doi.org/10.1109/ETFA.2018.8502545
70. Korsgaard, D., Bjørner, T., Sørensen, P.K., Burelli, P.: Creating user stereotypes for persona development from qualitative data through semi-automatic subspace clustering. User Model. User Adap. Inter. **30**, 81–125 (2020)
71. Lam, M.C., Muhammad Nizam, S.S., Arshad, H., Suwadi, N.A.: Tangible interaction technique with authoring capability for kitchen design. Multimed. Tools Appl. **82**(19), 30125–30150 (2023)
72. Lane, H.C., Ogan, A.E., Shute, V.: Intelligent educational games. In: Artificial Intelligence in Education, pp. 812–812. IOS Press (2009)
73. Lee, E.S.A., Danukarjanto, L., Sharmin, S., Hung, S.Y., Huang, S., Su, T.: Exploring student motivation in integration of soft skills training within three levels of computer science programs. In: Proceedings of the 55th ACM Technical Symposium on Computer Science Education, vol. 1, pp. 708–714 (2024)
74. Liapis, A., Guckelsberger, C., Zhu, J., Harteveld, C., Kriglstein, S., Denisova, A., Gow, J., Preuss, M.: Designing for playfulness in human-ai authoring tools. In: Proceedings of the 18th International Conference on the Foundations of Digital Games, pp. 1–4 (2023)

75. Lim, B., Wen, H.J.: Web services: An analysis of the technology, its benefits, and implementation difficulties. Inf. Syst. Manag. **20**(2) (2003)
76. Lyu, W., Liu, J.: Soft skills, hard skills: What matters most? Evidence from job postings. Appl. Energy **300**, 117307 (2021)
77. Ma, M., Oikonomou, A., Jain, L.C.: Serious Games and Edutainment Applications, vol. 504. Springer (2011)
78. de Macedo, G.T., Fontão, A., Gadelha, B.: Building soft skills through a role-play based approach for requirements engineering remote education. J. Braz. Comput. Soc. **30**(1), 1–16 (2024)
79. Mahadevan, A., Rivera, R., Najhawan, M., Saadat, S., Strehlow, M., Rao, G.R., Youm, J.: Assessing the efficacy of a novel massive open online soft skills course for south asian healthcare professionals. J. Med. Syst. **48**(1), 32 (2024)
80. Marques, A.B., Ferreira, B., Lopes, A., Silva, W.: Stimulating the development of soft skills in software engineering education through design thinking. In: SBES, pp. 690–699 (2020)
81. Martin, B.: Authoring educational games with greenmind. In: Intelligent Tutoring Systems: 9th International Conference, ITS 2008, Montreal, Canada, 23–27 June 2008 Proceedings, vol. 9, pp. 684–686. Springer (2008)
82. Masharova, T., Nimatulaev, M., Kharunzheva, E., Perevozchikova, M.: Formation of highly-demanded soft skills in the game environment of educational webquest. In: SLET-2020: International Scientific Conference on Innovative Approaches to the Application of Digital Technologies in Education, Stavropol, pp. 332–341 (2020)
83. Masneri, S., Dominguez, A., Mangina, E.: Workshop-authoring tools for xr applications in education. In: 2021 7th International Conference of the Immersive Learning Research Network (iLRN), pp. 1–2. IEEE (2021)
84. Massidda, M., Travaglini, L., Pescarin, S., et al.: Interactive digital narrative authoring tools and hybrid experiences in cultural heritage: an integrated review. In: GCH 2023-Eurographics Workshop on Graphics and Cultural Heritage (2023)
85. Mattos, D.P.D., Muchaluat-Saade, D.C., Ghinea, G.: Beyond multimedia authoring: on the need for mulsemedia authoring tools. ACM Comput. Surv. (CSUR) **54**(7), 1–31 (2021)
86. McGowan, N., López-Serrano, A., Burgos, D.: Serious games and soft skills in higher education: a case study of the design of compete! Electronics **12**(6), 1432 (2023)
87. Mehm, F., Göbel, S., Steinmetz, R.: An authoring tool for educational adventure games: concept, game models and authoring processes. Int. J. Game Based Learn. (IJGBL) **3**(1), 63–79 (2013)
88. Miljanovic, M.A., Bradbury, J.S.: Making serious programming games adaptive. In: Joint International Conference on Serious Games, pp. 253–259. Springer (2018)
89. Mitsea, E., Drigas, A., Mantas, P.: Soft skills & metacognition as inclusion amplifiers in the 21 st century. Int. J. Online Biomed. Eng. **17**(4) (2021)
90. Molek-Winiarska, D., Kawka, T.: Reducing work-related stress through soft-skills training intervention in the mining industry. Hum. Factors **66**(5), 1633–1649 (2024)
91. Mononen, A.: Perceptions of 21st-century digital skills and agency among design sprint participants in laurea uas, finland. J. Data Min. Digit, Hum (2024)
92. Mora-Cantallops, M.: Transhistorical perspective of the puzzle video game genre. In: Proceedings of the 13th International Conference on the Foundations of Digital Games, pp. 1–9 (2018)
93. Mougiakou, E., Virvou, M.: Based on gdpr privacy in uml: case of e-learning program. In: 2017 8th International Conference on Information, Intelligence, Systems & Applications (IISA), pp. 1–8. IEEE (2017)
94. Mougiakou, E., Papadimitriou, S., Virvou, M.: Intelligent tutoring systems and transparency: the case of children and adolescents. In: 2018 9th International Conference on Information, Intelligence, Systems and Applications (IISA), pp. 1–8. IEEE (2018)
95. Mougiakou, E., Papadimitriou, S., Virvou, M.: Automated decision making and personal data protection in intelligent tutoring systems: Design guidelines. In: Knowledge-Based Software Engineering: 2018: Proceedings of the 12th Joint Conference on Knowledge-Based Software Engineering (JCKBSE 2018) Corfu, Greece, vol. 12, pp. 231–241. Springer (2019)

96. Mougiakou, E., Papadimitriou, S., Virvou, M.: Synchronous and asynchronous learning methods under the light of general data protection regulation. In: 2020 11th International Conference on Information, Intelligence, Systems and Applications (IISA), pp. 1–7. IEEE (2020)
97. Moundridou, M., Virvou, M.: Authoring intelligent tutoring systems over the world wide web. Proceedings First International IEEE Symposium Intelligent Systems, IEEE **1**, 160–165 (2002)
98. Moundridou, M., Virvou, M.: Usability evaluation of a web-based authoring tool for building intelligent tutoring systems. Proceedings of the 10 th International (2003). Accessed http://citeseerx.ist.psu.edu/viewdoc/download
99. Mousavinasab, E., Zarifsanaiey, N., Niakan Kalhori, R.: S, Rakhshan M, Keikha L, Ghazi Saeedi M,: Intelligent tutoring systems: a systematic review of characteristics, applications, and evaluation methods. Interact. Learn. Environ. **29**(1), 142–163 (2021)
100. Müser, S., Maiero, J., Meyer, J., Hinkenjann, A.: Authoring tools for teaching in vr–an evaluation study. In: 2023 IEEE International Conference on Advanced Learning Technologies (ICALT), pp. 258–260. IEEE (2023)
101. Neto, L.V., Fontoura Junior, P.H., Bordini, R.A., Otsuka, J.L., Beder, D.M.: Design and implementation of an educational game considering issues for visually impaired people inclusion. Smart Learn. Environ. **7**(1), 4 (2020)
102. Noguez J, Espinosa, E.: Improving learning and soft skills using project oriented learning in software engineering courses. In: Proceedings of the 2nd International Workshop on Designing Computational Models of Collaborative Learning Interaction, Citeseer, pp. 83–88 (2004)
103. Novais, A.S., Matelli, J.A., Silva, M.B.: Fuzzy soft skills assessment through active learning sessions. Int. J. Artif. Intell. Educ. **34**(2), 416–451 (2024)
104. Ohlsson, S.: Constraint-based student modeling. In: Student modelling: The Key to Individualized Knowledge-Based Instruction, pp. 167–189. Springer (1994)
105. Pires de Oliveira, V., de Jesus Macedo, R., Vinicius de Freitas, F., Machado, A., Murari, T., Winkler, I.: Virtual reality authoring tools acceptance and use: an exploratory study with the utaut2 model. In: Proceedings of the 25th Symposium on Virtual and Augmented Reality, pp. 299–303 (2023)
106. Olney, A.M., Gilbert, S.B., Rivers, K.: Preface to the special issue on creating and improving adaptive learning: smart authoring tools and processes (2022)
107. Olsen, J.K., Belenky, D.M., Aleven, V., Rummel, N., Sewall, J., Ringenberg, M.: Authoring tools for collaborative intelligent tutoring system environments. In: Intelligent Tutoring Systems: 12th International Conference, ITS 2014, Honolulu, HI, USA, 5–9 June 2014. Proceedings, vol. 12, pp. 523–528. Springer (2014)
108. Orsborn, G.: Use of rapid e-learning authoring tools in higher education. Whitireia Nursing Health J. **24**, 47–51 (2017)
109. Panagoulias, D.P., Palamidas, F.A., Virvou, M., Tsihrintzis, G.A.: Evaluating the potential of llms and chatgpt on medical diagnosis and treatment. In: 2023 14th International Conference on Information, Intelligence, Systems & Applications (IISA), pp. 1–9. IEEE (2023)
110. Panagoulias, D.P., Virvou, M., Tsihrintzis, G.A.: A novel framework for artificial intelligence explainability via the technology acceptance model and rapid estimate of adult literacy in medicine using machine learning. Expert Syst. Appl. **248**, 123375 (2024)
111. Papadimitriou, S., Virvou, M.: Adaptivity in scenarios in an educational adventure game. In: 2017 8th International Conference on Information, Intelligence, Systems & Applications (IISA), pp. 1–6. IEEE (2017)
112. Papadimitriou, S., Chrysafiadi, K., Kamitsios, M., Virvou, M.: Fsp creator: A novel web service api creator of fuzzy students progress profile. In: 2018 9th International Conference on Information, Intelligence, Systems and Applications (IISA), pp. 1–7. IEEE (2018)
113. Papadimitriou, S., Chrysafiadi, K., Virvou, M.: Authoring technological and platform independent learning material and student's progress profile using web services. In: Knowledge-Based Software Engineering: 2018: Proceedings of the 12th Joint Conference on Knowledge-Based Software Engineering (JCKBSE 2018) Corfu, Greece, vol. 12, pp. 242–251. Springer (2019)

114. Papadimitriou, S., Chrysafiadi, K., Virvou, M.: Fuzzeg: Fuzzy logic for adaptive scenarios in an educational adventure game. Multimed. Tools Appl. **78**(22), 32023–32053 (2019). https://doi.org/10.1007/s11042-019-07955-w
115. Papadimitriou, S., Mougiakou, E., Virvou, M.: Smart educational games and consent under the scope of general data protection regulation. In: 2019 10th International Conference on Information, Intelligence, Systems and Applications (IISA), pp. 1–8. IEEE (2019)
116. Papadimitriou, S., Kamitsios, M., Chrysafiadi, K., Virvou, M.: Learn-and-play personalised reasoning from point-and-click to virtual reality mobile educational games. Intell. Decis. Technol. **15**(2), 321–332 (2021)
117. Papazoglou, M.: Web Services: Principles and Technology. Pearson Education (2008)
118. Paraskeva, F., Choustoulakis, E., Karampa, V., Lepenitsa, S., Nikoloudakis, D.: Investigating the impact of agile scrum processes on student engineers' 21st century (4cs) skills. In: International Workshop on Learning Technology for Education Challenges, pp. 96–109. Springer (2024)
119. Psarologou, A., Bourbakis, N., Virvou, M.: A mapping mechanism of nl sentences onto an spn state machine for understanding purposes. In: IISA 2014, The 5th International Conference on Information, Intelligence, Systems and Applications, pp. 321–324. IEEE (2014)
120. Putra, A.S., Novitasari, D., Asbari, M., Purwanto, A., Iskandar, J., Hutagalung, D., Cahyono, Y.: Examine relationship of soft skills, hard skills, innovation and performance: The mediation effect of organizational learning. Int. J. Sci. Manag. Stud. (IJSMS) **3**(3), 27–43 (2020)
121. Qaffas, A.: An operational study of video games' genres (2020)
122. Qizi, K.N.U.: Soft skills development in higher education. Univ. J. Educ. Res. **8**(5), 1916–1925 (2020)
123. Quintana, D., Rodríguez, A., Boada, I.: Authoring tools for procedural modeling of virtual reality-based rehabilitation exercises. IEEE Access **10**, 131567–131578 (2022)
124. Rahimi, A.R.: Beyond digital competence and language teaching skills: the bi-level factors associated with efl teachers' 21st-century digital competence to cultivate 21st-century digital skills. Educ. Inf. Technol. **29**(8), 9061–9089 (2024)
125. Richardson, L., Ruby, S.: RESTful Web Services. Inc, O'Reilly Media (2008)
126. Roig, J., Ribera, M., Alcaraz, R.: Authoring tools are critical to the accessibility of the documents: Recommendations for word processors, including interaction design. In: Proceedings of the 10th International Conference on Software Development and Technologies for Enhancing Accessibility and Fighting Info-Exclusion, pp. 160–165 (2022)
127. Rubab, S., Tang, J., Wu, Y.: Examining interaction techniques in data visualization authoring tools from the perspective of goals and human cognition: a survey. J. Vis. **24**, 397–418 (2021)
128. Salmen, F., Schroeder, U.: Evaluating Authoring Tools with the Explorable Authoring Requirements (2024). arXiv:2403.17714
129. Samala, A.D., Bojic, L., Vergara-Rodríguez, D., Klimova, B., Ranuharja, F.: Exploring the impact of gamification on 21st-century skills: Insights from dota 2. Int. J. Interact. Mob. Technol. 17(18) (2023)
130. dos Santos Silva, D.E., Sobrinho, M.C., Valentim, N.M.C.: Development of 21st-century skills and competencies in high school students through the interactive e-books creation. Revista Brasileira de Informática na Educação **31**, 971–1004 (2023)
131. Savvopoulos, A., Virvou, M., Sotiropoulos, D.N., Tsihrintzis, G.A.: Clustering for user modeling in recommender e-commerce application: a rup-based intelligent software life-cycle. In: Knowledge-Based Software Engineering, pp. 295–304. IOS Press (2008)
132. Schäfer, A.: Improving essential interactions for immersive virtual environments with novel hand gesture authoring tools. Ph.D thesis, Rheinland-Pfälzische Technische Universität Kaiserslautern-Landau (2023)
133. Schulz, B.: The Importance of Soft Skills: Education Beyond Academic Knowledge (2008)
134. Selvi, M., Çosan, A.Ö.: The effect of using educational games in teaching kingdoms of living things. Univ. J. Educ. Res. **6**(9), 2019–2028 (2018)
135. Serrano-Ausejo, E., Mårell-Olsson, E.: Opportunities and challenges of using immersive technologies to support students' spatial ability and 21st-century skills in k-12 education. Educ. Inf. Technol. **29**(5), 5571–5597 (2024)

136. Shakouri, G.H., Tavassoli, N.Y.: Implementation of a hybrid fuzzy system as a decision support process: a fahp-fmcdm-fis composition. Expert Syst. Appl. **39**(3), 3682–3691 (2012)
137. Shibolet, Y., Lombardo, V.: Resources for comparative analysis of idn authoring tools. In: International Conference on Interactive Digital Storytelling, pp. 513–528. Springer (2022)
138. Shiels, J., Tanner, M.: The perceived affordances and usage constraints of learning design authoring tools by south african high school teachers (2022)
139. Siddiq, F., Olofsson, A.D., Lindberg, J.O., Tomczyk, L.: What will be the new normal? digital competence and 21st-century skills: critical and emergent issues in education. Educ. Inf. Technol. **29**(6), 7697–7705 (2024)
140. Silva, L.F., Oliveira, P.S.G., Grander, G., Penha, R., Bizarrias, F.S.: Soft skills fuzzy topsis ranked multi-criteria to select project manager. Int. J. Inf. Decis. Sci. **16**(1), 19–45 (2024)
141. Silva, M., Roberto, R., Radu, I., Cavalcante, P., Schneider, B., Teichrieb, V.: Development of design principles for ar authoring tools for education based on teacher's perspectives. IEEE Trans. Learn, Technol (2023)
142. Solenthaler, B., Klingler, S., Käser, T., Gross, M.: Ten years of research on intelligent educational games for learning spelling and mathematics (2018). arXiv:1806.03257
143. Stankova, M., Tuparova, D., Mihova, P., Kamenski, T., Tuparov, G., Mehandzhiyska, K.: Educational computer games and social skills training. In: Handbook on Intelligent Techniques in the Educational Process: Recent Advances and Case Studies, vol. 1, pp. 361–392. Springer (2022)
144. Stathopoulou, I., Alepis, E., Tsihrintzis, G., Virvou, M.: On assisting a visual-facial affect recognition system with keyboard-stroke pattern. Research and Development in Intelligent Systems XXVI: Incorporating Applications and Innovations in Intelligent Systems XVII, p. 451 (2009)
145. Sucar, L.E., Noguez, J.: Student Modeling. A Practical Guide to Applications, pp. 173–185. Bayesian Networks (2008)
146. Supriana, I., Agustin, R.D., Bakar, M.A., Zin, N.A.M.: Serious games for effective learning. In: 2017 6th International Conference on Electrical Engineering and Informatics (ICEEI), pp. 1–6. IEEE (2017)
147. Takacs, B., Vincze, Z.: Deep authoring-an ai tool set for creating immersive multimedia experiences. Multimed. Tools Appl. **80**(20), 31105–31134 (2021)
148. Tang, K.N.: The importance of soft skills acquisition by teachers in higher education institutions. Kasetsart J. Soc. Sci. **41**(1), 22–27 (2020)
149. Tem, S., Kuroda, A., Tang, K.N.: The importance of soft skills development to enhance entrepreneurial capacity. Int. Educ. Res. **3**(3), p1–p1 (2020)
150. Tomasi, S., Liu, J., Sun, L.: Ai helps, doesn't it: the impact of ai on soft skills (2024)
151. Triantafyllou, A.M., Tsihrintzis, G.A., Virvou, M., Alepis, E.: A bimodal system for emotion recognition via computer of known or unknown persons in normal or fatigue situations. Advances in Core Computer Science-Based Technologies: Papers in Honor of Professor Nikolaos Alexandris, pp. 9–35 (2021)
152. Tsihrintzis, G., Toro, C., Rios, S., Howlett, R., Jain, L.: 27th kes International Conference on Knowledge-Based and Intelligent Information & Engineering Systems kes2023 (2023)
153. Tsihrintzis, G.A., Virvou, M., Phillips-Wren, G.: Surveys in artificial intelligence-based technologies. Intell. Decis. Technol. **13**(4), 393–394 (2019)
154. Tsihrintzis, G.A., Virvou, M., Sakkopoulos, E., Jain, L.C.: Applications of learning and analytics in intelligent systems. In: Applications of Learning and Analytics in Intelligent Systems, Machine Learning Paradigms, pp. 1–6 (2019)
155. Tsihrintzis, G.A., Virvou, M., Sakkopoulos, E., Jain, L.C.: Machine Learning Paradigms: Applications of Learning and Analytics in Intelligent Systems, vol. 1. Springer (2019)
156. Tsihrintzis, G.A., Virvou, M., Jain, L.C.: Advances in Machine Learning/Deep Learning-Based Technologies. Springer (2022)
157. Tsihrintzis, G.A., Virvou, M., Saruwatari, T.: Special Collection of Extended Selected Papers on "Novel Research Results Presented at the 14th International Joint Conference on Knowledge-Based Software Engineering (jckbse2022), 22–24 Aug 2022, Larnaca, Cyprus (2022). https://easyconferences.eu/jckbse2022/

158. Tsihrintzis, G.A., Phillips-Wren, G.E., Jain, L.C.: Intelligent decision technologies: an international journal (idt) heading into its 18th year: A short note from the editors-in-chief. Intell. Decis. Technol. **18**(1), 1–7 (2024)
159. Tsihrintzis, G.A., Virvou, M., Doukas, H., Jain, L.C.: Introduction to advances in artificial intelligence-empowered decision support systems. In: Advances in Artificial Intelligence-Empowered Decision Support Systems: Papers in Honour of Professor John Psarras, pp. 1–11. Springer (2024)
160. Tsiriga, V., Virvou, M.: Modelling the student to individualise tutoring in a web-based icall. Int. J. Contin. Eng. Educ. Life Long Learn. **13**(3–4), 350–365 (2003)
161. Vigneau, C., Mader, S., Rolland, C., Buendia, A.: Acquiring and assessing 21st century skills through game development: a case study with an escape game template. In: International Conference on Entertainment Computing, pp. 150–158. Springer (2023)
162. Virvou, M.: A human plausible reasoning theory in the context of an active help system for unix1 users. Ph.D thesis, University of Sussex, Falmer, East Sussex, UK (1993). https://ethos.bl.uk/OrderDetails.do?uin=uk.bl.ethos.787162
163. Virvou, M.: Automatic reasoning and help about human errors in using an operating system. Interact. Comput. **11**(5), 545–573 (1999)
164. Virvou, M.: A domain-independent reasoning mechanism for an its authoring tool. In: EdMedia+ Innovate Learning, Association for the Advancement of Computing in Education (AACE), pp. 1762–1764 (2000)
165. Virvou, M.: Modelling the knowledge and reasoning of users in a knowledge-based authoring tool. Int. J. Bus. Perform. Manag. **5**(2–3), 399–412 (2003)
166. Virvou, M.: On the evaluation of the combined role of virtual reality games and animated agents in edutainment. Intell. Comput. Graph. **2011**, 79–95 (2012)
167. Virvou, M.: A new era towards more engaging and human-like computer-based learning by combining personalisation and artificial intelligence techniques. In: Proceedings of the 23rd Annual ACM Conference on Innovation and Technology in Computer Science Education, pp. 2–3 (2018)
168. Virvou, M.: Artificial intelligence and user experience in reciprocity: Contributions and state of the art. Intell. Decis. Technol. **17**(Preprint), 1–53 (2023)
169. Virvou, M.: The emerging role of artificial intelligence in teaching 21st century skills in software engineering. In: Keynote 1 in Proceedings of the 35th IEEE Conference on Software Engineering Education and Training, CSEE&T 2023 (2023)
170. Virvou, M., Alepis, E.: Creating tutoring characters through a web-based authoring tool for educational software. In: SMC'03 Conference Proceedings. 2003 IEEE International Conference on Systems, Man and Cybernetics. Conference Theme-System Security and Assurance (Cat. No. 03CH37483), vol. 5, pp. 4884–4889. IEEE (2003)
171. Virvou, M., Alepis, E.: Mobile educational features in authoring tools for personalised tutoring. Comput. Educ. **44**(1), 53–68 (2005)
172. Virvou, M., Kabassi, K.: Reasoning about users' actions in a graphical user interface. Hum. Comput. Interact. **17**(4), 369–398 (2002)
173. Virvou, M., Manos, K.: Individualizing a cognitive model of students' memory in intelligent tutoring systems. In: Knowledge-Based Intelligent Information and Engineering Systems: 7th INternational Conference, KES 2003, Oxford, UK, Sept 2003. Proceedings, Part I, vol. 7, pp. 893–897. Springer (2003)
174. Virvou, M., Manos, K.: A simulated student-player in support of the authoring process in a knowledge-based authoring tool for educational games. In: Proceedings 3rd IEEE International Conference on Advanced Technologies, pp. 338–339. IEEE (2003)
175. Virvou, M., Maras, D.: Error diagnosis in an english tutor. In: Proceedings of the HCI International'99 (the 8th International Conference on Human-Computer Interaction) on Human-Computer Interaction: Communication, Cooperation, and Application Design, vol. 2, pp. 657–661 (1999)
176. Virvou, M., Moundridou, M.: An authoring tool for algebra-related domains. In: HCI (2), Citeseer, pp. 647–651 (1999)

177. Virvou, M., Moundridou, M.: A web-based authoring tool for algebra-related intelligent tutoring systems. J. Educ. Technol. Soc. **3**(2), 61–70 (2000)
178. Virvou, M., Stavrianou, A.: User modelling in a gui. In: Proceedings of HCI International (the 8th International Conference on Human-Computer Interaction) on Human-Computer Interaction: Ergonomics and User Interfaces, vol. I, pp. 261–265 (1999)
179. Virvou, M., Tsihrintzis, G.A.: Is chatgpt beneficial to education? A holistic evaluation framework based on intelligent tutoring systems. In: 2023 14th International Conference on Information, Intelligence, Systems & Applications (IISA), pp. 1–8. IEEE (2023)
180. Virvou, M., Tsihrintzis, G.A.: A novel trust state-chart model for requirements engineering of trustful ai-empowered software. In: 2023 14th International Conference on Information, Intelligence, Systems & Applications (IISA), pp. 1–6. IEEE (2023)
181. Virvou, M, Tsihrintzis, G.A.: Pre-made empowering artificial intelligence and chatgpt: the growing importance of human ai-experts. In: 2023 14th International Conference on Information, Intelligence, Systems & Applications (IISA), pp. 1–8. IEEE (2023)
182. Virvou, M., Manos, K., Katsionis, G., Tourtoglou, K.: Vr-integate: a knowledge-based authoring tool for virtual reality educational games. In: Proceedings of ED-MEDIA 2002 World Conferences on Educational Multimedia, Hypermedia and Telecommunications (2002)
183. Virvou, M., Lampropoulos, A.S., Tsihrintzis, G.A.: Inma: A knowledge-based authoring tool for music education. In: International Conference on Knowledge-Based and Intelligent Information and Engineering Systems, pp. 376–383. Springer (2006)
184. Virvou, M., Tsihrintzis, G., Alepis, E., Stathopoulou, I.O.: Designing a multi-modal affective knowledge-based user interface: combining empirical studies. In: Knowledge-Based Software Engineering, pp. 250–259. IOS Press (2008)
185. Virvou, M., Alepis, E., Sidiropoulos, S.C.: A learning analytics tool for supporting teacher decision. In: 2015 6th International Conference on Information, Intelligence, Systems and Applications (IISA), pp. 1–3. IEEE (2015)
186. Virvou, M., Alepis, E., Tsihrintzis, G.A., Jain, L.C.: Machine Learning Paradigms: Advances in Learning Analytics. Springer (2020)
187. Virvou, M., Tsihrintzis, G.A., Bourbakis, N.G., Jain, L.C.: Handbook on Artificial Intelligence-Empowered Applied Software Engineering: Smart Software Applications in Cyber-Physical Systems, vols. 2 and 3. Springer (2022)
188. Virvou, M., Tsihrintzis, G.A., Tsoukalas, L.H., Jain, L.C.: Introduction to advances in artificial intelligence-based technologies. In: Advances in Artificial Intelligence-based Technologies: Selected Papers in Honour of Professor Nikolaos G Bourbakis, vol. 1, pp. 1–5 (2022)
189. Virvou, M., Tsihrintzis, G.A., Sotiropouloss, D.N., Chrysafiadi, K., Sakkopoulos, E., Tsichrintzi, E.A.: Chatgpt in artificial intelligence-empowered e-learning for cultural heritage: The case of lyrics and poems. In: 2023 14th International Conference on Information, Intelligence, Systems & Applications (IISA), pp. 1–9. IEEE (2023)
190. Virvou, M., Tsihrintzis, G.A., Tsichrintzi, E.A.: Virtsi: A novel trust dynamics model enhancing artificial intelligence collaboration with human users-insights from a chatgpt evaluation study. Inf. Sci. **675**, 120759 (2024)
191. Wen, Z.: Designing educational video games and intelligent tutoring systems for drill-based training (2021)
192. Wiederhold, B.K.: Bridging tradition and technology: soft skills training for the modern day "house call" (2024)
193. Yang, C., Chiang, F.K., Cheng, Q., Ji, J.: Machine learning-based student modeling methodology for intelligent tutoring systems. J. Educ. Comput. Res. **59**(6), 1015–1035 (2021)
194. Yu, E.: (2023) Designing tools for 3d content authoring based on 3d sketching. Ph.D thesis, Université Côte d'Azur
195. Yuce, A., Abubakar, A.M., Ilkan, M.: Intelligent tutoring systems and learning performance: applying task-technology fit and is success model. Online Inf. Rev. **43**(4), 600–616 (2019)
196. Zahari, A.S., Rahim, L.A., Mehat, M.: A review of modelling languages for adventure educational games. In: 2016 3rd International Conference on Computer and Information Sciences (ICCOINS), pp. 495–500. IEEE (2016)

197. Zancanaro, M., Cappelletti, A., Signorini, C., Strapparava, C.: An authoring tool for intelligent educational games. In: International Conference on Virtual Storytelling, pp. 61–68. Springer (2001)
198. Zanfir-Fortuna, G.: (2016) Article 15 right of access by the data subject. In: The EU General Data Protection Regulation (GDPR), Oxford University Press
199. Zarei, N., Quek, F., Chu, S.L., Brown, S.A.: Towards designing enactment-scaffolded narrative authoring tools for elementary-school children. In: Proceedings of the 20th Annual ACM Interaction Design and Children Conference, pp. 387–395 (2021)

The manufacturer's authorised representative in the EU is Springer
Nature Customer Service Centre GmbH, Europaplatz 3, 69115 Heidelberg,
Germany. If you have any concerns regarding our products, please
contact ProductSafety@springernature.com

Printed and bound by CPI Group (UK) Ltd, Croydon, CR0 4YY
30/09/2026
02225562-0009